Lecture Notes in Computer Science 7071

Commenced Publication in 1973
Founding and Former Series Editors:
Gerhard Goos, Juris Hartmanis, and Jan van Leeuwen

Editorial Board

Bo-Yin Yang (Ed.)

Post-Quantum Cryptography

4th International Workshop, PQCrypto 2011
Taipei, Taiwan, November 29 – December 2, 2011
Proceedings

Springer

Volume Editor

Bo-Yin Yang
Academia Sinica
Institute of Information Science
128 Section 2 Academia Road, Taipei 115, Taiwan
E-mail: by@moscito.org

ISSN 0302-9743
e-ISSN 1611-3349
ISBN 978-3-642-25404-8
e-ISBN 978-3-642-25405-5
DOI 10.1007/978-3-642-25405-5
Springer Heidelberg Dordrecht London New York

Library of Congress Control Number: 2011940842

CR Subject Classification (1998): E.3, K.6.5, D.4.6, C.2, J.1, G.2.1

LNCS Sublibrary: SL 4 – Security and Cryptology

Typesetting: Camera-ready by author, data conversion by Scientific Publishing Services, Chennai, India

Printed on acid-free paper

Springer is part of Springer Science+Business Media (www.springer.com)

Foreword

With Shor's algorithm (Peter W. Shor, "Polynomial-Time Algorithms for Prime Factorization and Discrete Logarithms on a Quantum Computer", SIAM J. Sci. Statist. Comput. 41 (2): 303–332, 1999) and its first public instantiation in 2001, when Isaac Chuang and Neil Gershenfeld implemented Shor's algorithm on a 7-qubit quantum computer, it became common knowledge that RSA will crumble with the advent of large quantum computers. Follow-ups made it clear that discrete logarithm problems are equally as broken when thousands-of-qubits quantum computing became available.

A decade had passed and large quantum computers did not actually appear, but it seemed clear enough that the cryptographic research community should not await ostrich-like for the first public appearance of quantum computing to look for alternatives to RSA.

It was in this atmosphere that we saw the emergence, and in some cases renaissance, of "alternative" approaches to public-key cryptography that would survive quantum computers, for which the term "post-quantum cryptography" was affectionately coined.

Cryptographers were hard at work looking for new possibilities for public-key cryptosystems that could resist quantum computers, and currently there are four major families of post-quantum public-key cryptosystems: the code-based public-key cryptosystems, the hash-based public-key cryptosystems, the lattice-based public-key cryptosystems and the multivariate public-key cryptosystems. Many possibilities were proposed and quite a few were rejected. With the increase of research activity in post-quantum cryptography, it became clear that a venue is needed where ideas can be exchanged, results can be presented, and the newest developments can be made known to the world.

Thus was born the first Post-Quantum Cryptography, or PQCrypto, workshop in May 2006 in Leuven. This workshop did not have formal proceedings, and was only made possible with support of the European Union's Framework Program project ECRYPT. PQCrypto 2006 was such a success, however, that Post-Quantum Cryptography was encouraged to form a Steering Committee and run two more instances of these workshop in 2008 (October in Cincinnati, USA) and 2010 (May in Darmstadt, Germany).

The fourth event of this series, PQCrypto 2011, was organized in Taipei, Taiwan, by the Department of Electrical Engineering at the National Taiwan University during November 29–December 2, 2011. The Program Committee received 38 proposals of contributed talks from which 18 were selected. Each paper was thoroughly examined by several independent experts from the Program Committee and additional external reviewers. The papers along with the reviews were then scrutinized by the Program Committee members during a discussion phase after which recommendations were given to all authors. In several cases, we required the authors to work with a shepherd to ensure that the text was edited in accordance with the committee comments and a high standard of writing. Revised versions of the accepted contributions are published in these proceedings.

Thanks must go to all authors for submitting their quality research work to the conference. Even more deserving are the Program Committee and our external reviewers for their time and energy to ensure that a conference program and a volume of high scientific quality could be assembled.

I thank my fellow organizers: Chen-Mou Cheng, who made all the worldly arrangements, and Peter Schwabe, our capable indefatigable webmaster. We would also like to thank Springer, in particular Alfred Hofmann and Anna Kramer, for their support in publishing these proceedings.

September 2011 Bo-Yin Yang

Organization

PQCrypto 2011 was organized by the Department of Electrical Engineering at the National Taiwan University, Taipei, Taiwan; we thank Intel, the National Science Council of Taiwan, and Academia Sinica for sponsorship.

General Chair

Chen-Mou (Doug) Cheng	National Taiwan University, Taiwan

Program Chair

Bo-Yin Yang	Academia Sinica, Taiwan

Program Committee

Martin R. Albrecht	Université Pierre et Marie Curie, France
Paulo S.L.M. Barreto	Universidade de São Paulo, Brazil
Daniel J. Bernstein	University of Illinois at Chicago, USA
Johannes A. Buchmann	Technische Universität Darmstadt, Germany
Jintai Ding	University of Cincinnati, USA
Vivien Dubois	Direction générale de l'armement, France
Louis Goubin	Université de Versailles, France
Sean Hallgren	Pennsylvania State University, USA
Lars Knudsen	Danmarks Tekniske Universitet, Denmark
Tanja Lange	Technische Universiteit Eindhoven, The Netherlands
Richard Lindner	Technische Universität Darmstadt, Germany
Vadim Lyubashevsky	École Normale Supérieure Paris, France
Daniele Micciancio	University of California at San Diego, USA
Michele Mosca	University of Waterloo, Canada
Chris Peikert	Georgia Institute of Technology, USA
Christiane Peters	Technische Universiteit Eindhoven, The Netherlands
Bart Preneel	Katholieke Universiteit Leuven, Belgium
Nicolas Sendrier	INRIA Paris-Rocquencourt, France
Damien Stehlé	CNRS and École Normale Supérieure de Lyon, France
Jean-Pierre Tillich	INRIA Paris-Rocquencourt, France
Ralf-Philipp Weinmann	Université du Luxembourg, Luxembourg
Christopher Wolf	Ruhr-Universität Bochum, Germany

Webmaster

Peter Schwabe — National Taiwan University, Taiwan

External Reviewers

Romain Alleaume
Thierry Berger
Gaëtan Bisson
Stanislav Bulygin
Jean-Marc Couveignes
Christina Delfs
Kirsten Eisenträger
Pooya Farshim
Thomas Feller
Matthieu Finiasz
Philippe Gaborit
Steven Galbraith
Nicolas Gama
Ryan Henry
Jens Hermans
Stefan Heyse
Gerhard Hoffmann
Jeff Hoffstein
Andreas Hülsing
Po-Chun Kuo
Feng-Hao Liu
Pierre Loidreau
Alexander Meurer
Rafael Misoczki
Petros Mol
Michael Naehrig
Robert Niebuhr
Jacques Patarin
Kenny Paterson
Ludovic Perret
Edoardo Persichetti
Albrecht Petzoldt
Louis Salvail
Michael Schneider
Julien Schrek
Jieh-Ren Jarron Shih
Boris Skoric
Benjamin Smith
Douglas Stebila
Andreas Stein
Ron Steinfeld
Enrico Thomae
Valerie Gauthier Umana
Frederik Vercauteren
William Whyte

PQCrypto Steering Committee

Dan Bernstein	University of Illinois at Chicago, USA
Johannes Buchmann	Technische Universität Darmstadt, Germany
Claude Crépeau	McGill University, Canada
Jintai Ding	University of Cincinnati, USA
Philippe Gaborit	Université de Limoges, France
Tanja Lange	Technische Universiteit Eindhoven, The Netherlands
Daniele Micciancio	University of California at San Diego, USA
Werner Schindler	BSI, Germany
Nicolas Sendrier	INRIA, France
Shigeo Tsujii	Chuo University, Japan
Bo-Yin Yang	Academia Sinica, Taiwan

Sponsoring Institutions

Institute of Information Science, Academia Sinica
Center of Information Technology and Innovation, Academia Sinica
The Intel Connected Context Computing Center (at National Taiwan University)

Table of Contents

General Fault Attacks on Multivariate Public Key Cryptosystems

Yasufumi Hashimoto[1], Tsuyoshi Takagi[2], and Kouichi Sakurai[3]

[1] Department of Mathematical Sciences, University of the Ryukyus
[2] Institute of Mathematics for Industry, Kyushu University
[3] Department of Informatics, Kyushu University,
Institute of Systems, Information Technologies and Nanotechnologies

Abstract. The multivariate public key cryptosystem (MPKC), which is based on the problem of solving a set of multivariate systems of quadratic equations over a finite field, is expected to be secure against quantum attacks. Although there are several existing schemes in MPKC that survived known attacks and are much faster than RSA and ECC, there have been few discussions on security against physical attacks, aside from the work of Okeya *et al.* (2005) on side-channel attacks against Sflash. In this study, we describe general fault attacks on MPKCs including Big Field type (*e.g.* Matsumoto-Imai, HFE and Sflash) and Stepwise Triangular System (STS) type (*e.g.* UOV, Rainbow and TTM/TTS). For both types, recovering (parts of) the secret keys S, T with our fault attacks becomes more efficient than doing without them. Especially, on the Big Field type, only *single* fault is sufficient to recover the secret keys.

Keywords: post-quantum cryptography, multivariate public-key cryptosystems, fault attacks.

1 Introduction

It is well known that, if a large scale quantum computer is realized, RSA and elliptic curve cryptosystems (ECC) can be broken by Shor's algorithm [43], and that post-quantum cryptography is now one of the most avidly studied areas in cryptology. Lattice-based cryptosystems, code-based cryptosystems and the multivariate public key cryptosystem (MPKC) are leading candidates for the post quantum cryptosystems.

The cryptosystem studied in the present paper is MPKC, which is based on the problem of solving a set of multivariate systems of quadratic equations over a finite field. The MPKC schemes were first proposed by Matsumoto and Imai [34] and Tsujii *et al.* [44] in the 1980's and have since been extensively developed. Although some of these schemes have already been broken (*e.g.* Matsumoto-Imai's and Tsujii's schemes were broken by Patarin [39], and Hasegawa and Kaneko [28] respectively), others, such as (variants of) HFE and Rainbow, have survived known attacks like the Gröbner basis attacks [23,24], the rank attacks [32,46] and the differential attacks [41,26,22]. Another attractive advantage of

B.-Y. Yang (Ed.): PQCrypto 2011, LNCS 7071, pp. 1–18, 2011.

MPKC is its efficiency. Chen *et al.* [10] presented several MPKC implementations that are more efficient than RSA and ECC on modern x86 CPUs. Before MPKCs can be implemented for practical use, it is necessary to check their security against physical attacks. However, at this time, there have been few such works, aside from the one on a side channel attack against Sflash by Okeya *et al.* [37]. On the other hand, there have been many quite recent studies [36,2,9] on physical attacks against lattice- and code-based cryptosystems.

In this paper, we propose general fault attacks on MPKCs, and discuss the security of MPKCs against the proposed fault attacks comprehensively. A fault attack is a physical attack, first introduced by Boneh *et al.* [7], that causes faults on the parameters in a target device. This type of attacks has been studied with regard to its influence on RSA [30], ECC [6,12] and Pairing [38], but to the best of our knowledge this is the first work that deals with fault attacks on MPKC.

1.1 The Proposed Fault Attacks

Two different fault attacks on MPKCs are discussed in this paper. The first is an attack in which the attacker causes a fault to change coefficients of unknown terms in the central quadratic map G. For the Big Field type scheme (*e.g.* MI [34], HFE [40], Sflash [1] and lIC [21]), the attacker can simplify the target problem to an easier one (*e.g.* HFE to MI) and can recover the secret affine transforms S, T by only single fault and sufficiently many pairs of messages and signatures given by the faulty central map. For the Stepwise Triangular System (STS) type scheme (*e.g.* Tsujii's scheme [44], Shamir's scheme [42], UOV [31], Rainbow [20] and TTM/TTS [35,46]), the attacker can recover a part of the secret affine transform T on the quadratic forms directly from a pair of message and signature given by the faulty central map. On our attacks in practice, the fault can not always change the coefficients in G, since there are three possible system parameters (G, S, or T). However, the success probability of changing the central map G by one fault attack is high enough for attackers (see Table 2), and we are able to distinguish whether the location of the fault is in G or not.

The other fault attack is an attack in which the attacker causes faults such that the parameters r chosen randomly in the process of signature generation are (partially) fixed to the same values. For the "minus" variation of the Big Field type signature scheme, the attacker can simplify the "minus" to the original scheme. For the STS type signature schemes, the attacker can recover a part of the secret affine transform S on the variables.

In the side-channel attack on Sflash by Okeya *et al.* [37], the attacker reduces Sflash to MI by finding the secret key Δ using the side channel attack, and generates a dummy signature of MI by Patarin's attack [39]. On the other hand in our fault attacks, we do not use the secret information discovered by the side-channel attacks but use pairs of messages and signatures given by the faulty secret information to recover the secret affine maps S, T. Though these fault attacks do not necessarily break the schemes directly, partial information of secret keys recovered by the fault attacks is usually critical in terms of preserving the security against known attacks. In other words, the fault attacks weaken the security of the MPKC schemes. It is therefore crucial for the practical implementation of MPKC to determine how to protect the schemes against fault attacks.

2 Multivariate Public Key Cryptosystems

Let $n, m \geq 1$ be integers. For a power of prime q, denote by k a finite field of order q. In an MPKC, the public key is given by a quadratic map $F : k^n \to k^m$, namely $F(x) = (f_1(x), \cdots, f_m(x))^t$ for $x \in k^n$ is described by

$$f_l(x) = \sum_{1 \leq i \leq j \leq n} a_{ij}^{(l)} x_i x_j + \sum_{1 \leq i \leq n} b_i^{(l)} x_i + c^{(l)}, \tag{1}$$

where $a_{ij}^{(l)}, b_i^{(l)}, c^{(l)} \in k$ and $1 \leq l \leq m$. For most schemes in MPKC, the trapdoors are established as follows.

Let S and T be invertible affine transforms of k^n and k^m respectively, namely, for $x \in k^n$ and $y \in k^m$,

$$S(x) = S_1 x + s_2, \quad T(y) = T_1 y + t_2, \tag{2}$$

where S_1 and T_1 are linear transforms and $s_2 \in k^n$ and $t_2 \in k^m$ are constants. Denote by $G : k^n \to k^m$ a quadratic map with a computationally feasible inversion and call G by the central map of the corresponding scheme.

The secret keys are S and T and the public key is $F := T \circ G \circ S$.

$$F : k^n \xrightarrow{S} k^n \xrightarrow{G} k^m \xrightarrow{T} k^m \tag{3}$$

The encryption for a message $x \in k^n$ is computed by $y = F(x)$, and the decryption for the cipher text $y \in k^m$ is recovered by

$$x = S^{-1}(G^{-1}(T^{-1}(y))). \tag{4}$$

In the case of using F for a signature scheme, equation (4) becomes a signature generation function. Some ephemeral random values $r \in k^u$ $(u < m)$ are usually used to generate the signature x of message $y \in k^{m-u}$, namely

$$x = S^{-1}(G^{-1}(T^{-1}(y, r))) \tag{5}$$

The major two types of the constructions G are described in the following subsection.

2.1 Basic Constructions of G

2.1.1 Big Field (BF) Type

The first is called the "Big Field" type; the central map G is given by polynomials over an extension of the original field k. The Matsumoto-Imai (MI) [34], HFE [39] and lIC [21] belong to this type.

Let N be a divisor of $\gcd(n, m)$ and K an algebraic extension of k with $[K; k] = N$. The central map G is given by $G = \psi \circ \mathcal{G} \circ \phi$, where $\phi : k^n \to K^{n/N}$ and $\psi : K^{m/N} \to k^m$ are one-to-one maps and $\mathcal{G} : K^{n/N} \to K^{m/N}$ is a map given by polynomials over K.

$$G : k^n \xrightarrow{\phi} K^{n/N} \xrightarrow{\mathcal{G}} K^{m/N} \xrightarrow{\psi} k^m \tag{6}$$

This type was first introduced by Matsumoto and Imai [34]. In their original scheme, $n = m = N$ and

$$\mathcal{G}(X) := X^{q^i+1}, \tag{7}$$

where i is an integer with $\gcd(q^i + 1, q^n - 1) = 1$. The decryption is computed by $Y^\theta \equiv G^{-1}(Y) = X$ with $\theta \equiv (q^i + 1)^{-1} \bmod q^n - 1$. Unfortunately, Patarin [39] developed a linearization attack that could find the message.

The hidden field equation (HFE) is a modification of MI proposed by Patarin [40]. In this scheme, $\mathcal{G}$ is given by

$$\mathcal{G}(X) = \sum_{0 \le i \le j < d} \alpha_{ij} X^{q^i+q^j} + \sum_{0 \le i < d} \beta_i X^{q^i} + \gamma, \tag{8}$$

where $d \ge 1$ is an integer and $\alpha_{ij}, \beta_i, \gamma \in K$. The inversion of $\mathcal{G}$ is computed by the Berlekamp algorithm whose complexity depends on the degree of $\mathcal{G}$. Because the degree of $\mathcal{G}$ is at most $2q^{d-1}$, the number d cannot be too large. Kipnis-Shamir [32] developed an attack to recover the secret keys S and T on HFE by the MinRank attacks (see also [25]). The complexity of these attacks depends on d, namely, if d is small, the attack will find them effectively. Alternatively, Faugère-Joux [24] developed attacks that find the message by the Gröbner basis algorithm [23]. While estimations of the complexity of the Gröbner basis algorithm is not easy, the message of the HFE with a small d seems to be found efficiently.

The MI and HFE are given by a univariate $\mathcal{G}$. The lIC [21] and lHFE [11] are derived from $\mathcal{G}$ of l variables $(X_1, \cdots, X_l)$. For example, in standard 3IC, $n = m$ is a multiple of 3, $N = n/3$ and $\mathcal{G}(X_1, X_2, X_3) = (X_1X_2, X_2X_3, X_3X_1)$. Both the encryption and the decryption of lIC are faster than MI and HFE. However, Fouque *et al.* [27] found that the Gröbner basis attack recovers the message efficiently. The lHFE is derived from more complicated polynomials. Though Chen *et al.* [11] claimed that this scheme is also efficient, Bettale *et al.* [5] recently pointed out that it is less secure than HFE with the equal-sized keys.

2.1.2 Stepwise Triangular System (STS) Type

This type was first introduced independently of each other by Tsujii *et al.* [44] and Shamir [42]. The basic idea is as follows.

Let $r \ge 1$ be an integer and $\{n_1, \cdots, n_r\}$ and $\{m_1, \cdots, m_r\}$ series of integers with $1 \le n_1 < n_2 < \cdots < n_r = n$ and $1 \le m_1 < m_2 < \cdots < m_r = m$ respectively. The central map

$$G(x) = (g_1(x), \cdots, g_m(x))^t \tag{9}$$

is described by

$$g_l(x) = \sum_{n_{u-1}+1 \le i \le j \le n} a_{ij}^{(l)} x_i x_j + \sum_{n_{u-1}+1 \le i \le n} b_i^{(l)} x_i + c^{(l)}, \tag{10}$$

for $m_{u-1}+1 \le l \le m_u$ ($n_0 = m_0 = 0$). For the inversion, first solve the equations

$$y_1 = g_1(x), \quad \cdots, \quad y_{n_1} = g_{n_1}(x).$$

By the construction of G, the equations above are of variables $x_{n_{r-1}+1}, \cdots, x_n$. If one finds a solution $x_{n_{r-1}+1}, \cdots, x_n$, substitute them into other equations $y_{m_1+1} = g_{m_1+1}(x), \cdots, y_m = g_m(x)$. Next, solve the equations $y_{m_1+1} = g_{m_1+1}(x)$, $\cdots, y_{m_2} = g_{m_2}(x)$ and substitute the solution $x_{n_{r-2}+1}, \cdots, x_{n_{r-1}}$ into other equations $y_{m_2+1} = g_{m_2+1}(x), \cdots, y_m = g_m(x)$. Continuing such steps, enables the inversion of G. In most cases, the ranks of the coefficient matrices of $g_l(x)$ are small. We know that the rank attacks [46] recover a part of T when $n - n_{r-1}$ or n_1 is small, so the values $n - n_{r-1}$ and n_1 should be made large enough.

In the original scheme of Tsujii *et al.* [44], $n = m = r$, $n_l = m_l = l$ and

$$g_l(x) = x_l \times (x_1, \cdots, x_{l-1}\text{-linear}) + (x_1, \cdots, x_{l-1}\text{-quadratic}).$$

Shamir's signature scheme [42] is quite similar. For both schemes, attacks to recover the secret keys had already been proposed [28,13].

The central map G of UOV [31] is as follows.

$$\begin{aligned} g_l(x) &= \sum_{1 \le i \le m} (x_{m+1}, \cdots, x_n\text{-linear}) x_j + (x_{m+1}, \cdots, x_n\text{-quadratic}) \\ &= x^t \begin{pmatrix} 0_m & * \\ * & * \end{pmatrix} x + (\text{liner form}). \end{aligned}$$

In the process of signature generation, first choose constants $r = (r_1, \cdots, r_{n-m})^t \in k^{n-m}$ ($u = n - m$) randomly and substitute them with

$$x_{m+1} = r_1, \quad \cdots, \quad x_n = r_{n-m}. \tag{11}$$

Then $g_1(x), \cdots, g_m(x)$ are linear forms of $x_1, \cdots, x_m$ and are inverted by the elimination. Kipnis-Shamir proposed an attack [33,31] to find (a part of) the secret key S on UOV with the complexity is $O(q^{n-2m} m^4)$, which means that n must be sufficiently larger than $2m$ on UOV.

The Rainbow [20] is a multi-layer UOV, with a central map given by

$$g_l(x) = x^t \begin{pmatrix} 0_{n_{u-1}} & 0 & 0 \\ 0 & 0_{n_u - n_{u-1}} & * \\ 0 & * & *_{n-n_u} \end{pmatrix} x + (\text{linear})$$

for $m_{u-1}+1 \le l \le m_u$. For the signature generation, first choose random values $r = (r_1, \cdots, r_{n-m})^t \in k^{n-m}$ and input $x_{m+1} = r_1, \cdots, x_n = r_{n-m}$. Then, one can invert G in a similar way to the UOV case. Because F is given by

$$f_l(x) = x^t S_1^t \begin{pmatrix} 0_{n_1} & * \\ * & *_{n-n_1} \end{pmatrix} S_1 x + (\text{linear}),$$

Kipnis-Shamir's attack on UOV [33,31] also finds (a part) of S with the complexity $O(q^{n-2n_1}n_1^4)$.

In TTM [35] and TTS [46], G is given by a convolution of two (or more) STS type special invertible non-linear maps. The current examples of TTS (*e.g.* [46]) are special (sparse) versions of Rainbow. The advantages of TTS compared to Rainbow are its efficiency, faster signature generation and smaller sized G.

2.2 Variations of Basic G

Several variations have been proposed to enhance the security of the basic schemes given in the previous subsection. We now describe the major ones.

2.2.1 "Minus(−)" and "Plus(+)"

The "minus" method is given by removing several polynomials in G. If $G(x) := (g_1(x), \cdots, g_m(x))^t$, the central map $G_- : k^n \to k^{n-u}$ $(u < m)$ of "minus" is

$$G_-(x) := (g_1(x), \cdots, g_{m-u}(x)), \tag{12}$$

namely the polynomials $g_{m-u+1}(x), \cdots, g_m(x)$ are hidden in the "minus". This is used for signature schemes. The signature generation process is as follows. For a message $y \in k^{m-u}$, choose $r \in k^u$ randomly. The signature is then generated by

$$x = S^{-1}(G^{-1}(T^{-1}(y), r)).$$

This is usually used in the BF type, because the "minus" of STS is also described as the STS type. MI−, HFE− and lIC− [41,21] are examples of the "minus" of BF type. Note that Sflash [1] is a further modification of MI−. By removing u polynomials in G, they prevents the attacks on the original schemes. However, a differential attack recover the hidden polynomials in MI− and Sflash [26,22].

The "plus" method is given by adding several polynomials, namely the central map of "plus" is $G_+ = (g_1(x), \cdots, g_m(x), h_1(x), \cdots, h_{u_1}(x))$ where $u_1 \geq 1$ is a small integer and h_l is a randomly chosen quadratic forms. The decryption is about q^r times slower than the original one. Note that the security of MI± (the "plus" of MI−) is still open (further discussed by Patarin *et al.* [41]).

2.2.2 "Vinegar"

The "vinegar" method is given by adding several variables. For the original scheme $G(x) := (g_1(x), \cdots, g_m(x))$ with

$$g_l(x) := \sum_{1\leq i\leq j\leq n} a_{ij}x_ix_j + \sum_{1\leq i\leq n} b_ix_i + c,$$

the "vinegar" $G_v(x) := (g_1(\tilde{x}), \cdots, g_m(\tilde{x}))$ is given by

$$g_l(x) := \sum_{1\leq i\leq j\leq n} a_{ij}x_ix_j + \sum_{1\leq i\leq n} v_i^{(l)}(x_{n+1}, \cdots, x_{n+u})x_i + w_i^{(l)}(x_{n+1}, \cdots, x_{n+u}),$$

where $u \geq 1$, $x_{n+1}, \cdots, x_{n+u}$ are additional variables, $\tilde{x} := (x_1, \cdots, x_{n+u})^t$, $v_i^{(l)}$ is a linear form and $w_i^{(l)}$ is a quadratic form. For the signature generation, first choose $r = (r_1, \cdots, r_u) \in k^u$ randomly and substitute

$$x_{n+1} = r_1, \quad \cdots, \quad x_{n+u} = r_u. \tag{13}$$

Then G_v becomes the original G and the signature can be generated. The scheme HFEv- (Quartz) [40,31] is the "minus" of HFEv (the "vinegar" of HFE-). It is known that the secret keys can be recovered if the u is small [14,19].

2.2.3 Other Randomizations

The internal perturbation (IP) [16] is a randomization of G by adding the "perturbing" quadratic polynomials to G. This makes the decryption much slower than the original scheme. Ding *et al.* [18] observed that the IP improves the security against the Gröbner basis attack. However, the differential attack [26] removes the perturbation on PMI (the IP of MI) and then PMI is reduced to Matsumoto-Imai's scheme. The "plus" of PMI (PMI+) has previously been proposed to prevent the differential attack [17].

The piece in hand (PH) [45] has been proposed to randomize G. Tsujii *et al.* [45] have observed that PH improves the security against the Gröbner basis attacks.

2.3 Major Attacks

While it is not easy to list all the attacks on MPKC, we give an overview of several major attacks.

2.3.1 Direct Attacks

The direct attack is to find a solution $x \in k^n$ of the equation $y = F(x)$ directly. The Gröbner basis algorithms [23], the XL algorithms [15], and the fast exhaustive searches [8] are the major approaches. Though their precise complexity estimations are difficult [3], it is known that the Gröbner basis algorithm can effectively recover the message of the HFE if $\mathcal{G}$ is simple [24].

2.3.2 Rank Attacks

The min-rank attack is based on the "min-rank problem". This problem is, in general, difficult to be solved. However, if the corresponding matrix is a linear combination of other matrices $B_1, \cdots, B_m$ and one (or more) of them are of small rank, this problem can be solved efficiently.

In most STS type schemes (except UOV), the minimal rank of the coefficient matrices in G is $n - n_{r-1}$ (see (10)). Thus, if $n - n_{r-1}$ is small, the partial information of T can be found efficiently. In HFE (and MI), $\mathcal{G}$ is described by a quadratic form of $(X, X^q, X^{q^2}, \cdots, X^{q^{n-1}})^t$ over K and its coefficient matrix is of rank (at most) d. Then min-rank attack will find partial information of T when d is small. Note that, for most schemes, S can be recovered if T is recovered.

(This paper does not describe the complexity estimations, but detailed discussions of the min-rank attack are available elsewhere [32,25]).

On the other hand, the high-rank attack is used when the gap between the highest and second highest rank is small. For the STS type scheme, the high-rank attack recovers (a part of) T effectively when n_1 in (10) is small [13,46].

2.3.3 Differential Attacks

The differential attack is an attack using the differential $F(x+t) - F(x) - F(t)$, where F is the public key and $x, t \in k^n$. The differential is a linear map and its kernel or rank will give secret information. In fact, it is known that dummy signatures of MI− [41], PMI [26] and Sflash [22] can be generated by the differential attacks.

2.3.4 Individual Attacks

In addition to the general attacks above, several attacks on individual schemes can be used on other similar schemes. For example, Kipnis-Shamir's attack on UOV [33,31], which recovers a part of S from the public key F, is used when the public key F is given by

$$f_l(x) = x^t S_1^t \begin{pmatrix} 0_o & * \\ * & *_{n-o} \end{pmatrix} S_1 x + (\text{linear form of } x),$$

where $1 \leq o \ll n$. Rainbow and TTS are the examples of such schemes. Note that this attack finds M such that

$$S_1 \begin{pmatrix} I_o & 0 \\ M & I_{n-o} \end{pmatrix} = \begin{pmatrix} *_o & * \\ 0 & *_{n-o} \end{pmatrix}$$

with the complexity $O(q^{n-2o}o^4)$. Such partial information M of S is important, since

$$f_l\left(\begin{pmatrix} I_o & 0 \\ M & I_{n-o} \end{pmatrix} x\right) = x^t \begin{pmatrix} 0_o & * \\ * & *_{n-o} \end{pmatrix} x + (\text{linear}). \qquad (14)$$

The quadratic form above is also of UOV type, and thus a signature x for any given message y can be generated. In Rainbow and TTS, (14) does not generate a signature directly, however if such an M is found, recovering (a part of) T becomes much easier than doing so with only the original F, and ensuring the security of Rainbow becomes as easy as when Rainbow was a smaller size.

3 The Proposed Fault Attacks on MPKCs

In this section, we discuss the fault attacks to be used on MPKCs.

3.1 Attack Model

The entries of the affine maps S, T in (2) and the coefficients of the central map G are stored in the device as the fixed parameters used in MPKCs. Cipher texts of MPKCs are decrypted using secret keys S, T and G by equation (4), and signatures are generated using S, T, G and random ephemeral values $r = (r_1, \cdots, r_u)$ by equation (5). We deal with the following two attacks on MPKCs:

(Fault attacks on G) the attacker causes a fault to change a coefficient of the central map G,
(Fault attacks on r) the attacker causes a fault such that several random ephemeral values of r are fixed to the same values.

3.2 Fault Attacks on G

Assume that the coefficient α_{ij} in (6) for a BF type or $a_{ij}^{(l)}$ in (10) for an STS type is changed to α'_{ij} or $(a_{ij}^{(l)})'$ by the fault, and α'_{ij} or $(a_{ij}^{(l)})'$ cannot be recovered to α_{ij} or $a_{ij}^{(l)}$ again. For the correct (not faulty) central map G and the public key $F = T \circ G \circ S$ from equation (3), denote by G' the faulty central map, $F' := T \circ G' \circ S$ and

$$\Delta F := F' - F = T_1 \circ (G' - G) \circ S,$$

where T_1 is given in (2). The basic approach of our fault attack is as follows.

Step 1. Cause a fault on G and make G'.
Step 2. Decrypt randomly chosen messages $y^{(1)}, \cdots, y^{(N)} \in k^m$ $(N \geq 1)$ by the faulty map G' using equation (4);

$$x^{(l)} := S^{-1}(G'^{-1}(T^{-1}(y^{(l)}))).$$

Step 3. Encrypt $x^{(1)}, \cdots, x^{(N)} \in k^n$ by the correct (not faulty) public key F in equation (3);

$$z^{(l)} := F(x^{(l)}).$$

Step 4. Put

$$\delta^{(l)} := y^{(l)} - z^{(l)}.$$

Find (partial information of) S and T by the pairs $\{(x^{(l)}, \delta^{(l)})\}_{1 \leq l \leq N}$.

Since $y^{(l)} = F'(x^{(l)})$ and $z^{(l)} := F(x^{(l)})$, we have

$$\delta^{(l)} = \Delta F(x^{(l)}) = T_1 \circ (G' - G) \circ S(x^{(l)}).$$

Many coefficients of both G and G' are the same, so then $G - G'$ is a sparse polynomial. From the sparseness of $G - G'$, we try to recover S and T. The details of recovering S and T are described in the following subsections.

3.2.1 Fault Attack on G for Big Field Type

We propose the fault attack on HFE, which is a typical BF-type model. Let

$$\Delta F(x) = (\Delta f_1(x), \cdots, \Delta f_m(x))^t$$

and each equation $\Delta f_l(x)$ be defined similar to equation (1) as

$$\Delta f_l(x) := \sum_{1\le i\le j\le n} a_{ij}^{(l)} x_i x_j + \sum_i b_i^{(l)} x_i + c^{(l)},$$

where $a_{ij}^{(l)}, b_i^{(l)}, c^{(l)} \in k$. At this point, the coefficients $a_{ij}^{(l)}, b_i^{(l)}, c^{(l)}$ are unknown. To find them, substitute the pairs $(x^{(l)}, \delta^{(l)})$ into the above and construct a system of linear equations of unknowns $a_{ij}^{(l)}, b_i^{(l)}, c^{(l)}$. Since the total number of $a_{ij}^{(l)}, b_i^{(l)}, c^{(l)}$ is $(n+1)(n+2)/2$ for a fixed l, we can solve such equations and find $a_{ij}^{(l)}, b_i^{(l)}, c^{(l)}$ with $(n+1)(n+2)/2$ pairs of $(x^{(l)}, \delta^{(l)})$. Remark that a univariate polynomial equation $\mathcal{G}'(X) = Y^{(l)}$, where $Y^{(l)} \in K$ is derived from $y^{(l)} \in k^n$, does not always have a solution, and the probability that it has no solutions is around $0.33 \sim 0.5$ [29] (about $0.368\cdots \sim 1/e$ for large q, n). Thus the attacker has to compute more $\mathcal{G}'^{-1}(Y^{(l)})$ in practice.

Next, we explain how to find S and T by ΔF. Assume that the fault changes α_{ij} to α'_{ij}. Then the central map of ΔF is derived from

$$(\mathcal{G}' - \mathcal{G})(X) = (\alpha_{ij} - \alpha'_{ij}) X^{q^i + q^j},$$

which is similar to $\mathcal{G}(X)$ in MI. Since the rank of the coefficient matrix of $(\mathcal{G}' - \mathcal{G})(X)$ as a quadratic form of $(X, X^q, \cdots, X^{q^{n-1}})^t$ is 2, Kipnis-Shamir's attack [32,25] (the rank attack) for rank 2 will find (a part of) S and T.

If there is more than one G fault – namely several coefficients are changed – $\Delta F(x)$ is no longer the case above and $\Delta F(x)$ is reduced to a public key of HFE with a smaller u. For example, when two coefficients are changed, ΔF is a public key of HFE derived from $\mathcal{G}$ with two terms. The attacker can then find S and T by Kipnis-Shamir's attack [32,25] for rank 4 at most, which is much smaller than the rank of the coefficient matrix for $\mathcal{G}$ of the HFE without the fault. Since, if the rank is smaller, Kipnis-Shamir's attack recover S and T with less complexity (see [32,25] for the complexity estimations), our fault attack reduces the complexity of recovering S and T.

3.2.2 Fault Attack on G for STS Type

We propose the fault attack on the STS type, which has a central map given by (10). Note that, different to the attack on the BF type, we need to cause a fault in several times for STS type.

First, assume that the coefficient $a_{i_1 j_1}^{(l_1)}$ is changed to $(a_{i_1 j_1}^{(l_1)})'$ by the fault. In this case, it is easy to see that

$$(G - G')(x) = \Big(\overbrace{0, \cdots, 0}^{l_1 - 1}, (a_{i_1 j_1}^{(l_1)} - (a_{i_1 j_1}^{(l_1)})') x_{i_1} x_{j_1}, 0, \cdots, 0\Big)^t,$$

where $x = (x_1, \cdots, x_n)^t$. We then see that

$$\delta^{(1)} = \Delta F(x^{(1)}) = T_1 \circ (G - G') \circ S(x^{(1)}) = c T_1 (0, \cdots, 0, \overset{l_1}{\check{1}}, 0, \cdots, 0)^t,$$

Table 1. The number of faults and messages to recover the secret key in our fault attacks on G

	Big Field (§3.2.1)	STS (§3.2.2)
#Fault	1	$n-1$
$\#(x,\delta)$	$\frac{1}{2}(n+1)(n+2)$	1
Recovering	parts of S,T	a part of T

where c is a constant. This means that $\delta^{(1)} = (\delta_1^{(1)}, \cdots, \delta_m^{(1)})^t$ coincides with a constant multiple of the l_1-th column vector $(t_{1l_1}, \cdots, t_{ml_1})^t$ in T_1. Thus, taking the transform

$$T^{(1)} := \left(\begin{array}{c|c} 1 & 0 \\ \hline -\delta_2^{(1)}/\delta_1^{(1)} & \\ \vdots & I_{m-1} \\ -\delta_m^{(1)}/\delta_1^{(1)} & \end{array}\right), \quad \text{we have} \quad T^{(1)}T_1 = \begin{pmatrix} * & \overset{l_1}{*} & * \\ * & 0 & * \\ \vdots & \vdots & \vdots \\ * & 0 & * \end{pmatrix}. \qquad (15)$$

Next, cause another fault on G', and get a pair $(x^{(2)}, \delta^{(2)})$ by Step 1-4 with $(G')' = G''$, namely, $x^{(2)} := S^{-1}(G''^{-1}(T^{-1}(y^{(2)})))$, $\delta^{(2)} := F''(x^{(2)}) - F(x^{(2)})$. Assume that the fault changes the coefficient $a_{i_2j_2}^{(l_2)}$ to $(a_{i_2j_2}^{(l_2)})'$. Then we see that

$$\delta^{(2)} = T_1 \circ (G - G'') \circ S(x^{(2)}) = T_1\Big((0, \cdots, 0, \overset{l_2}{\check{c}_2}, 0, \cdots, 0, \overset{l_1}{\check{c}_1}, 0, \cdots, 0)^t\Big),$$

where c_1 and c_2 are constants. Then, similar to (15), the attacker can reduce the elements in the l_2-th column in T_1.

Repeating such processes $n-1$ times, one can reduce T_1 to a permutation of a triangle matrix. Recall that the purpose of the rank attack is to find (a part of) T. Thus the fault attack reduces the parameters in T to be found by the rank attack.

Table 1 shows a comparison of the fault attack results against the Big Field type in section 3.2.1 and the STS type in section 3.2.2. #Fault is the number of faults required for the attack and $\#(x,\delta)$ is the number of pairs $(x^{(l)}, \delta^{(l)})$ given in Step 4 for *each fault*. The proposed fault attack can recover parts of the secret keys S and T for the Big Field type and a part of T in equation (2) for the STS type, respectively.

3.2.3 Success Probability and Distinguishing the Faulty Place

Our fault attacks on G succeed if the fault changes a parameter in G. However, there are fixed system parameters other than G, which are in S and T. If the fault changes entries in S or T, our fault attack fails to recover the desired secret information. In this subsection, we discuss the probability that a parameter in G is changed by a random fault, and how to distinguish which map (G, S or T) is faulty.

Success probability. The numbers of the entries (over k) in S and T are $n(n+1)$ and $m(m+1)$ respectively, and the total number of the coefficients

Table 2. Success probability of our proposed fault attacks on some MPKCs

Scheme	q	n	m	S	G	T
Quarz(2,103,129,3,4) [14]	2	107	100	0.38	0.29	0.33
4HFE(31,10) [11]	31	40	40	0.37	0.26	0.37
Rainbow(31,24,20,20) [10]	31	64	40	0.07	0.90	0.03
Rainbow(256,18,12,12) [10]	256	42	24	0.10	0.87	0.03

(over k) in m quadratic forms of n variables is $m(n+1)(n+2)/2$. In general, the total number of the coefficients (over k) in G is less than $m(n+1)(n+2)/2$, since G is a special quadratic map. Such total numbers depend on the scheme (namely G), so it is not easy to discuss the general situations. We select several example schemes, Quartz (HFEv−) [14], 4HFE [11] and Rainbows [10], and calculated the number of k-elements in S, G and T. The results are shown in Table 2. In "S", we describe the ratio of the number of entries in S over the total sum of the k-elements in S, G and T. The numbers in "G" and "T" are the same.

In Quartz and 4HFE (which are BF type), the number of k-elements in G is less than those in S and T, since, if a BF type G is constructed contains a large number of coefficients, the decryption process becomes much complex. However, the probability around 25% ∼ 30% is large enough for attackers. On the other hand, in two Rainbow examples (which are STS type), the number of k-elements in G is much larger than that in S and T. For such schemes, there is quite a large probability that G is faulty.

Distinguishing the faulty place. Our fault attack succeeds if the fault changes a coefficient in G, and it does not if an entry in S or T is changed. We now discuss how to distinguish the faulty place, G, S or T.

First, assume that the (l_1, l_2)-entry in T is changed by the fault. Then

$$\delta^{(i)} = (T - T') \circ G \circ S(x^{(i)}) = (0, \cdots, 0, \overset{l_1}{*}, 0, \cdots, 0)^t. \tag{16}$$

This shows that, if only one entry in $\delta^{(i)}$ is non-zero and the others are all zero, T is faulty with high probability.

Next, assume that the (l_1, l_2)-entry in S is faulty and denote by

$$\delta^{(i)} = T \circ G \circ S(x^{(i)}) - T \circ G \circ S'(x^{(i)}) = (\delta_1^{(i)}, \cdots, \delta_m^{(i)})^t,\ x^{(i)} = (x_1^{(i)}, \cdots, x_n^{(i)})^t$$

This shows that

$$\delta_j^{(i)} = x_{l_2}^{(i)} \times (\text{linear form of } x_1^{(i)}, \cdots, x_n^{(i)}). \tag{17}$$

Based on this, we can check whether S is faulty as follows.

Prepare N pairs of $(x^{(i)}, \delta^{(i)})$ with $N \geq n + 2$. Put

$$y_j^{(l)} = x_v \left(\tilde{a}_1^{(j,l)} x_1 + \cdots + \tilde{a}_n^{(j,l)} x_n + \tilde{b}^{(j,l)} \right), \tag{18}$$

where $1 \leq l \leq n$, $1 \leq j \leq m$ and $a_i^{(j,l)} \in k$. Substitute $x = x^{(i)}$ and $y_j^{(l)} = \delta_j^{(i)}$ into (18). Then N linear equations of $n+1$ variables $(\tilde{a}_1^{(j,l)}, \cdots, \tilde{a}_n^{(j,l)}, \tilde{b}^{(j,l)})$ appear. According to (17), we see that, if such equations have no common solutions for any l, then S is not faulty.

3.3 Fault Attacks on r

In the signature schemes, ephemeral random values $r_1, \cdots, r_u \in k$ are used for the signature generation in equation (5). In our fault attack on r, we assume that the attacker can fix a part of such $r_1, \cdots, r_u \in k$ by a fault. Without the loss of generality, suppose that $(r_{u-u_1+1}, \cdots, r_u)$ is fixed for $u_1 \leq u$. The basic approach of the fault attack is as follows.

Step 1. Cause a fault in which $(r_{u-u_1+1}, \cdots, r_u)$ is fixed in every signature generation process. Suppose that $(r_{u-u_1+1}, \cdots, r_u) = (\tilde{r}_{u-u_1+1}, \cdots, \tilde{r}_u)$ where $\tilde{r}_{u-u_1+1}, \cdots, \tilde{r}_u \in k$ are fixed (unknown) values.
Step 2. For given messages $y^{(1)}, \cdots, y^{(N)}$, generate the corresponding signature $x^{(1)}, \cdots, x^{(N)}$ with $r = (r_1, \cdots, r_{u-u_1}, \tilde{r}_{u-u_1+1}, \cdots, \tilde{r}_u)$.
Step 3. Recover (parts of) S and T by using $\{(x^{(l)}, y^{(l)})\}_{1 \leq l \leq N}$.

Next, we will describe precisely how to recover the S and T details.

3.3.1 Fault Attack on r for ("Minus" of) BF Type

In the basic BF type, *e.g.* MI and HFE, random parameters are not used. In the "minus" and the "vinegar" variation, however, ephemeral random values are used for the signature generation. Since the fault attack on r for the vinegar is similar to that on the STS type given later, we now explain the fault attack on r for the minus of the BF type.

Assume that $r_{u-u_1+1}, \cdots, r_u$ are fixed to be $(\tilde{r}_{u-u_1+1}, \cdots, \tilde{r}_u)$ $(1 \leq u_1 \leq u)$. The signatures $x^{(1)}, \cdots, x^{(N)}$ are then the solutions of

$$g_{m-u_1+1}(S(x)) = \tilde{r}_{u-u_1+1}, \qquad \cdots, \qquad g_m(S(x)) = \tilde{r}_u, \tag{19}$$

where g_l is the hidden polynomial. Taking differences of (19), we have

$$g_l(S(x^{(l)})) - g_l(S(x^{(1)})) = 0, \quad (m - u_1 + 1 \leq l \leq m). \tag{20}$$

Similar to the case of the fault attack on G for the BF type, we see that, if $N > (n+1)(n+2)/2$, we can recover the central polynomials $g_{m-u_1+1}(S(x)), \cdots, g_m(S(x))$.

Recall that the "minus" hides the central polynomials $g_{m-u+1}(x), \cdots, g_m(x)$ of the original scheme and, if the number u of the hidden polynomials is small, the hidden polynomials can be discovered. Our fault attack finds several hidden polynomials, namely the number of hidden polynomials is reduced. Our fault attack therefore reduces the security of the "minus" variations. In particular, if $u_1 = u$, all of the hidden central polynomials are recovered. In this case, the "minus" is completely reduced to the original scheme (*e.g.* MI− → MI, HFE− → HFE) and the attacks on the original scheme are then used directly.

3.3.2 Fault Attack on r for STS Type (and the "Vinegar")

In most STS type signature schemes and "vinegar variations, the signature $x \in k^n$ for a message $y \in k^m$ and ephemeral random values $r = (r_1, \cdots, r_u)^t \in k^u$ in (5) are given by

$$x = S^{-1}(G^{-1}(T^{-1}(y, r))) = S^{-1}\begin{pmatrix} z \\ r \end{pmatrix},$$

where $z \in k^{n-u}$, namely the lower u entries of $S(x)$ coincide with r. In fact, the signatures of UOV, Rainbow, TTS and HFEv are all given in this way. We now propose the fault attack on r by using this property.

Assume that $(r_{u-u_1+1}, \cdots, r_u)$ is fixed to be $(\tilde{r}_{u-u_1+1}, \cdots, \tilde{r}_u)$ $(u \leq u_1)$. Let $r^{(l)} = (r_1^{(l)}, \cdots, r_{u-u_1}^{(l)}, \tilde{r}_{u-u_1+1}, \cdots, \tilde{r}_u)^t = (\eta^{(l)t}, \tilde{r}^t)^t$ where $\eta^{(l)} \in k^{u-u_1}$ are the ephemeral random values corresponding to the signature $x^{(l)}$ given in Step 2. Then we have

$$S_1 x^{(l)} + s_2 = \begin{pmatrix} z^{(l)} \\ \eta^{(l)} \\ \tilde{r} \end{pmatrix}, \qquad (21)$$

where S_1, s_2 are given in (2). Divide $x^{(l)}, S_1$ and s_2 by

$$x^{(l)} = \begin{pmatrix} x^{(l,1)} \\ x^{(l,2)} \end{pmatrix}, \quad S_1 = \begin{pmatrix} A & B \\ C & D \end{pmatrix}, \quad s_2 = \begin{pmatrix} s_{21} \\ s_{22} \end{pmatrix}$$

with $x^{(l,1)}, s_{21} \in k^{n-u_1}$, $x^{(l,2)}, s_{22} \in k^{u_1}$, $A \in k^{(n-u_1)\times(n-u_1)}$, $B \in k^{(n-u_1)\times u_1}$, $C \in k^{u_1\times(n-u_1)}$ and $D \in k^{u_1 \times u_1}$, and assume that D is invertible. Then the lower u_1 entries in (21) are $Cx^{(l,1)} + Dx^{(l,2)} + s_{22} = \tilde{r}$. Since s_{22} and $\tilde{r}$ are fixed values, we can take

$$C\tilde{x}^{(l,1)} + D\tilde{x}^{(l,2)} = 0,$$

where $\tilde{x}^{(l,1)} := x^{(l,1)} - x^{(1,1)}$ and $\tilde{x}^{(l,2)} := x^{(l,2)} - x^{(1,2)}$ for $2 \leq l \leq N$. We can thus recover $D^{-1}C$ by

$$D^{-1}C = -X_2 X_1^{-1}, \qquad (22)$$

where $X_1 := (\tilde{x}^{(1,1)}, \cdots, \tilde{x}^{(n-u_1,1)})$, $X_2 := (\tilde{x}^{(1,2)}, \cdots, \tilde{x}^{(n-u_1,2)})$. Since

$$S\begin{pmatrix} I_{n-u_1} & 0 \\ -D^{-1}C & I_{u_1} \end{pmatrix} = \begin{pmatrix} *_{n-u_1} & * \\ 0 & *_{u_1} \end{pmatrix},$$

we see that our fault attack reduces the complexity $O(q^{n-2o}o^4)$ of Kipnis-Shamir's attack on UOV [33,31] to $O(q^{n-2o-u_1}o^4)$. The fault attack on r then weakens the security of UOV like schemes (UOV, Rainbow and TTS) on Kipnis-Shamir's attack to find (a part of) S.

Table 3. The number of messages to recover the central map or secret key in our fault attacks on r

	Big Field (§3.3.1)	STS (§3.3.2)
$\#(x,y)$	$\frac{1}{2}(n+1)(n+2)$	$n-u_1+1$
Recovering	hidden $g_{m-u_1+1},\cdots,g_m$	a part of S

Even if the scheme is not UOV like, the information $D^{-1}C$ is necessary to prevent the rank attacks. In fact, when the rank for the rank attack is R, our fault attack on r reduces the rank for the rank attack to $R-u_1$. This weakens the security against the rank attacks.

In "vinegar", the information $D^{-1}C$ is enough to discover the original scheme. Thus, the fault attack reduces the "vinegar" with u random values to the "vinegar" with $u-u_1$ random values. In particular, if $u=u_1$, the polynomials in the original scheme are recovered and then the attacks against the original scheme can be used directly.

Table 3 shows a comparison of the fault attack results on r against ("minus" of) the Big Field type in section 3.3.1 and the STS type (and "vinegar") in section 3.3.2. $\#(x,y)$ is the number of pairs given in Step 2. The proposed fault attack can recover central polynomials $g_{m-u_1+1}(x),\cdots,g_m(x)$ in G hidden to generate the "minus" for the Big Field type and the secret key T in equation (2) for the STS type.

3.4 Countermeasures

In this section, we explain naive countermeasures against our fault attacks.

Fault attacks on G. Recall that our fault attack on G requires several signatures x derived from randomly chosen messages y and the faulty central map G'. The basic strategy to prevent the fault attack is to check whether G is faulty and, if so, to not generate the signature. For example, prepare c_G the sum of the coefficients of the polynomials in G and check whether c_G coincides with the sum of the coefficients in the central map before the signature generation process. If it does not, reject the given message for the signature generation. Our fault attack will not work, because the signatures cannot be generated by the faulty central map G'.

Fault attack on r. Recall that the fault attack on r require the signature x derived from several randomly chosen messages y and (partially) fixed random ephemeral values r. The basic strategy to prevent this attack is to recall the random ephemeral values r chosen in the past several signature generations and if there are (partial) coincidences of r, to stop the next signature generation process. Our fault attack will not work because a sufficient number of signatures with a fixed r cannot be given.

4 Conclusion

The present paper proposed general fault attacks on multivariate public key cryptosystems and discussed the security of MPKC against the proposed fault attacks comprehensively. The proposed fault attacks reduce the complexity of finding the secret keys S and T of the underlying schemes by causing faults on the central map G or faults on the ephemeral random values r.

Our approach to fault attacks can be applied to other physical attacks (*e.g.* side-channel attacks). Investigating the security against such attacks is also crucial to ensure the practical implementations of MPKC's schemes. Finally, applying the fault attacks in this paper to QUAD [4], which is a stream cipher based on multivariate quadratic forms, is an interesting open problem.

Acknowledgment. We would like to thank Jintai Ding, Kazuo Sakiyama and Junko Takahashi for giving us some comments on the draft of this paper. The first and the third authors are partially supported by JST Strategic Japanese-Indian Cooperative Programme on multidisciplinary Research Field, which combines Information and Communications Technology with Other Fields.

References

1. Akkar, M.L., Courtois, N., Goubin, L., Duteuil, R.: A Fast and Secure Implementation of Sflash. In: Desmedt, Y.G. (ed.) PKC 2003. LNCS, vol. 2567, pp. 267–278. Springer, Heidelberg (2002)
2. Avanzi, R.M., Hoerder, S., Page, D., Tunstall, M.: Side-Channel Attacks on the McEliece and Niederreiter Public-Key Cryptosystems (2010), `http://eprint.iacr.org/2010/479`
3. Bardet, M., Faugère, J.C., Salvy, B., Yang, B.Y.: Asymptotic Expansion of the Degree of Regularity for Semi-Regular Systems of Equations. In: MEGA 2005 (2005)
4. Berbain, C., Gilbert, H., Patarin, J.: QUAD: A Practical Stream Cipher with Provable Security. In: Vaudenay, S. (ed.) EUROCRYPT 2006. LNCS, vol. 4004, pp. 109–128. Springer, Heidelberg (2006)
5. Bettale, L., Faugere, J.C., Perret, L.: Cryptanalysis of Multivariate and Odd-Characteristic HFE Variants. In: Catalano, D., Fazio, N., Gennaro, R., Nicolosi, A. (eds.) PKC 2011. LNCS, vol. 6571, pp. 441–458. Springer, Heidelberg (2011)
6. Biehl, I., Meyer, B., Müller, V.: Differential Fault Attacks on Elliptic Curve Cryptosystems. In: Bellare, M. (ed.) CRYPTO 2000. LNCS, vol. 1880, pp. 131–146. Springer, Heidelberg (2000)
7. Boneh, D., DeMillo, R.A., Lipton, R.J.: On the Importance of Checking Cryptographic Protocols for Faults. In: Fumy, W. (ed.) EUROCRYPT 1997. LNCS, vol. 1233, pp. 37–51. Springer, Heidelberg (1997)
8. Bouillaguet, C., Chen, H.C., Cheng, C.M., Chou, T., Niederhagen, R., Shamir, A., Yang, B.Y.: Fast Exhaustive Search for Polynomial Systems in F_2. In: Mangard, S., Standaert, F.-X. (eds.) CHES 2010. LNCS, vol. 6225, pp. 203–218. Springer, Heidelberg (2010)
9. Cayrel, P.L., Dusart, P.: Fault injection's sensitivity of the McEliece PKC. In: Proc. of 5th International Conference on Future Information Technology, pp. 1–6 (2010)
10. Chen, A.I.-T., Chen, M.-S., Chen, T.-R., Cheng, C.-M., Ding, J., Kuo, E.L.-H., Lee, F.Y.-S., Yang, B.-Y.: SSE Implementation of Multivariate PKCs on Modern x86 CPUs. In: Clavier, C., Gaj, K. (eds.) CHES 2009. LNCS, vol. 5747, pp. 33–48. Springer, Heidelberg (2009)

11. Chen, C.H.O., Chen, M.S., Ding, J., Werner, F., Yang, B.Y.: Odd-char multivariate Hidden Field Equations (2008), http://eprint.iacr.org/2008/543
12. Ciet, M., Joye, M.: Elliptic curve cryptosystems in the presence of permanent and transient faults. Designs, Codes and Cryptography 36, 33–43 (2005)
13. Coppersmith, D., Stern, J., Vaudenay, S.: Attacks on the Birational Permutation Signature Schemes. In: Stinson, D.R. (ed.) CRYPTO 1993. LNCS, vol. 773, pp. 435–443. Springer, Heidelberg (1994)
14. Courtois, N.T., Daum, M., Felke, P.: On the Security of HFE, HFEv- and Quartz. In: Desmedt, Y.G. (ed.) PKC 2003. LNCS, vol. 2567, pp. 337–350. Springer, Heidelberg (2002)
15. Courtois, N.T., Klimov, A.B., Patarin, J., Shamir, A.: Efficient Algorithms for Solving Overdefined Systems of Multivariate Polynomial Equations. In: Preneel, B. (ed.) EUROCRYPT 2000. LNCS, vol. 1807, pp. 392–407. Springer, Heidelberg (2000)
16. Ding, J.: A New Variant of the Matsumoto-Imai Cryptosystem through Perturbation. In: Bao, F., Deng, R., Zhou, J. (eds.) PKC 2004. LNCS, vol. 2947, pp. 305–318. Springer, Heidelberg (2004)
17. Ding, J., Gower, J.E.: Inoculating Multivariate Schemes Against Differential Attacks. In: Yung, M., Dodis, Y., Kiayias, A., Malkin, T. (eds.) PKC 2006. LNCS, vol. 3958, pp. 290–301. Springer, Heidelberg (2006)
18. Ding, J., Gower, J.E., Schmidt, D., Wolf, C., Yin, Z.: Complexity Estimates for the F_4 Attack on the Perturbed Matsumoto-imai Cryptosystem. In: Smart, N.P. (ed.) Cryptography and Coding 2005. LNCS, vol. 3796, pp. 262–277. Springer, Heidelberg (2005)
19. Ding, J., Schmidt, D.: Cryptanalysis of HFEv and Internal Perturbation of HFE. In: Vaudenay, S. (ed.) PKC 2005. LNCS, vol. 3386, pp. 288–301. Springer, Heidelberg (2005)
20. Ding, J., Schmidt, D.: Rainbow, a New Multivariate Polynomial Signature Scheme. In: Ioannidis, J., Keromytis, A.D., Yung, M. (eds.) ACNS 2005. LNCS, vol. 3531, pp. 164–175. Springer, Heidelberg (2005)
21. Ding, J., Wolf, C., Yang, B.-Y.: $l-$invertible Cycles for Multivariate Quadratic (MQ) Public Key Cryptography. In: Okamoto, T., Wang, X. (eds.) PKC 2007. LNCS, vol. 4450, pp. 266–281. Springer, Heidelberg (2007)
22. Dubois, V., Fouque, P.-A., Shamir, A., Stern, J.: Practical Cryptanalysis of SFLASH. In: Menezes, A. (ed.) CRYPTO 2007. LNCS, vol. 4622, pp. 1–12. Springer, Heidelberg (2007)
23. Faugère, J.C.: A new efficient algorithm for computing Grobner bases (F_4). Journal of Pure and Applied Algebra 139, 61–88 (1999)
24. Faugère, J.-C., Joux, A.: Algebraic Cryptanalysis of Hidden Field Equation (HFE) Cryptosystems using Gröbner Bases. In: Boneh, D. (ed.) CRYPTO 2003. LNCS, vol. 2729, pp. 44–60. Springer, Heidelberg (2003)
25. Faugère, J.-C., Levy-dit-Vehel, F., Perret, L.: Cryptanalysis of Minrank. In: Wagner, D. (ed.) CRYPTO 2008. LNCS, vol. 5157, pp. 280–296. Springer, Heidelberg (2008)
26. Fouque, P.-A., Granboulan, L., Stern, J.: Differential Cryptanalysis for Multivariate Schemes. In: Cramer, R. (ed.) EUROCRYPT 2005. LNCS, vol. 3494, pp. 341–353. Springer, Heidelberg (2005)
27. Fouque, P.-A., Macario-Rat, G., Perret, L., Stern, J.: Total Break of the ℓ-IC Signature Scheme. In: Cramer, R. (ed.) PKC 2008. LNCS, vol. 4939, pp. 1–17. Springer, Heidelberg (2008)

28. Hasegawa, S., Kaneko, T.: An attacking method for a public-key cryptosystem based on the difficulty of solving a system of non-linear equations (in Japanese). In: Proc. 10th SITA, vol. JA5-3 (1987)
29. Jiang, X., Hu, L., Ding, J., Sun, S.: On the Kipnis-Shamir method solving the MinRank problem. In: Proc. IWSEC 2010 – Short Papers, pp. 1–13 (2010)
30. Joye, M., Lenstra, A.K., Quisquater, J.J.: Chinese Remaindering Based Cryptosystems in the Presence of Faults. J. Cryptology 12, 241–245 (1999)
31. Kipnis, A., Patarin, J., Goubin, L.: Unbalanced Oil and Vinegar Signature Schemes. In: Stern, J. (ed.) EUROCRYPT 1999. LNCS, vol. 1592, pp. 206–222. Springer, Heidelberg (1999)
32. Kipnis, A., Shamir, A.: Cryptanalysis of the HFE Public Key Cryptosystem by Relinearization. In: Wiener, M. (ed.) CRYPTO 1999. LNCS, vol. 1666, pp. 19–30. Springer, Heidelberg (1999)
33. Kipnis, A., Shamir, A.: Cryptanalysis of the Oil & Vinegar Signature Scheme. In: Krawczyk, H. (ed.) CRYPTO 1998. LNCS, vol. 1462, pp. 257–267. Springer, Heidelberg (1998)
34. Matsumoto, T., Imai, H.: Public Quadratic Polynomial-Tuples for Efficient Signature-Verification and Message-Encryption. In: Günther, C.G. (ed.) EUROCRYPT 1988. LNCS, vol. 330, pp. 419–453. Springer, Heidelberg (1988)
35. Moh, T.: A public key system with signature and master key functions. Communications in Algebra 27, 2207–2222 (1999)
36. Strenzke, F., Tews, E., Molter, H.G., Overbeck, R., Shoufan, A.: Side Channels in the McEliece PKC. In: Buchmann, J., Ding, J. (eds.) PQCrypto 2008. LNCS, vol. 5299, pp. 216–229. Springer, Heidelberg (2008)
37. Okeya, K., Takagi, T., Vuillaume, C.: On the Importance of Protecting Δ in SFLASH against Side Channel Attacks. IEICE Trans. 88-A, 123–131 (2005)
38. Page, D., Vercauteren, F.: A Fault Attack on Pairing-Based Cryptography. IEEE Transactions on Computers 55, 1075–1080 (2006)
39. Patarin, J.: Cryptanalysis of the Matsumoto and Imai Public Key Scheme of Eurocrypt '88. In: Coppersmith, D. (ed.) CRYPTO 1995. LNCS, vol. 963, pp. 248–261. Springer, Heidelberg (1995)
40. Patarin, J.: Hidden Fields Equations (HFE) and Isomorphisms of Polynomials (IP): Two New Families of Asymmetric Algorithms. In: Maurer, U.M. (ed.) EUROCRYPT 1996. LNCS, vol. 1070, pp. 33–48. Springer, Heidelberg (1996)
41. Patarin, J., Goubin, L., Courtois, N.T.: $C*-+$ and HM: Variations around Two Schemes of T. Matsumoto and H. Imai. In: Ohta, K., Pei, D. (eds.) ASIACRYPT 1998. LNCS, vol. 1514, pp. 35–50. Springer, Heidelberg (1998)
42. Shamir, A.: Efficient Signature Schemes Based on Birational Permutations. In: Stinson, D.R. (ed.) CRYPTO 1993. LNCS, vol. 773, pp. 1–12. Springer, Heidelberg (1994)
43. Shor, P.W.: Polynomial-Time Algorithms for Prime Factorization and Discrete Logarithms on a Quantum Computer. SIAM J. Computing 26, 1484–1509 (1997)
44. Tsujii, S., Kurosawa, K., Itoh, T., Fujioka, A., Matsumoto, T.: A public-key cryptosystem based on the difficulty of solving a system of non-linear equations. IEICE Trans. Inf. & Syst. (Japanese Edition) J69-D, 1963–1970 (1986)
45. Tsujii, S., Tadaki, K., Fujita, R.: Proposal for Piece in Hand Matrix: General Concept for Enhancing Security of Multivariate Public Key Cryptosystems. IEICE Trans. 90-A, 992–999 (2007)
46. Yang, B.-Y., Chen, J.-M.: Building Secure Tame-like Multivariate Public-key Cryptosystems: The New TTS. In: Boyd, C., González Nieto, J.M. (eds.) ACISP 2005. LNCS, vol. 3574, pp. 518–531. Springer, Heidelberg (2005)

Towards Quantum-Resistant Cryptosystems from Supersingular Elliptic Curve Isogenies

David Jao[1] and Luca De Feo[2]

[1] Department of Combinatorics and Optimization
University of Waterloo, Waterloo, Ontario, N2L 3G1, Canada
djao@math.uwaterloo.ca
[2] Laboratoire PRiSM
Université de Versailles, 78035 Versailles, France
http://www.prism.uvsq.fr/~dfl

Abstract. We present new candidates for quantum-resistant public-key cryptosystems based on the conjectured difficulty of finding isogenies between supersingular elliptic curves. The main technical idea in our scheme is that we transmit the images of torsion bases under the isogeny in order to allow the two parties to arrive at a common shared key despite the noncommutativity of the endomorphism ring. Our work is motivated by the recent development of a subexponential-time quantum algorithm for constructing isogenies between ordinary elliptic curves. In the supersingular case, by contrast, the fastest known quantum attack remains exponential, since the noncommutativity of the endomorphism ring means that the approach used in the ordinary case does not apply. We give a precise formulation of the necessary computational assumption along with a discussion of its validity. In addition, we present implementation results showing that our protocols are multiple orders of magnitude faster than previous isogeny-based cryptosystems over ordinary curves.

Keywords: elliptic curves, isogenies, quantum-resistant public-key cryptosystems.

1 Introduction

The Diffie-Hellman scheme is a fundamental protocol for public-key exchange between two parties. Its original definition over finite fields is based on the hardness of computing the map $g, g^a, g^b \mapsto g^{ab}$ for $g \in \mathbb{F}_p^*$, while its elliptic curve analogue depends on the difficulty of computing $P, aP, bP \mapsto abP$ for points P on an elliptic curve. Recently, Stolbunov [19] proposed a Diffie-Hellman type system based on the difficulty of computing isogenies between ordinary elliptic curves, with the stated aim of obtaining quantum-resistant cryptographic protocols. The fastest known (classical) probabilistic algorithm for solving this problem is the algorithm of Galbraith and Stolbunov [9], based on the algorithm of Galbraith, Hess, and Smart [8]. This algorithm is exponential, with a worst-case running time of $O(\sqrt[4]{q})$. However, on a quantum computer, recent work of

B.-Y. Yang (Ed.): PQCrypto 2011, LNCS 7071, pp. 19–34, 2011.

Childs et al. [5] has shown that the private keys in Stolbunov's system can be recovered in subexponential time. Moreover, even if we only consider classical attacks in assessing security levels, Stolbunov's scheme requires 229 seconds (even with precomputation) to perform a single key exchange operation at the 128-bit security level on a desktop PC [19, Table 1].

In this work we present isogeny-based cryptosystems that address both the performance and security drawbacks of Stolbunov's system. Our scheme achieves performance on the order of one second (cf. Section 6) at the 128-bit security level (as measured against the fastest known quantum attacks) using desktop PCs, making it far faster than Stolbunov's scheme. In terms of security, our scheme is not vulnerable to the algorithm of Childs et al. [5], nor to any algorithm of this type, since it is not based on a group action. The fastest known attacks against our scheme, even on quantum computers, require fully exponential time. Our scheme involves a new computational assumption upon which its quantum resistance is based, and like all new computational assumptions, further study and the passage of time is needed for validation. Nevertheless, we believe our proposal represents a promising candidate for quantum-resistant isogeny-based public-key cryptography.

Our proposal uses isogenies between *supersingular* elliptic curves rather than ordinary elliptic curves. The main technical difficulty is that, in the supersingular case, the endomorphism ring is noncommutative, whereas Diffie-Hellman type protocols require commutativity. We show how to overcome this obstacle by providing the outputs of the isogeny on certain points as auxiliary input to the protocol. To the best of our knowledge, nothing similar to this idea has ever previously appeared in the literature. Providing this auxiliary input does not seem to make the problem of finding isogenies any easier; see Section 5.2 for a full discussion. The multiple orders of magnitude of performance gains in our scheme arise from the fact that supersingular isogeny graphs are much faster to navigate than ordinary graphs. In Section 5.1 we provide formal statements of the hardness assumptions and security reductions for our system. Finally, in Section 6 we present implementation results confirming the correctness and performance of our protocol.

2 Isogenies

Let E_1 and E_2 be elliptic curves defined over a finite field $\mathbb{F}_q$. An isogeny $\phi : E_1 \to E_2$ defined over $\mathbb{F}_q$ is a non-constant rational map defined over $\mathbb{F}_q$ which is also a group homomorphism from $E_1(\mathbb{F}_q)$ to $E_2(\mathbb{F}_q)$ [15, III.4]. The degree of an isogeny is its degree as a rational map. For separable isogenies, to have degree ℓ means to have kernel of size ℓ. Every isogeny of degree greater than 1 can be factored into a composition of isogenies of prime degree over $\bar{\mathbb{F}}_q$ [6].

An endomorphism of an elliptic curve E defined over $\mathbb{F}_q$ is an isogeny $E \to E$ defined over $\mathbb{F}_{q^m}$ for some m. The set of endomorphisms of E together with the zero map forms a ring under the operations of pointwise addition and composition; this ring is called the endomorphism ring of E and denoted $\mathrm{End}(E)$. The ring End(E) is isomorphic either to an order in a quaternion algebra or to an order in an imaginary quadratic field [15, V.3.1]; in the first case we say E is supersingular and in the second case we say E is ordinary.

Two elliptic curves E_1 and E_2 defined over $\mathbb{F}_q$ are said to be isogenous over $\mathbb{F}_q$ if there exists an isogeny $\phi\colon E_1 \to E_2$ defined over $\mathbb{F}_q$. A theorem of Tate states that two curves E_1 and E_2 are isogenous over $\mathbb{F}_q$ if and only if $\#E_1(\mathbb{F}_q) = \#E_2(\mathbb{F}_q)$ [21, §3]. Since every isogeny has a dual isogeny [15, III.6.1], the property of being isogenous over $\mathbb{F}_q$ is an equivalence relation on the finite set of $\bar{\mathbb{F}}_q$-isomorphism classes of elliptic curves defined over $\mathbb{F}_q$. Accordingly, we define an isogeny class to be an equivalence class of elliptic curves, taken up to $\bar{\mathbb{F}}_q$-isomorphism, under this equivalence relation.

The ℓ-torsion group of E, denoted $E[\ell]$, is the set of all points $P \in E(\bar{\mathbb{F}}_q)$ such that ℓP is the identity. For ℓ such that $p \nmid \ell$, we have $E[\ell] \cong \mathbb{Z}/\ell\mathbb{Z} \oplus \mathbb{Z}/\ell\mathbb{Z}$.

Curves in the same isogeny class are either all supersingular or all ordinary. Traditionally, most elliptic curve cryptography uses ordinary curves; however, for this work we will be interested in supersingular curves. We assume for the remainder of this paper that we are in the supersingular case.

Supersingular curves are all defined over $\mathbb{F}_{p^2}$, and for every prime $\ell \nmid p$, there exist $\ell + 1$ isogenies (counting multiplicities) of degree ℓ originating from any given such supersingular curve. Given an elliptic curve E and a finite subgroup Φ of E, there is up to isomorphism a unique isogeny $E \to E'$ having kernel Φ [15, III.4.12]. Hence we can identify an isogeny by specifying its kernel, and conversely given a kernel subgroup the corresponding isogeny can be computed using Vélu's formulas [24]. Typically, this correspondence is of little use, since the kernel, or any representation thereof, is usually as unwieldy as the isogeny itself. However, in the special case of kernels generated by $\mathbb{F}_{p^2}$-rational points of smooth order, specifying a generator of the kernel allows for compact representation and efficient computation of the corresponding isogeny, as we demonstrate below.

3 Public-Key Cryptosystems Based on Supersingular Curves

In this section we present a key-exchange protocol and a public-key cryptosystem analogous to those of [14,19], using supersingular elliptic curves. Since the discrete logarithm problem is unimportant when elliptic curves are used in an isogeny-based system, we propose using supersingular curves of smooth order to improve performance. In the supersingular setting, it is easy to construct curves of smooth order, and using a smooth order curve will give a large number of isogenies that are fast to compute. Specifically, we fix $\mathbb{F}_q = \mathbb{F}_{p^2}$ as the field of definition, where p is a prime of the form $\ell_A^{e_A} \ell_B^{e_B} \cdot f \pm 1$. Here ℓ_A and ℓ_B are small primes, and f is a cofactor such that p is prime. Alice and Bob will each take a random walk on a different isogeny graph; Alice will use the graph consisting of isogenies of degrees ℓ_A, and Bob will use the graph of degree ℓ_B isogenies. The main technical modification is that, since ideal classes no longer commute (or indeed even multiply together) in the supersingular case, extra

$\mathcal{A}$		$\mathcal{B}$
Input: A, B, sID		**Input:** B
$m_A, n_A \in_R \mathbb{Z}/\ell_A^{e_A}\mathbb{Z}$		$m_B, n_B \in_R \mathbb{Z}/\ell_B^{e_B}\mathbb{Z}$
$\phi_A := E_0/\langle [m_A]P_A + [n_A]Q_A \rangle$		$\phi_B := E_0/\langle [m_B]P_B + [n_B]Q_B \rangle$
	A,sID, $\phi_A(P_B)$, $\phi_A(Q_B)$, E_A $\longrightarrow$	
	$\longleftarrow$ B,sID, $\phi_B(P_A)$, $\phi_B(Q_A)$, E_B	
$E_{AB} :=$ $E_B/\langle [m_A]\phi_B(P_A)+[n_A]\phi_B(Q_A)\rangle$		$E_{BA} :=$ $E_A/\langle [m_B]\phi_A(P_B)+[n_B]\phi_A(Q_B)\rangle$
Output: $j(E_{AB})$, sID		**Output:** $j(E_{BA})$, sID

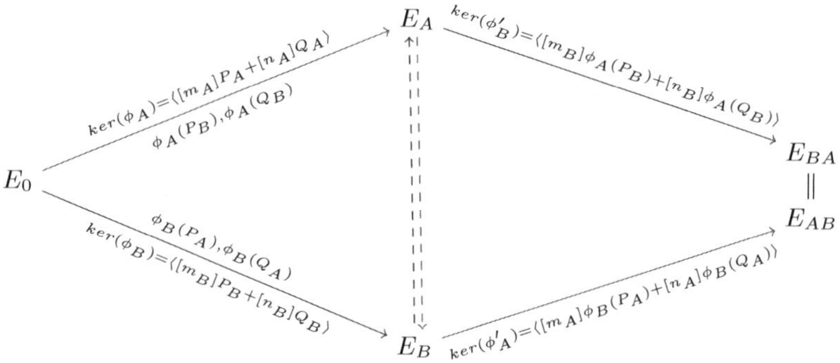

Fig. 1. Key-exchange protocol using isogenies on supersingular curves

information must be communicated as part of the protocol in order to ensure that both parties arrive at the same common value. This is in contrast to the ordinary case [19], where the existence of an abelian class group allows for the straightforward creation of a Diffie-Hellman type system.

3.1 Key Exchange

We fix as public parameters a supersingular curve E_0 defined over $\mathbb{F}_{p^2}$, and bases $\{P_A, Q_A\}$ and $\{P_B, Q_B\}$ which generate $E_0[\ell_A^{e_A}]$ and $E_0[\ell_B^{e_B}]$ respectively, so that $\langle P_A, Q_A\rangle = E_0[\ell_A^{e_A}]$ and $\langle P_B, Q_B\rangle = E_0[\ell_B^{e_B}]$. Alice chooses two random elements $m_A, n_A \in_R \mathbb{Z}/\ell_A^{e_A}\mathbb{Z}$, not both divisible by ℓ_A, and computes an isogeny $\phi_A\colon E_0 \to E_A$ with kernel $K_A := \langle [m_A]P_A + [n_A]Q_A\rangle$. Alice also computes the image $\{\phi_A(P_B), \phi_A(Q_B)\} \subset E_A$ of the basis $\{P_B, Q_B\}$ for $E_0[\ell_B^{e_B}]$ under her secret isogeny ϕ_A, and sends these points to Bob together with E_A. Similarly, Bob selects random elements $m_B, n_B \in_R \mathbb{Z}/\ell_B^{e_B}\mathbb{Z}$ and computes an isogeny $\phi_B\colon E_0 \to E_B$ having kernel $K_B := \langle [m_B]P_B + [n_B]Q_B\rangle$, along with the points $\{\phi_B(P_A), \phi_B(Q_A)\}$. Upon receipt of E_B and $\phi_B(P_A), \phi_B(Q_A) \in E_B$

from Bob, Alice computes an isogeny $\phi'_A \colon E_B \to E_{AB}$ having kernel equal to $\langle [m_A]\phi_B(P_A) + [n_A]\phi_B(Q_A)\rangle$; Bob proceeds *mutatis mutandis*. Alice and Bob can then use the common j-invariant of

$$E_{AB} = \phi'_B(\phi_A(E_0)) = \phi'_A(\phi_B(E_0)) = E_0/\langle [m_A]P_A+[n_A]Q_A, [m_B]P_B+[n_B]Q_B\rangle$$

to form a secret shared key. For specific details of how each of the above computations can be performed efficiently, we refer the reader to Section 4.

The full protocol is given in Figure 1. We denote by A and B the identifiers of Alice and Bob, and use sID to denote the unique session identifier.

3.2 Public-Key Encryption

The key-exchange protocol of Section 3.1 can easily be adapted to yield a public-key cryptosystem, in much the same way that Elgamal encryption follows from Diffie-Hellman. We briefly give the details here. All notation is the same as above. Stolbunov [19] uses a similar construction, upon which ours is based.

Setup: Choose $p = \ell_A^{e_A}\ell_B^{e_B} \cdot f \pm 1$, E_0, $\{P_A, Q_A\}$, $\{P_B, Q_B\}$ as above. Let $\mathcal{H} = \{H_k : k \in K\}$ be a hash function family indexed by a finite set K, where each H_k is a function from $\mathbb{F}_{p^2}$ to the message space $\{0,1\}^w$.

Key generation. Choose two random elements $m_A, n_A \in_R \mathbb{Z}/\ell_A^{e_A}\mathbb{Z}$, not both divisible by ℓ_A. Compute $E_A, \phi_A(P_B), \phi_A(Q_B)$ as above, and choose a random element $k \in_R K$. The public key is $(E_A, \phi_A(P_B), \phi_A(Q_B), k)$ and the private key is (m_A, n_A, k).

Encryption. Given a public key $(E_A, \phi_A(P_B), \phi_A(Q_B), k)$ and a message $m \in \{0,1\}^w$, choose two random elements $m_B, n_B \in_R \mathbb{Z}/\ell_B^{e_B}\mathbb{Z}$, not both divisible by ℓ_B, and compute

$$\begin{aligned} h &= H_k(j(E_{AB})), \\ c &= h \oplus m. \end{aligned}$$

The ciphertext is $(E_B, \phi_B(P_A), \phi_B(Q_A), c)$.

Decryption. Given a ciphertext $(E_B, \phi_B(P_A), \phi_B(Q_A), c)$ and a private key (m_A, n_A, k), compute the j-invariant $j(E_{AB})$ and set

$$\begin{aligned} h &= H_k(j(E_{AB})), \\ m &= h \oplus c. \end{aligned}$$

The plaintext is m.

4 Algorithmic Aspects

We now give specific algorithms to implement the abovementioned steps efficiently.

4.1 Parameter Generation

For any fixed choice of $\ell_A^{e_A}$ and $\ell_B^{e_B}$, one can easily test random values of f (of any desired cryptographic size) until a value is found for which $p = \ell_A^{e_A}\ell_B^{e_B} \cdot f - 1$ or $p = \ell_A^{e_A}\ell_B^{e_B} \cdot f + 1$ is prime; the prime number theorem in arithmetic progressions (specifically, the effective version of Lagarias and Odlyzko [11]) provides a sufficient lower bound for the density of such primes.

Once the prime $p = \ell_A^{e_A}\ell_B^{e_B} \cdot f \pm 1$ is known, Bröker [2] has shown that it is easy to find a supersingular curve E over $\mathbb{F}_{p^2}$ having cardinality $(p \mp 1)^2 = (\ell_A^{e_A}\ell_B^{e_B} \cdot f)^2$. Starting from E, one can select a random supersingular curve E_0 over $\mathbb{F}_{p^2}$ by means of random walks on the isogeny graph; alternatively, one can simply take $E_0 = E$. In either case, E_0 has group structure $(\mathbb{Z}/(p \mp 1)\mathbb{Z})^2$. To find a basis for $E_0[\ell_A^{e_A}]$, choose a random point $P \in_R E_0(\mathbb{F}_{p^2})$ and multiply it by $(\ell_B^{e_B} \cdot f)^2$ to obtain a point P' of order dividing $\ell_A^{e_A}$. With high probability, P' will have order exactly $\ell_A^{e_A}$; one can of course check this by multiplying P' by powers of ℓ_A. If the check succeeds, then set $P_A = P'$; otherwise try again with another P. A second point Q_A of order $\ell_A^{e_A}$ can be obtained in the same way. To check whether Q_A is independent of P_A, simply compute the Weil pairing $e(P_A, Q_A)$ in $E[\ell_A^{e_A}]$ and check that the result has order $\ell_A^{e_A}$; as before, this happens with high probability, and if not, just choose another point Q_A. Note that the choice of basis has no effect on the security of the scheme, since one can convert from one basis to another using extended discrete logarithms, which are easy to compute in $E_0[\ell_A^{e_A}]$ by [22].

4.2 Key Exchange

It remains to describe how Alice and Bob can compute isogenies of a given kernel. We show how to compute $\phi_A \colon E_0 \to E_A$ where $E_A = E_0/\langle [m_A]P_A + [n_A]Q_A\rangle$; the same procedure suffices to compute all the other isogenies mentioned. The computation is performed using a version of Hensel lifting modulo ℓ_A. Let $R_0 := [m_A]P_A + [n_A]Q_A$. The order of R_0 is $\ell_A^{e_A}$. For $0 \le i < e_A$, let

$$E_{i+1} = E_i/\langle \ell_A^{e_A-i-1}R_i\rangle, \qquad \phi_i : E_i \to E_{i+1}, \qquad R_{i+1} = \phi_i(R_i),$$

where ϕ_i is a degree ℓ_A isogeny from E_i to E_{i+1}. Then $E_A = E_{e_A}$ and $\phi_A = \phi_{e_A-1} \circ \cdots \circ \phi_0$.

Figure 2 gives two algorithms for this task. They both compute iteratively $(R_i, \ell_A^{e_A-i-1}R_i, \phi_i, E_{i+1})$ for $i < e_A$, but they differ in the strategy. The first one, which we will refer to as *multiplication-oriented*, computes at each iteration $\ell_A^{e_A-i-1}R_i$ from R_i using point addition (or duplication, or triplication). The second one, which we call *isogeny-oriented*, first forms the list $(\ell_A^j R_0)_{j<e_A}$ using point addition, then at each iteration computes the list $(\ell_A^j R_{i+1})_{j<e_A-i-1}$ by evaluating $\phi_i(\ell_A^j R_i)$ for each j. Observe that Alice and Bob can use one algorithm or the other independently.

A quick analysis shows that both algorithms require $O(\log^2 p)$ operations in $\mathbb{F}_p$. The major cost in the multiplication-based one is scalar point multiplication; this costs $O(e_A \log_2 \ell_A)$ double-and-adds at each iteration and is repeated

Multiplication based

Input: E_0, R_0
1: **for** $0 \le i < e_A$ **do**
2: $P_i \leftarrow \ell_A^{e_a-i-1} R_0$;
3: Compute $\phi_i : E_i \to E_i/\langle P_i \rangle$;
4: $E_{i+1} \leftarrow E_i/\langle P_i \rangle$;
5: $R_{i+1} \leftarrow \phi_i(R_i)$;
6: **end for**
Output: E_{e_A}

Isogeny based

Input: E_0, R_0
1: $Q_0 \leftarrow R_0$;
2: **for** $0 \le j < e_A - 1$ **do**
3: $Q_{j+1} \leftarrow \ell_A Q_j$;
4: **end for**
5: **for** $0 \le i < e_A$ **do**
6: Compute $\phi_i : E_i \to E_i/\langle Q_{e_A-1} \rangle$;
7: $E_{i+1} \leftarrow E_i/\langle Q_{e_A-1} \rangle$;
8: **for** $i \le j < e_A - 1$ **do**
9: $Q_{j+1} \leftarrow \phi_i(Q_j)$;
10: **end for**
11: **end for**
Output: E_{e_A}

Fig. 2. Key exchange algorithms

$e_A \sim \log_{\ell_A} \sqrt{p}$ times. The major cost in the isogeny-based algorithm is the isogeny evaluation at step 8; each evaluation costs $O(\ell_A)$ operations and there are $\frac{1}{2}e_A(e_A - 1)$ of them. By forming the ratio of these quantities, we obtain $O(\log_2 \ell_A/\ell_A)$, so we see that the multiplication-based algorithm is preferable as ℓ_A grows—but we cannot grow ℓ_A indefinitely, because eventually Step 3 becomes the dominant cost. Our implementation, described in Section 6, supports the isogeny-oriented approach for $\ell_A = 2, 3$ and the multiplication-oriented approach for $\ell_A > 2$.

4.3 Isogenies of Montgomery Curves

Independently of which method is chosen, it is important to use pick models for elliptic curves that offer the fastest isogeny evaluation performance. The literature on efficient formulas for evaluating small degree isogenies is much less extensive than for point multiplication. In this section we provide explicit and efficient formulas for evaluating isogenies using curves in Montgomery form.

Each of our curves has group structure $(\mathbb{Z}/(p \mp 1)\mathbb{Z})^2$ and its twist has group structure $(\mathbb{Z}/(p \pm 1)\mathbb{Z})^2$. Hence either the curve or its twist has a point of order 4. Consequently, we can write our curves in Montgomery form as follows:

$$E \,:\, B^2y^2 = x^3 + Ax^2 + x \qquad (1)$$

Montgomery curves have very efficient arithmetic on their Kummer line (i.e. by representing points by the coordinates $(X : Z)$ where $x = X/Z$) [12]. The Kummer line identifies P with $-P$, and thus it is not possible to add two distinct points; however it is still possible to compute any scalar multiple of a point. Also observe that since P and $-P$ generate the same subgroup, isogenies can be defined and evaluated correctly on the Kummer line. The goal of this section is to give explicit and efficient formulas for such isogenies.

Let E be the Montgomery curve defined by Eq. (1). It has a point of order two in point $P_2 = (0,0)$, and a point of order four in $P_4 = (1, \sqrt{(A+2)/B})$—eventually defined over a quadratic extension—such that $[2]P_4 = P_2$. Montgomery curves have twists of the form $\tilde{y} = \sqrt{c}y$; these are isomorphisms when c is a square. The change of coordinates $\tilde{x} = x/B, \tilde{y} = y/B$ brings the curve E to the Weierstrass form

$$\tilde{E} : \tilde{y}^2 = \tilde{x}^3 + \frac{A}{B}\tilde{x}^2 + \frac{1}{B^2}\tilde{x},$$

and the point P_4 to $P_4' = (1/B, \ldots)$. Inversely, given a Weierstrass curve $\tilde{E}$ with equation $\tilde{y}^2 = \tilde{x}^3 + a\tilde{x}^2 + b\tilde{x}$, together with a point $P_4 = (1/\beta, \ldots)$—with its ordinate possibly lying in a quadratic extension—such that $[2]P_4 = (0,0)$, the change of variables $\tilde{x} = x/\beta, \tilde{y} = y/\beta$ brings $\tilde{E}$ to the Montgomery form $\beta y^2 = x^3 + a\beta x^2 + x$.

Let G be a subgroup of the Montgomery curve E of odd cardinality ℓ and let h be the degree $(\ell-1)/2$ polynomial vanishing on the abscissas of G. With a twist $y = \tilde{y}/\sqrt{B}$, we can put E in the form $\tilde{y}^2 = \tilde{x}^3 + A\tilde{x}^2 + \tilde{x}$, and this doesn't change the abscissas of G or the polynomial h. Now we can use Vélu's formulas to compute an isogeny having G for kernel: this gives an isogeny ϕ and a curve F such that

$$\begin{aligned} F &: y^2 = x^3 + a_2x^2 + a_4x + a_6, \\ \phi &: E \to F, \\ (x,y) &\mapsto \left(\frac{g(x)}{h(x)^2}, y\sqrt{B}\left(\frac{g(x)}{h(x)^2}\right)'\right). \end{aligned}$$

Because ℓ is odd, the point $(0,0)$ of E is sent to a point of order two in F. A closer look at Vélu's formulas (see Eq. (3) below) shows that $\phi(0,0) = (p_{-1} - p_1, 0)$, where p_1 is the sum of the abscissas of G and p_{-1} is the sum of their inverses. By the change of variables $\tilde{x} = x - p_{-1} + p_1$, we bring F to the form $\tilde{F} : \tilde{y}^2 = \tilde{x}^3 + a\tilde{x}^2 + b\tilde{x}$. Now $\phi(P_4)$ is a point of order four (possibly in a quadratic extension). Its abscissa in $\tilde{F}$ is rational and is given by $1/\beta = g(1)/h(1) - p_{-1} + p_1$, so we apply the change of variables $\tilde{x} = \bar{x}/\beta, \tilde{y} = \bar{x}/\beta$ to obtain a Montgomery curve. Finally, we have to twist back the image curve to obtain a curve isogenous over the base field: the twist $\bar{y} = y\sqrt{B}$ cancels with the first one and leaves us with square-root-free formulas. The image curve is

$$B\beta y^2 = x^3 + a\beta x^2 + x. \tag{2}$$

To efficiently evaluate these isogenies (either on the full curve or on the Kummer line) we use [1, Proposition 4.1], which says:

$$\frac{g}{h} = \ell x + p_1 - 2(3x^2 + 2Ax + 1)\frac{h'}{h} - 4(x^3 + Ax^2 + x)\left(\frac{h'}{h}\right)'. \tag{3}$$

To evaluate at a point (x_0, y_0), we compute $h(x_0), h'(x_0), h''(x_0), h'''(x_0)$; applying Horner's rule, this costs $\sim 2\ell$ multiplications using affine coordinates, or $\sim 3\ell$ using projective coordinates. Then we inject these values in Eq. (3) and in its derivative to evaluate the isogeny, this only takes a constant number of multiplications (plus one inversion in affine coordinates). Finally, the image of (x_0, y_0) is given by $(\beta(g(x_0)/h(x_0) - p_{-1} + p_1), \beta y_0 (g/h)'(x_0))$. Note that if the y-coordinate is not needed[1], we can avoid computing $h'''(x_0)$, thus saving $\sim \ell/2$ multiplications. Of course, for specific small ℓ, such as $\ell = 3, 5$, it is more convenient to write down the isogeny explicitly in terms of the kernel points and find optimized formulas.

When $\ell = 2$, things are more complex, but in our specific case we can easily deal with it. The isogeny of E vanishing $(0,0)$ is readily seen as being

$$F \,:\, y^2 = x^3 + \frac{A+6}{B}x^2 + 4\frac{A+2}{B^2}x, \tag{4}$$

$$\phi : E \to F,$$

$$(x,y) \mapsto \left(\frac{1}{B}\frac{(x-1)^2}{x}, \frac{1}{B}\left(y - \frac{y}{x^2}\right)\right). \tag{5}$$

If a point P_8 satisfying $[4]P_8 = (0,0)$ is known, then $\sqrt{A+2}$ can be computed from the abscissa of $\phi(P_8)$, and F can be put in Montgomery form as before. The isogeny vanishing on a generic point of order two $P_2 \neq (0,0)$ can be easily computed when a point P_4 satisfying $[2]P_4 = P_2$ is known: change coordinates to bring P_2 in $(0,0)$, then use the abscissa of P_4 to express the resulting curve in Montgomery form (this is the same technique as above, taking $\ell = 1$); notice that this step needs to be done at most once per key exchange. When points of order 8 or 4 are not available, as in the last few steps of our setting, ordinary Weierstrass forms yield formulas that require a few extra multiplications.

We conclude this section with operation counts for the key exchange algorithms. We write I, M, S for the costs of one inversion, multiplication and squaring in $\mathbb{F}_{p^2}$ respectively, and we make the assumption $S \leq M$; we count multiplication by constants as normal multiplications. For simplicity, we only list quadratic terms in e_A.

Multiplication-based. If P is a point on the Kummer line, computing P times an n-bit integer costs $(7M + 4S)\log_2 n$ (see [12]). Thus the cumulative cost of Step 2 is

$$\sum_{i=1}^{e_A - 1} (7M + 4S)\log_2 \ell_A^i \sim \frac{1}{2}(7M + 4S)(\log_2 \sqrt{p})^2 \log_{\ell_A} 2.$$

Doubling a point on the Kummer line only costs $3M + 2S$, and thus the cost for $\ell_A = 2$ drops down to $\frac{1}{2}(3M + 2S)(\log_2 \sqrt{p})^2$.

Isogeny-based. The only quadratic term in e_A appears at Step 8. Since we do not need the y coordinate of the points involved in this step, we only need the

[1] While x coordinates are enough to compute Vélu's isogenies and the image curve, this forces the other party to use y-coordinate-free formulas for point multiplication.

Table 1. Comparative costs for the multiplication and isogeny based algorithms using projective coordinates. The entries must be multiplied by $\frac{1}{2}(\log_2 \sqrt{p})^2$ to obtain the full cost.

ℓ_A	2	3	5	11	19
$\log_{\ell_A} 2$	1	0.63	0.43	0.29	0.23
Isogeny	$2M + S$	$4.0M + 0.8S$	$1.7M + 0.5S$	$2.0M + 0.2S$	$2.4M + 0.2S$
Multiplication	$3M + 2S$	$4.4M + 2.5S$	$3.0M + 1.7S$	$2.0M + 1.1S$	$1.6M + 0.9S$

values $h(x_0), h'(x_0), h''(x_0)$ in order to apply Eq. (3). We let $s = (\ell_A - 1)/2$ be the degree of h. In affine coordinates, since h is monic, Horner's rule requires $(3s - 6)M$, except when $s = 1, 2$. Then, to compute $\beta(g(x_0)/h(x_0) - p_{-1} + p_1)$ we need $I + 8M + 2S$. For $\ell_A = 3$ the total count drops to $I + 6M + 2S$, and for $\ell_A = 5$ it is $I + 8M + 2S$.

In projective coordinates, we first compute $Z, \ldots, Z^s$ at a cost of $(s - 1)M$. Then, if $h = \sum_i h_i X^{s-i} Z^i$, we compute the monomials $h_i Z^i$ using sM. Finally we compute h, h', h'' using three applications of Horner's rule, costing again $(3s - 6)M$ when $s \neq 1, 2$. The final computation requires $11M + 3S$. For $\ell_A = 3$ the total count is $10M + 2S$, and for $\ell_A = 5$ it is $14M + 3S$.

The difference between the affine and the projective formulas is $I - 2(s - 1)M - S$, so the choice between the two must be done according to the ratio I/M.

Finally for $\ell_A = 2$, assuming a point of order 8 on the domain curve is known (which will always be the case, except in the last two iterations), evaluating the x part of Eq. 4 in projective coordinates and bringing the result back to a Montgomery curve costs $2M + S$.

There are $e_A(e_A - 1)$ isogeny evaluations in the algorithm, so, assuming that N is the cost of doing one evaluation, the total cost is about $\frac{1}{2}e_A^2 N = \frac{1}{2}N(\log_2 \sqrt{p})^2(\log_{\ell_A} 2)^2$. We summarize the main costs of the two algorithms in Table 1.

5 Security

5.1 Complexity Assumptions and Security Proofs

As before, let p be a prime of the form $\ell_A^{e_A} \ell_B^{e_B} \cdot f \pm 1$, and fix a supersingular curve E_0 over $\mathbb{F}_{p^2}$ together with bases $\{P_A, Q_A\}$ and $\{P_B, Q_B\}$ of $E_0[\ell_A^{e_A}]$ and $E_0[\ell_B^{e_B}]$ respectively. In analogy with the case of isogenies over ordinary elliptic curves, we define the following computational problems, adapted for the supersingular case:

Problem 5.1 (Supersingular Isogeny (SSI) problem). Let $\phi_A \colon E_0 \to E_A$ be an isogeny whose kernel is $\langle [m_A]P_A + [n_A]Q_A \rangle$, where m_A and n_A are chosen at random from $\mathbb{Z}/\ell_A^{e_A}\mathbb{Z}$ and not both divisible by ℓ_A. Given E_A and the values $\phi_A(P_B)$, $\phi_A(Q_B)$, find a generator R_A of $\langle [m_A]P_A + [n_A]Q_A \rangle$.

We remark that given a generator $R_A = [m_A]P_A + [n_A]Q_A$, it is easy to solve for (m_A, n_A), since E_0 has smooth order and thus extended discrete logarithms are easy in E_0 [22].

Problem 5.2 (Supersingular Computational Diffie-Hellman (SSCDH) problem). Let $\phi_A\colon E_0 \to E_A$ be an isogeny whose kernel is $\langle [m_A]P_A + [n_A]Q_A \rangle$, and let $\phi_B\colon E_0 \to E_B$ be an isogeny whose kernel is $\langle [m_B]P_B + [n_B]Q_B \rangle$, where m_A, n_A (respectively m_B, n_B) are chosen at random from $\mathbb{Z}/\ell_A^{e_A}\mathbb{Z}$ (respectively $\mathbb{Z}/\ell_B^{e_B}\mathbb{Z}$) and not both divisible by ℓ_A (respectively ℓ_B). Given the curves E_A, E_B and the points $\phi_A(P_B)$, $\phi_A(Q_B)$, $\phi_B(P_A)$, $\phi_B(Q_A)$, find the j-invariant of $E_0/\langle [m_A]P_A + [n_A]Q_A, [m_B]P_B + [n_B]Q_B \rangle$.

Problem 5.3 (Supersingular Decision Diffie-Hellman (SSDDH) problem). Given a tuple sampled with probability 1/2 from one of the following two distributions:

- $(E_A, E_B, \phi_A(P_B), \phi_A(Q_B), \phi_B(P_A), \phi_B(Q_A), E_{AB})$, where E_A, E_B, $\phi_A(P_B)$, $\phi_A(Q_B)$, $\phi_B(P_A)$, $\phi_B(Q_A)$ are as in the SSCDH problem and
$$E_{AB} \cong E_0/\langle [m_A]P_A + [n_A]Q_A, [m_B]P_B + [n_B]Q_B \rangle,$$
- $(E_A, E_B, \phi_A(P_B), \phi_A(Q_B), \phi_B(P_A), \phi_B(Q_A), E_C)$, where E_A, E_B, $\phi_A(P_B)$, $\phi_A(Q_B)$, $\phi_B(P_A)$, $\phi_B(Q_A)$ are as in the SSCDH problem and
$$E_C \cong E_0/\langle [m'_A]P_A + [n'_A]Q_A, [m'_B]P_B + [n'_B]Q_B \rangle,$$
where m'_A, n'_A (respectively m'_B, n'_B) are chosen at random from $\mathbb{Z}/\ell_A^{e_A}\mathbb{Z}$ (respectively $\mathbb{Z}/\ell_B^{e_B}\mathbb{Z}$) and not both divisible by ℓ_A (respectively ℓ_B),

determine from which distribution the triple is sampled.

We conjecture that these problems are computationally infeasible, in the sense that for any polynomial-time solver algorithm, the advantage of the algorithm is a negligible function of the security parameter $\log p$. The resulting security assumptions are referred to as the SSI, SSCDH, and SSDDH assumptions, respectively. Using the methods of Stolbunov [18], it is a routine exercise to prove that the protocols of Section 3 are secure under SSDDH:

Theorem 5.4. *Under the SSDDH assumption, the key-agreement protocol of Section 3.1 is session-key secure in the authenticated-links adversarial model of Canetti and Krawczyk [3].*

Theorem 5.5. *If the SSDDH assumption holds, and the hash function family $\mathcal{H}$ is entropy-smoothing, then the public-key cryptosystem of Section 3.2 is IND-CPA.*

Remark 5.6. As in the ordinary case [14,19], our protocols do not provide authentication. One possible workaround for the time being is to use classical public-key authentication schemes in conjunction with the standard observation [16, §6.2] that the authentication step only needs to be secure against the adversary at the time of the initial connection.

5.2 Hardness of the Underlying Assumptions

Given an SSI (respectively, SSCDH) solver, it is trivial to solve SSCDH (respectively, SSDDH). There is of course no known reduction in the other direction, and given that the corresponding question of equivalence for discrete logarithms and Diffie-Hellman has not yet been completely resolved in all cases, it is reasonable to assume that the question of equivalence of SSI, SSCDH, and SSDDH is at least hard to resolve. For the purposes of this discussion, we will presume that SSI is equivalent to SSDDH.

In the context of cryptography, the problem of computing an isogeny between isogenous supersingular curves was first considered by Galbraith [7] in 1999. The first published cryptographic primitive based on supersingular isogeny graphs is the hash function proposal of Charles et al. [4], which remains unbroken to date (the cryptanalysis of [13] applies only to the LPS graph-based hash function from [4], and not to the supersingular isogeny graph-based hash functions). The fastest known algorithm for finding isogenies between supersingular curves in general takes $O(\sqrt{p}\log^2 p)$ time [4, §5.3.1]; however our problem is less general because the degree of the isogeny is known in advance and is smooth. Since we are the first to propose using isogenies of this type, there is no existing literature addressing the security of the isogenies of the special form that we propose.

There is an easy exponential attack against our cryptosystem that improves upon exhaustive search. To find an isogeny of degree $\ell_A^{e_A}$ between E and E_A, an attacker builds two trees of all curves isogenous to E (respectively, E_A) via isogenies of degree $\ell_A^{e_A/2}$. Once the trees are built, the attacker tries to find a curve lying in both trees. Since the degree of the isogeny ϕ_A is $\sim \sqrt{p}$ (much shorter than the size of the isogeny graph), it is unlikely that there will be more than one isogeny path—and thus more than one match—from E to E_A. Given two functions $f : A \to C$ and $g : B \to C$ with domain of equal size, finding a pair (a, b) such that $f(a) = g(b)$ is known as the *claw problem* in complexity theory. The claw problem can obviously be solved in $O(|A|+|B|)$ time and $O(|A|)$ space on a classical computer by building a hash table holding $f(a)$ for any $a \in A$ and looking for hits for $g(b)$ where $b \in B$. This gives a $O(\ell_A^{e_A/2}) = O(\sqrt[4]{p})$ classical attack against our cryptosystem. With a quantum computer, one can do better using the algorithm in [20], which has complexity $O(\sqrt[3]{|A||B|})$, thus giving an $O(\ell_A^{e_A/3}) = O(\sqrt[6]{p})$ quantum attack against our cryptosystem. These complexities are optimal for a black-box claw attack [25].

We consider the question of whether the auxiliary data points $\phi_A(P_B)$ and $\phi_A(Q_B)$ might assist an adversary in determining ϕ_A. Since (P_B, Q_B) forms a basis for $E_0[\ell_B^{e_B}]$, the values $\phi_A(P_B)$ and $\phi_A(Q_B)$ allow the adversary to compute ϕ_A on all of $E_0[\ell_B^{e_B}]$. This is because any element of $E_0[\ell_B^{e_B}]$ is a (known) linear combination of P_B and Q_B (known since extended discrete logarithms are easy [22]). However, there does not appear to be any way to use this capability to determine ϕ_A. Even on a quantum computer, where finding abelian hidden subgroups is easy, there is no hidden subgroup to find, since ϕ_A has degree $\ell_A^{e_A}$, and thus does not annihilate any point in $E_0[\ell_B^{e_B}]$ other than the identity. Of course, if one could evaluate ϕ_A on arbitrary points of $E_0[\ell_A^{e_A}]$, then a quantum computer could easily break the scheme, and indeed in this case the scheme is also easily broken classically by using a few calls to the oracle to compute a generator of the kernel of the dual isogeny $\hat{\phi}_A$. However, it does not seem possible to translate the values of ϕ_A on $E_0[\ell_B^{e_B}]$ into values on $E_0[\ell_A^{e_A}]$.

Finally, we discuss the possibility of adapting the quantum algorithm of Childs et al. [5] for the ordinary case to the supersingular case. For both ordinary and supersingular curves, there is a natural bijection between isogenies (up to isomorphism) and (left) ideal classes in the endomorphism ring. The algorithm of Childs et al. depends crucially on the fact that the ideal classes in the ordinary case form an abelian group. In the supersingular case, the endomorphism ring is a maximal order in a noncommutative quaternion algebra, and the left ideal classes do not form a group at all (multiplication is not well defined). Thus we believe that no reasonable variant of this strategy would apply to supersingular curves.

6 Implementation Results and Example

We have implemented our cryptosystem in the computer algebra system Sage [17] using a mixed C/Cython/Sage architecture. This allows us to access the large palette of number theoretic algorithms distributed with Sage, while still permitting very efficient code in C/Cython for the critical parts such as the algorithms of Section 4.2. The source code can be downloaded from the second author's web page.

Arithmetic in $\mathbb{F}_{p^2}$ is written in C. We use the library GMP for arithmetic modulo p. The field $\mathbb{F}_{p^2}$ is implemented as $\mathbb{F}_{p^2}[X]/(X^2+1)$ (this requires $p = 3 \bmod 4$); using this representation, one multiplication in $\mathbb{F}_{p^2}$ requires three multiplications ($3M$) in $\mathbb{F}_p$, one $\mathbb{F}_{p^2}$ squaring requires two multiplications ($2M$) in $\mathbb{F}_p$, and one $\mathbb{F}_{p^2}$ inversion requires one inversion, two squarings, and two multiplications ($I + 2S + 2M$) in $\mathbb{F}_p$. Our experiments show that, for the sizes we are interested in, $I = 10M$ and $S = 0.8M$.

We implemented the isogeny-based key exchange algorithm for $\ell = 2, 3$ and the multiplication-based algorithm for $\ell > 2$. The main loop is implemented in Cython, while the isogeny evaluation and the Montgomery ladder formulas are written in C.

Finally, the parameter generation is implemented in plain Sage. Because Sage is a collection of many open source mathematical systems, various of its subsystems are involved in this last part. Of these, Pari [23] plays an important role because it is used to compute Hilbert class polynomials and to factor polynomials over finite fields.

All tests ran on a 2.4 GHz Opteron running in 64-bit mode. The results are summarized in Table 2. At the quantum 128-bit security level (768-bit p), our numbers improve upon Stolbunov's reported performance figures [19, Table 1] by over two orders of magnitude (.758 seconds vs. 229 seconds). This is the highest security level appearing in [19, Table 1], so comparisons at higher levels are difficult. Nevertheless, it seems safe to assume that the improvement is even greater at the 256-bit security level. Our results demonstrate that the proposed scheme is practical.

Table 2. Benchmarks for various group sizes and structures

	Alice		Bob	
	round 1	round 2	round 1	round 2
$2^{253}3^{161}7-1$	365 ms	363 ms	318 ms	314 ms
$5^{110}7^{91}284-1$	419 ms	374 ms	369 ms	326 ms
$11^{74}13^{69}384-1$	332 ms	283 ms	321 ms	272 ms
$17^{62}19^{60}210+1$	330 ms	274 ms	331 ms	276 ms
$23^{56}29^{52}286+1$	339 ms	274 ms	347 ms	277 ms
$31^{51}41^{47}564-1$	355 ms	279 ms	381 ms	294 ms
$2^{384}3^{242}8-1$	1160 ms	1160 ms	986 ms	973 ms
$5^{165}7^{137}2968-1$	1050 ms	972 ms	916 ms	843 ms
$11^{111}13^{104}78+1$	790 ms	710 ms	771 ms	688 ms
$17^{94}19^{90}116-1$	761 ms	673 ms	750 ms	661 ms
$23^{85}29^{79}132-1$	755 ms	652 ms	758 ms	647 ms
$31^{77}41^{72}166+1$	772 ms	643 ms	824 ms	682 ms
$2^{512}3^{323}799-1$	2570 ms	2550 ms	2170 ms	2150 ms
$23^{113}29^{105}1004-1$	1480 ms	1330 ms	1470 ms	1300 ms

6.1 Example

As a convenience, we provide an example calculation of a key-exchange transaction. Let $\ell_A = 2$, $\ell_B = 3$, $e_A = 63$, $e_B = 41$, and $f = 11$. We use the starting curve $E_0 : y^2 = x^3 + x$. For the torsion bases, we use

$$P_A = (2374093068336250774107936421407893885897i + 2524646701852396349308425328218203569693,$$
$$1944869260414574206229153243510104781725i + 1309099413211767078055232768460483417201)$$
$$P_B = (1556716033657530876728525059284431761206i + 1747407329595165241335131647929866065215,$$
$$3456956202852028835529419995475915388483i + 1975912874247458572654720717155755005566)$$

and $Q_A = \psi(P_A), Q_B = \psi(P_B)$, where $i = \sqrt{-1}$ in $\mathbb{F}_{p^2}$ and $\psi(x, y) = (-x, iy)$ is a distortion map [10]. The secret values are

$$m_A = 2575042839726612324, \quad n_A = 8801426132580632841,$$
$$m_B = 4558164392438856871, \quad n_B = 20473135767366569910$$

The isogeny $\phi_A \colon E_0 \to E_A$ is specified by its kernel, and thus the curve E_A is only well defined up to isomorphism; its exact value may vary depending on the implementation. In our case, the curve is $E_A : y^2 = x^3 + ax + b$ where

$$a = 428128245356224894562061821180718114127i + 2147708009907711790134986624604674525769$$
$$b = 3230359267202197460304331835170424053093i + 1577264336482370197045362359104894884862$$

and the values of $\phi_A(P_B)$ and $\phi_A(Q_B)$ are

$$\phi_A(P_B) = (1216243037955078292900974859441066026976i + 1666291136804738684832637187674330905572,$$
$$3132921609453998361853372941893500107923i + 2823164938573549485619800034616855236 6)$$
$$\phi_A(Q_B) = (2039728694420930519155732965018291910660i + 2422092614322988112492931615528155727388,$$
$$1688115812694355145549889238510457034272i + 1379185984608240638912948890349738467536)$$

Similarly, in our implementation $E_B : y^2 = x^3 + ax + b$ is the curve with

$$a = 2574722398094022968578313861884608943122i + 4645075571495590621841741325716474277 22$$
$$b = 2863478907513088792144998311229772886197i + 1767078036714109405796777065089868386753$$

The common j-invariant of $E_{AB} \cong E_{BA}$, computed by both Alice and Bob, is equal to

$$j(E_{AB}) = 14371454943626551191684828087021114413744i + 8334980967783864529517222853105920563 51.$$

7 Conclusion

We propose a new family of conjecturally quantum-resistant cryptographic protocols for key exchange and public-key cryptosystems using isogenies between supersingular elliptic curves of smooth order. In order to compensate for the noncommutative endomorphism rings that arise in this setting, we introduce the idea of providing the images of torsion bases as part of the protocol. Against the fastest known attacks, the resulting scheme improves upon all previous isogeny-based schemes by orders of magnitude in performance at conventional security levels, making it the first practical isogeny-based public-key cryptosystem. Unlike prior such schemes, our proposal admits no known subexponential-time attacks even in the quantum setting.

Acknowledgements. We thank Andrew M. Childs, Alfred Menezes, Vladimir Soukharev, and the anonymous reviewers for helpful comments and suggestions. This work is supported in part by NSERC CRD Grant CRDPJ 405857-10.

References

1. Bostan, A., Morain, F., Salvy, B., Schost, É.: Fast algorithms for computing isogenies between elliptic curves. Math. Comp. 77(263), 1755–1778 (2008)
2. Bröker, R.: Constructing supersingular elliptic curves. J. Comb. Number Theory 1(3), 269–273 (2009)
3. Canetti, R., Krawczyk, H.: Analysis of Key-Exchange Protocols and their Use for Building Secure Channels. In: Pfitzmann, B. (ed.) EUROCRYPT 2001. LNCS, vol. 2045, pp. 453–474. Springer, Heidelberg (2001)
4. Charles, D., Lauter, K., Goren, E.: Cryptographic hash functions from expander graphs. Journal of Cryptology 22, 93–113 (2009)
5. Childs, A., Jao, D., Soukharev, V.: Constructing elliptic curve isogenies in quantum subexponential time (2010), `http://arxiv.org/abs/1012.4019/`
6. Couveignes, J.: Hard homogeneous spaces (2006), `http://eprint.iacr.org/2006/291/`
7. Galbraith, S.: Constructing isogenies between elliptic curves over finite fields. LMS J. Comput. Math. 2, 118–138 (1999)
8. Galbraith, S.D., Hess, F., Smart, N.P.: Extending the GHS Weil Descent Attack. In: Knudsen, L.R. (ed.) EUROCRYPT 2002. LNCS, vol. 2332, pp. 29–44. Springer, Heidelberg (2002)
9. Galbraith, S., Stolbunov, A.: Improved algorithm for the isogeny problem for ordinary elliptic curves (2011), `http://arxiv.org/abs/1105.6331/`
10. Joux, A.: The Weil and Tate Pairings as Building Blocks for Public Key Cryptosystems. In: Fieker, C., Kohel, D.R. (eds.) ANTS 2002. LNCS, vol. 2369, pp. 20–32. Springer, Heidelberg (2002)

11. Lagarias, J., Odlyzko, A.: Effective versions of the Chebotarev density theorem. In: Proc. Sympos. on Algebraic Number Fields: L-functions and Galois Properties, Univ. Durham, Durham, 1975, pp. 409–464. Academic Press, London (1977)
12. Montgomery, P.: Speeding the Pollard and elliptic curve methods of factorization. Mathematics of Computation 48(177), 243–264 (1987)
13. Petit, C., Lauter, K., Quisquater, J.-J.: Full Cryptanalysis of LPS and Morgenstern Hash Functions. In: Ostrovsky, R., De Prisco, R., Visconti, I. (eds.) SCN 2008. LNCS, vol. 5229, pp. 263–277. Springer, Heidelberg (2008)
14. Rostovtsev, A., Stolbunov, A.: Public-key cryptosystem based on isogenies (2006), `http://eprint.iacr.org/2006/145/`
15. Silverman, J.: The arithmetic of elliptic curves. Graduate Texts in Mathematics, vol. 106. Springer, New York (1992); Corrected reprint of the 1986 original
16. Stebila, D., Mosca, M., Lütkenhaus, N.: The Case for Quantum Key Distribution. In: Sergienko, A., Pascazio, S., Villoresi, P. (eds.) QuantumComm 2009. LNICS, vol. 36, pp. 283–296. Springer, Heidelberg (2010)
17. Stein, W., et al.: Sage Mathematics Software (Version 4.6.2). The Sage Development Team (2011), `http://www.sagemath.org`
18. Stolbunov, A.: Reductionist security arguments for public-key cryptographic schemes based on group action. In: Mjølsnes, S.F. (ed.) Norsk informasjonssikkerhetskonferanse (NISK), pp. 97–109 (2009)
19. Stolbunov, A.: Constructing public-key cryptographic schemes based on class group action on a set of isogenous elliptic curves. Adv. Math. Commun. 4(2), 215–235 (2010)
20. Tani, S.: Claw Finding Algorithms Using Quantum Walk. arXiv:0708.2584 (March 2008)
21. Tate, J.: Endomorphisms of abelian varieties over finite fields. Invent. Math. 2, 134–144 (1966)
22. Teske, E.: The Pohlig-Hellman method generalized for group structure computation. Journal of Symbolic Computation 27(6), 521–534 (1999)
23. The PARI Group, Bordeaux. PARI/GP, version 2.4.3 (2008) `http://pari.math.u-bordeaux.fr/`
24. Vélu, J.: Isogénies entre courbes elliptiques. C. R. Acad. Sci. Paris Sér. A-B 273, A238–A241 (1971)
25. Zhang, S.: Promised and Distributed Quantum Search. In: Wang, L. (ed.) COCOON 2005. LNCS, vol. 3595, pp. 430–439. Springer, Heidelberg (2005)

Full Cryptanalysis of the Chen Identification Protocol

Philippe Gaborit[1], Julien Schrek[1], and Gilles Zémor[2]

[1] Université de Limoges, XLIM-DMI,
123, Av. Albert Thomas
87060 Limoges Cedex, France
{gaborit,julien.schrek}@unilim.fr

[2] Institut Mathématiques de Bordeaux UMR 5251 - Université Bordeaux I,
351 cours de la Libération, 33451, Talence, France
{gilles.zemor}@math.u-bordeaux.fr

Abstract. In 1995, K. Chen proposed a 5-pass zero-knowledge identification protocol based on the rank distance. The protocol is a 5-pass protocol with cheating probability $\frac{1}{2}$ in the spirit of Shamir's PKP protocol and Stern's SD protocol, but it has the additional property of avoiding the use of a hash function. This latter feature is very interesting from a low-cost cryptography perspective, but it also raises the suspicion of being too good to be true.

The contribution of this paper is twofold, first we show that the protocol's proof of zero-knowledge is flawed and we describe how to fully break the protocol in two different ways and in time polynomial in the size of the parameters. Secondly we propose a new zero-knowledge identification protocol for rank distance, for which we give a rigorous proof of zero-knowledge: however the proof requires the use of a hash function. The parameters of the new protocol are substantially improved compared to those of Chen's original protocol.

Keywords: Chen protocol, zero-knowledge, cryptanalysis, rank metric.

1 Introduction

Today, identification protocols are essential for the security of computer networks and smart cards. Zero-Knowledge is a useful tool to prove that a protocol can be reused without loss of security. Most practical zero-knowledge protocols are based on number theory and it is worth searching for alternatives for at least two reasons: first, number theory based protocols are often costly in terms of computational complexity and second their security will collapse if a workable quantum computer comes to exist.

Over the years several protocols have been proposed that rely on NP-Complete problem and based on linear functions [11], [13], [14]. Although recent advances have appeared, e.g. [8], that improve some of these protocols they are still mostly unsatisfactory because of their key size and/or their communication cost, moreover they intrinsically rely on a hash function which can be seen as a drawback

B.-Y. Yang (Ed.): PQCrypto 2011, LNCS 7071, pp. 35–50, 2011.

from the point of view of low cost cryptography. In this context K. Chen proposed in 1995 [5] an efficient zero-knowledge authentication protocol *without using a hash function* and based on the syndrome decoding problem for the rank metric.

The Chen protocol is in the spirit of Shamir's PKP protocol, but the hard underlying problem is, as in the case of Stern's SD protocol, a syndrome decoding problem, though for the rank distance rather than the Hamming distance.

The syndrome decoding problem for the rank metric is less known and less studied than it's Hamming distance counterpart but it is believed to be hard by the community. In particular the complexity of the best known attacks is more than exponential in the size of the parameters, which enables one to consider small parameters and hence obtain rather small key size. This latter aspect and the fact that the Chen protocol does not use any hash function makes it a very appealing alternative to more classical protocols.

The proposed parameters for the Chen protocol have already been attacked twice. First by Chabaud and Stern at AsiaCrypt'96 in [4] and later by Ourivski and Johansson in [10]: however the attacks proposed in these two papers were not specific to the Chen protocol. In fact they improved the algorithms for solving the generic syndrome decoding problem for the rank distance and as a by-product broke some of the proposed parameters for the Chen protocol. In practice, since the algorithms for the rank distance syndrome decoding problem remains largely exponential, a small increase of the parameters for the Chen protocol permits to resist these attacks while preserving relatively small key sizes. Overall the Chen protocol was still unbroken in itself and was still attractive for applications.

Our Contribution: The main result of the present paper is a total break of the Chen zero-knowledge identification protocol in two different ways. These attacks are made possible by the fact that the zero-knowledge proof of the Chen protocol is flawed. These attacks are polynomial (more precisely cubic) in the parameter size of the code and therefore lead to a total break of the protocol. The first attack relies on an unsatisfactory way to mask a word by a simple right matrix multiplication in the protocol. The second attack uses the fact that no hash function is used in the commitment step of the protocol, and uses the commitment to derive information on the secret key.

In a second part of the paper we propose a new protocol which permits the use of the rank metric for a zero-knowledge identification scheme. Our new protocol is a 3-pass protocol for which we add a hash function for the commitment step and a correct way to mask a word. The protocol we propose is based on Stern's SD protocol for the Hamming distance, adapted to the context of the rank metric. We then give a correct zero-knowledge proof for our new protocol whose security relies in part on the security of the hash function as in the case of Stern's SD and Shamir's PKP protocols. Finally, a really good feature of the use of a hash function is that it enables us to dramatically improve the soundness proof and to significantly recuce parameter sizes. These features should make the new protocol a strong candidate for low-cost cryptography.

The paper is organized as follows: Section 2 and 3 recall the main properties of the rank distance and Chen's protocol. In Section 4 we explain the two attacks that we propose and which break the original Chen protocol. Section 5 proposes a repaired version of the protocol: the new version is immune to our attacks, but at the cost of introducing a hash function. Finally Section 6 considers parameters for the repaired protocol.

2 Basic Facts on Rank Distance

The rank distance was introduced in coding theory by Gabidulin in 1985 in [7]. Since the decoding problem is seemingly more difficult for the rank distance than for the Hamming distance, codes for the rank metric have been variously proposed as alternatives to ordinary error-correcting codes when they play a role in a cryptographic protocol. In the following we recall basic facts on this metric. We refer the reader to P. Loidreau's paper [9] for more details on the rank distance and its use in cryptography.

2.1 Definitions and Notation

Notation :
Let q be a power of a prime p, m an integer and let V_n be a n dimensional vector space over the finite field GF(q^m). Let $\beta = (\beta_1, \ldots, \beta_m)$ be a basis of GF(q^m) over GF(q).
Let $\mathcal{F}_i$ be the map from GF(q^m) to GF(q) where $\mathcal{F}_i(x)$ is the i-th coordinate of x in the basis β.
To any $v = (v_1, \ldots, v_n)$ in V_n we associate the matrix $\overline{v} \in \mathcal{M}_{m,n}(\mathrm{GF}(q))$ in which $\overline{v}_{i,j} = \mathcal{F}_i(v_j)$.
The rank weight of a vector v can be defined as the rank of the associated matrix $\overline{v}$. If we name this value rank(v) we can have a distance between two vectors x, y using the formula $\mathrm{rd}(x, y) = \mathrm{rank}(x - y)$.

We now introduce an equivalence relation $\sim$ between two vectors x and y of V_n:

Definition 1. *For $x, y \in V_n$ we say that x and y are equivalent (denoted by $x \sim y$) if and only if their coordinates generate the same vector space.*

We remark that this definition is independent of the basis and that it is possible to have $\mathrm{rank}(x) = \mathrm{rank}(y)$ without $x \sim y$.

We can see that $x \sim y$ is equivalent to the existence of a $n \times n$ invertible matrix P over GF(q) which satisfy $xP = y$.

2.2 Properties

The isometry group for the rank distance is computed in [1] and is composed of two families: right multiplication by an invertible matrix over GF(q) and multiplication of columns by a non-zero element of GF(q^m). With these two families it is not possible to transform a word with fixed rank to any word with same rank. The fact that isometries may transform a word of given (rank or Hamming) weight to any other word with the same weight is very useful, it is for instance the case for the Hamming weight with permutation. We now introduce the product "$*$" in order to obtain a similar propriety with the rank metric.

For a given basis β, we denote Φ_β the inverse of the function $V_n \to \mathcal{M}_{m,n}(\mathrm{GF}(q))$: $x \to \overline{x}$ computed with the basis β.

Definition 2 (product). *Let Q be in $\mathcal{M}_{m,m}(\mathrm{GF}(q))$, $v \in V_n$ and β a basis. We define the product $Q * v$ such that $Q * v = \Phi_\beta(Q\overline{v})$, where $\overline{v}$ is constructed from the basis β.*

The following property of the product is useful.

Proposition 1. *For any $x \in V_n$, $P \in \mathcal{M}_{n,n}(\mathrm{GF}(q))$ and $Q \in \mathcal{M}_{m,m}(\mathrm{GF}(q))$, we have $(Q * x)P = Q * (xP)$.*

Proof. It is clear that $(Q\overline{x})P = Q(\overline{x}P)$. To conclude we just have to notice that $\overline{x}P = \overline{xP}$.

Moreover the product has the two important straightforward following properties (rank preservation and linearity over the base field $GF(q)$):

Proposition 2. *For any $x \in V_n$ and $Q \in GL_m(q)$, $\mathrm{rank}(Q * x) = \mathrm{rank}(x)$ and for any $a, b \in GF(q)$ and $x, y \in V_n$: $aQ * x + bQ * y = Q * (ax + by)$.*

Finally the following property shows that the "$*$" permits to associate a word with a given rank to any word with the same rank.

Proposition 3. *For any $x, y \in V_n$ and $\mathrm{rank}(x) = \mathrm{rank}(y)$, it is possible to find $P \in \mathcal{M}_{n,n}(\mathrm{GF}(q))$ and $Q \in \mathcal{M}_{m,m}(\mathrm{GF}(q))$ such that $x = Q * yP$.*

Proof. Let us denote by r the rank of x and y. Notice that r is then less or equal to m and n. To prove the property we just have to prove that there is $Q \in \mathcal{M}_{m,m}(\mathrm{GF}(q))$ which satisfies $x \sim Q * y$. As we said previously, this is equivalent to the existence of a matrix P which satisfies $Q*yP = x$. Let $y_1, \ldots, y_n$ be the coordinates of y. We reorder them to obtain $y_{\sigma(1)}, \ldots, y_{\sigma(r)}, \ldots, y_{\sigma(n)}$ with $y_{\sigma(1)}, \ldots, y_{\sigma(r)}$ linearly independent. We extend the family $y_{\sigma(1)}, \ldots, y_{\sigma(r)}$ to make a basis γ. We do the same with $x_1, \ldots, x_n$ to make γ', an extended basis of the family (x_i). Let Q be the matrix which transforms γ into γ'; $Q * y$ is now a vector with all its coordinates in γ' and moreover, all its coordinates are in $x_{\sigma'(1)}, \ldots, x_{\sigma'(r)}$.

2.3 Codes for the Rank Distance

We refer to [9] for more details on codes for the rank distance.

A code C of length n and dimension k over $\mathrm{GF}(q^m)$ is a subspace of dimension k of $\mathrm{GF}(q^m)$. The minimum rank distance of the code C is the miminum rank of non-zero vectors of the code.

It is known from [9] that these codes satisfy a Gilbert-Varshamov-like bound and that random codes attain this bound with a very high probability. In the following we will use this result to set up our parameters.

2.4 Rank Distance and Cryptography

The security of the Chen protocol is based on the rank version of an NP-hard problem, namely the syndrome decoding problem for the Hamming distance [2].

Syndrome Decoding Problem
Let H be a $((n-k) \times n)$ matrix over GF(q^m) with $k \leq n$, $i \in$ GF$(q^m)^k$ and ω an integer. The problem is to find s such that $wt(s) \leq \omega$ and $Hs^t = i$ where wt denotes the Hamming weight.

The problem is considered hard in general, especially when the matrix H is chosen at random. The best known algorithms for solving this problem are all exponential in ω, a recent survey on this complexity can be found in [6].

The previous problem can be naturally extended to the rank distance:

Syndrome Decoding Problem for the Rank Distance. Let H be a $((n-k) \times n)$ matrix over GF(q^m) with $k \leq n$, $i \in$ GF$(q^m)^k$ and r an integer. The problem is to find s such that rank$(s) = r$ and $Hs^t = i$.

In that case it is not proven that the problem is NP-hard, but the relation with the Hamming case and the fact that the best known algorithms are all exponential makes this problem difficult in practice and the problem is generally believed to be hard.

There are two main approaches to this problem in the case of the rank matrix: Chabaud and Stern proposed an algorithm to solve the problem in $O((nr+m)^3 q^{(m-r)(r-1)})$(see [4]) Ourivski and Johannson proposed two algorithms, the first uses a basis enumeration and is in $O((k+r)^3 q^{(m-r)(r-1)+2})$, the second uses a coordinate enumeration and is in $O((k+r)^3 q^{(m-r)(k+1)})$(see [10]).

3 The Chen Protocol

The Chen identification protocol is a 5-pass zero-knowledge protocol with a cheating probability of 1/2. This protocol has the attractive feature that it does not use any hash function, as the original Fiat-Shamir protocol, but unlike other zero-knowledge protocols based on linear functions. Indeed, the PKP protocol and Stern's SD protocol both need a hash function. Chen's protocol works as follows. Let H be a random matrix over GF(q^m), s a word of low rank weight and $i = Hs^t$. If Alice knows the secret s then i is called her identity. The aim of the protocol is for Alice to prove that she knows s and for Bob to know whether Alice knows s. The protocol is split into six steps and uses the word s of rank r as private key and a random matrix H and the syndrome $i = Hs^t$ as public key.

The difficulty of the protocol relies on the difficulty of solving the syndrome decoding problem for the rank distance. In the original paper, Chen proved that to satisfy the soundness property one needs to choose the rank weight of the

1. [First commitment step]
   The prover chooses $x \in V_n$ and $P \in \mathrm{GL}_n(\mathrm{GF}(q))$.
   He sends $c = HP^t x^t$ and $c' = Hx^t$.
2. [First challenge step]
   The verifier sends a random $\lambda \in \mathrm{GF}(q^m)$.
3. [Second commitment step]
   The prover computes $w = x + \lambda s P^{-1}$ and sends it.
4. [Second challenge step]
   The verifier sends at random $b \in \{0, 1\}$.
5. [Answer step]
   The prover sends P if $b = 0$ or x if $b = 1$.
6. [Verification step]
   If $b = 1$ and $\lambda \neq 0$, the verifier checks c' and if $\mathrm{rank}(w - x) = r$.
   If $b = 1$ and $\lambda = 0$, the verifier checks c' and if $\mathrm{rank}(w - x) = 0$.
   If $b = 0$, the verifier checks if $HP^t w^t = c + \lambda i$.

Fig. 1. Chen's protocol

secret not more than $\frac{d}{3}$ where d is the minimum rank distance of the code whose parity-check matrix is H (d corresponds to the Gilbert-Varshamov-like bound). Notice that the rank weight of the secret was increased to $\frac{d}{2}$ by Chabaud-Stern in [4]. Later (in section 6) we will see how it is possible to push it up to d with a new protocol.

4 Cryptanalysis

In this section we first explain why there are flaws in the zero-knowledge proof of the Chen protocol and then we propose two full cryptanalysis by passive attacks on the protocol which uses these flaws. The attacker will just need access to the public data exchanged during the protocol to retrieve the secret key of the protocol. The first attack relies on the fact that in the protocol, the masking of the secret at Step 3 by a right multiplication by an invertible matrix is not enough. This attack can be easily countered by modifying the masking procedure. The second attack involves a recovery of the secret by linear algebra from leaking information when $b = 0$ and does not seem to be reparable without the introduction of hash functions.

4.1 Flaws in Chen's Zero-Knowledge Proof

The two attacks rely on a flaw in Chen's proof of zero-knowledge. Proofs of zero-knowledge are generally made with a simulator of the scheme. The simulator proves that someone can create a scheme indistinguishable from a real execution in reasonable time without knowing the private key. With this assumption, it is clear that the view of the execution of the scheme is not needed to attack it. The simulator given by Chen in his paper [5] is in fact false since the zero-knowledge proof omits the last sending of the scheme. Hence the incompleteness of the simulator may imply attacks using a potential leak of information of the protocol. We present here two attacks based on using that leak of information. Indeed, we prove that Hx^t gives information about x (by linearity) and sP gives information about s (the underlying vector space of sP is the same as the underlying vector space of s), with the two following attacks.

4.2 The Support Attack

High level overview. In the protocol the matrix P is used to mask the secret s into a potential secret sP like it is done in PKP [11] or in SD [13], P taking the place of the permutation in these protocols. However in Chen's protocol the masking by P is weak and leaks information. To understand this point, consider the case of Stern's SD protocol: a permutation of coordinates has the property that it can transform any codeword with a given Hamming weight into *any* codeword with the same weight. This point is crucial for indistinguishability in the zero-knowledge proof of Stern SD protocol. In the case of Chen's protocol, the multiplication by the matrix P is the equivalent notion to that of the permutation for the Hamming distance in Stern's SD protocol, however this transformation does not possess the same feature as the permutation in the Hamming distance case. Indeed this transformation does not permit to associate a given word with a given rank distance, to *any* word with the same rank distance, it only gives a subset of codewords with the same rank distance. In practice it leads to a leak of information.

Indeed a right multiplication of s by P does not change the basis generated by the coordinates of s and therefore all elements of the form sP generate the same vector space. Therefore the masking sP is not general enough and gives information on the secret s by revealing the vector space generated by its coordinates.

If we compare this attack with the Chabaud-Stern attack of [4], we derive a vector space basis which contains the secret s when their attack (which is more general on the rank distance) consists in an exponential search for such a vector space basis. Our attack therefore avoids the exponential search of a basis and reads it directly from sP, leaving only the linear algebra part of the attack.

Description. Each time a prover wants to be identified with the protocol and his key s, he has to pass a sequence of challenges. The attack presented in this paragraph works when the challenge is the one represented by the value $b = 1$ in the description of the Chen protocol. This challenge occurs so many times that the attack can be executed in a passive way instead of an active way. We consider an execution of the protocol when the prover has to send the value x (case $b = 1$). In this case the value sP^{-1} can be computed with the formula $\lambda^{-1}(w - x)$. The attack uses the fact that there is a way to retrieve s from sP^{-1} with a very low complexity.

In the Basic Facts section we saw that $s \sim sP^{-1}$, so their coordinates generate the same vector space E over GF(q). The specific rank r of s implies that E is a vector space of dimension r. The coordinates of sP^{-1} generate the vector space E so we can construct a basis of E over GF(q) with them.

Let us denote $\gamma = (\gamma_1, \ldots, \gamma_r)$ this basis, with $\gamma \in \mathrm{GF}(q^m)^r$. We can express each coordinate s_j, j from 1 to n, of s into this basis and obtain $s_j = \sum_{k=1}^{r} a_{k,j}\gamma_k)$ with $a_{k,j} \in \mathrm{GF(q)}$ for k between 1 and r and j between 1 and n. We construct a basis of $\mathrm{GF}(q^m)$ over $\mathrm{GF}(q)$ such that its first r vectors are equal to $\gamma_1, \ldots, \gamma_r$. We call it $\gamma' = (\gamma_1, \ldots, \gamma_r, \gamma_{r+1}, \ldots, \gamma_m)$ with $\gamma_l \in \mathrm{GF}(q^m)$ for l from 1 to m. Writing the equation $Hs^t = i$ into $\mathrm{GF}(q)$ with the basis γ', permits to obtain $(n-k) \times m$ equations on $\mathrm{GF}(q)$ and $n \times r$ unknowns (the $a_{k,j}$). In the parameters proposed for the Chen protocol we always have $n \times r \leq (n-k) \times m$, so the system is directly solvable by Gaussian elimination and one can recover the secret s.

Attack's complexity. For this algorithm we have to generate a basis of cardinality r from n vectors, complete this basis and invert an $n \times r$ matrix over $\mathrm{GF}(q)$. The only cost we have to focus on is the matrix inversion, a $\mathrm{O}(N^3)$ algorithm, with $N = n \times r$ equal to 128 as proposed in [4], this attack cost about 2^{21} operations in $\mathrm{GF}(q)$. Our approach , which is specific to this protocol, permits to avoid the (exponential) search for a basis in which one can express the secret s. As soon as the previous inequality is satisfied, which is the case for all proposed parameters (the original Chen parameters and Chabaud and Stern parameters) the protocol can be broken in polynomial time at the cost of a Gaussian elimination for a matrix of size $n \times r$. We checked on examples that the attack worked in practice with this complexity.

We now describe our second attack.

4.3 Linear Attack

High level overview. This attack relies on the fact that no hash function is used in the commitment step, which makes it possible to use information from each commitment.

Suppose a passive attacker can observe an exchange between a prover and a verifier with $b = 0$ in the protocol. In this case, the values HP^tx^t and Hx^t correspond to a system of $2k$ equations in the n coordinates of x. In the Chen parameters, where n is equal to $2k$, it is possible to retrieve x in only one iteration of the protocol. With the knowledge of P (in the $b = 0$ case) and x, the secret s can be deduced from $w = x + \lambda sP^{-1}$. In this paragraph we show how to deduce the secret s for any type of possible parameters. Indeed, each occurrence of the protocol for the $b = 0$ case has the possibility to give new linear equations in the variables which compose the secret key s. More specifically, every time the prover sends a new P, (say P_j), there are as many new equations as the number of rows of HP_j^t independent from the rows of HP_k^t for $k \leq j$ and H.

Description. In order to prove his identity to a verifier, a prover runs the protocol several times. For every execution of the protocol, the prover sends two possible values (see step 5 of the description of the Chen protocol). The linear attack works when the prover has to send the variable named P (see description of the protocol). Since the protocol is based on a choice between two challenges, we can consider that it happens often. We call this challenge the challenge "$b = 0$". Let us denote by x, P, w, λ, c', the variables used in a challenge "$b = 0$" encountered, corresponding to the variables with same names as in the description of the Chen protocol. Remember that for the challenge "$b = 0$" the variable x is unknown and the variables P, w, λ and c' are known. We will prove that other occurrences of the challenge "$b = 0$" give linear equations in the coordinates of s.

The definition of $Hs^t = i$ gives equations in the coordinates of s but not enough to solve a system. To add more equations to the system we use those given by the challenge "$b = 0$" :

$$w = x + \lambda s P^{-1}$$

$$c' = xH^t$$

We obtain :

$$Hw^t = c' + \lambda H(sP^{-1})^t$$

In the challenge "$b = 0$" the only unknown here is s, this gives new equations in the coordinates of s as long as these equations are linearly independent. Supposing that we have $n - 1$ equations in our system, which is the worst case, the probability for uniformly generated equations to be dependent of the system is equal to $\frac{1}{q^m}$. The fact that P is uniformly generated implies that HP^{-t} forms uniformly generated equations. The probability of obtaining linearly independent equations is therefore very high.

The attack is finished when the number of equations is sufficient to solve the system.

Example[Active attack] The previous passive attack can easily be turned into an active attack. An attacker can choose to send only the value 0 in the fourth step of the protocol and obtain information leakage which eventually gives the previous attack.

The attack is based on the fact that no hash function is used in the commitment which makes a leakage of information possible.

Complexity of the attack and implementation. A square matrix in GF(q^m) has a good probability of being invertible, since the cardinality of GF(q^m) is far from 2. The only cost we have to focus on is the matrix inversion which is cubic in O(N^3) with a simple algorithm and can be decreased with fast multiplication. With n about equal to 32 as proposed in [5], the complexity of the attack is about $W \times (2^5)^3$ with W the cost of a multiplication in GF(q^m). An implementation of this attack using the mathematics software sage [12] found 2^{21} secrets s in 38013.98 seconds and it happened only once during the 2^{21} tests that the matrix was not invertible. It makes the attack very fast and is coherent with the expected computational complexity.

5 Countermeasures

5.1 Defense against the Support Attack

The support attack uses the fact that $s \sim sP$. A simple repair is to multiply s by a matrix Q in $GL_m(q)$ using the product $*$ and a matrix P in $GL_n(\mathrm{GF}(q))$ instead of just using the matrix P. The fact that every vector can be turned into any vector with the same rank implies that a support attack is not possible. The change affects also c' which has to be $c' = H(Q * xP)^t$, w becomes $x + Q^{-1}sP^{-1}$ and if $(b = 0)$ the prover has to send Q and P instead of P. We notice that the verification has no problem because we have $(Q * x)P = Q * (xP)$. However this repair does not affect the linear attack.

5.2 Defense against the Linear Attack

The problem of the Chen protocol exploited in the linear attack is that Hx^t is a linear equation which simplifies other linear equations. A simple way to repair the linear attack is to replace $c' = Hx^t$ by a non linear equation. We use here $c = hash(x)$ where $hash$ is a hash function to be sure that no information can be obtained from c.

6 A New Protocol

We saw in previous sections that the zero-knowledge proof of the Chen protocol was incomplete and hence opened the door to potential attacks. We saw that such attacks could be made possible through the exploitation of a bad masking and the fact that no hash function was used in the commitment step. We hence propose a new protocol for which a correct zero-knowledge proof is possible. It implies in particular the use of a correct masking (see Proposition 3) and a hash function for commitments. The protocol can be seen as an adaptation of the Stern SD protocol in a context of rank distance, we prove that our protocol benefits from the same feature as the Stern SD protocol: a 3-pass zero-knowledge identification protocol with a 2/3 cheating probability. However in terms of communication cost it is better than Stern's SD protocol.

6.1 Description of the Protocol

The protocol proposed here is a rank metric adaptation of the Stern protocol [13] in which the masking of a codeword by a permutation is replaced by the masking $x \to Q * xP$ which has the same property in terms of rank distance as a permutation for a codeword with Hamming distance, since it can transform any given x with given rank to *any* element with the same rank. This property (which is not obtained if one only applies a right multiplication by P) is crucial for the zero-knowledge proof of the protocol. It also contains a hash function for the commitment.

In the following the notation $(a|b)$ corresponds to the concatenation of a and b. The notation $hash(a)$ is the hash value of a.

A given basis β is fixed and known in advance for the '*' product (see Section 2.2).

For the protocol a public $k \times n$ matrix over $GF(q^m)$ H is fixed. The **secret key** is a vector s of $V_n(= (GF(q^m)^n)$ with rank r. The **public key** consists of the matrix H, the syndrome $i = Hx^t$ and the rank r of s. The protocol is described in Fig. 2. For the protocol the small base field is $GF(2)$, (ie: $q = 2$).

1. [Commitment step] The prover PR chooses $x \in V_n$, $P \in GL_n(\mathrm{GF}(q))$ and $Q \in GL_m(q)$. He sends c_1, c_2, c_3 such that :

   $$c_1 = hash(Q|P|Hx^t), c_2 = hash(Q * xP), c_3 = hash(Q * (x+s)P)$$

2. [Challenge step] The verifier V sends $b \in \{0,1,2\}$ to P.
3. [Answer step] there are three possibilities :
   - if $b = 0$, PR reveals x and $(Q|P)$
   - if $b = 1$, PR reveals $x + s$ and $(Q|P)$
   - if $b = 2$, PR reveals $Q * xP$ and $Q * sP$
4. [Verification step] there are three possibilities :
   - if $b = 0$, V checks c_1 and c_2.
   - if $b = 1$, V checks c_1 and c_3.
   - if $b = 2$, V checks c_2 and c_3 and that $rank(Q * sP) = r$.

Fig. 2. Repaired protocol

Verification Step

Here we explain the verification step. There are three cases in the verification, $b = 0, 1, 2$.

In the case $b = 0$, V receives x and $(Q|P)$, he computes Hx^t and $Q * xP$ and checks the hash values $c_1 = hash(Q|P|Hx^t)$ and $c_2 = hash(Q * xP)$.

In the case $b = 1$, V receives $x+s$ and $(Q|P)$, he computes $Hx^t = H(x+s)^t - i$ since $i = Hs^t$ and $Q*(x+s)P$ which permits to check c_1 and c_3. In the case $b = 2$, V receives $Q * xP$ and $Q * sP$, he computes the hash values $c_2 = hash(Q * xP)$ and $c_3 = hash(Q * xP + Q * sP) = hash(Q * (x+s)P)$. He also checks that $rank(Q * sP) = r$.

6.2 Zero-Knowledge Properties

We saw that the Chen protocol has flaws in its zero-knowledge proof which implied a total break of the system: deriving correct proofs is therefore important. In the case of the new protocol, the zero-knowledge proof described hereafter is based on the zero-knowledge proof of the Stern SD protocol, adapted to the rank metric context. Crucial is the 'equivalent' notion of a permutation for the Hamming distance.

Completeness. An honest prover who knows s will be able to construct c_1, c_2 and c_3. He will always be identified by a verifier because the verifications match with the data c_1, c_2 and c_3.

Soundness. The proof exposed here is not the same as for the Chen protocol [5], because of the use of hash functions. In the Chen protocol the secret key s is taken of weight under $\frac{d}{3}$ ($\frac{d}{2}$ in Chabaud-Stern). The use of a hash function permits us to reach d (it dramatically improves the parameters which can be taken in the new protocol).

We prove the following theorem:

Theorem 1. *If a prover PR is able to correctly answer all three challenges then either he is able to find a collision for the hash function, or he has access to the secret key s.*

Proof:
If someone can answer in the three cases $b = 0, 1, 2$ for a chosen triple (c_1, c_2, c_3), then he knows:

– x_1 and (Q_1, P_1) such that :

$$c_1 = hash(Q_1|P_1|Hx_1^t), c_2 = hash(Q_1 * x_1 P_1)$$

– w and $(Q_2|P_2)$ such that :

$$c_1 = hash(Q_2|P_2|Hw^t - i), c_3 = hash(Q_2 * wP_2)$$

– x_2 and z such that :

$$c_2 = hash(x_2), c_3 = hash(x_2 + z), rank(z) = r$$

So either the attacker finds a collision for the hash function or all the respective pre-images of c_1, c_2 and c_3 are equal and we have :

$$Q_1 = Q_2, P_1 = P_2, Q_1 * x_1 P_1 = x_2, Q_2 * wP_2 = x_2 + z$$

and

$$Hx_1^t = Hw^t - i$$

From which we deduce that

$$Q_1 * wP_1 = x_2 + z = Q_1 * x_1 P_1 + z$$

and then

$$w - x_1 = Q_1^{-1} * zP_1^{-1}$$

and now:

$$i = Hw^t - Hx_1^t = H(w - x_1)^t = H(Q_1^{-1} * zP_1^{-1})^t,$$

with $rank(z) = r$.

Since $rank(Q_1^{-1} * zP_1^{-1}) = rank(z)$, the attacker knows a solution of the difficult problem which describes the secret key s (and which is unique for r less than the Gilbert-Varshamov-like bound). He is therefore able to reconstruct the secret s □

A corollary of the theorem is that the probability of success when the secret key s is not known is less or equal to $\frac{2}{3}$. Now it is possible to anticipate (as for the Stern SD scheme) any two challenges among the three:

- for $b = 0$ or 1, a cheater takes random P,Q and x and constructs z which satisfies $Hz^t = Hx^t + i$ (notice that since there is no weight constraint finding a z is easy). He takes c_1 and c_2 as in the protocol and $c_3 = hash(Q * zP)$. For $b = 0$ the cheater reveals x and $(P|Q)$ and for $b = 1$ he reveals P, Q and z. The verification follows.
- for $b = 0$ or 2, a cheater takes random P,Q and x and constructs z random of rank weight r. He takes c_1 and c_2 as in the protocol and $c_3 = hash(Q * (x + z)P)$. For $b = 0$ the cheater reveals x and $(P|Q)$ and for $b = 2$ he reveals $Q * xP$ and $(Q * zP)$. The verification works.
- for $b = 1$ or 2, a cheater takes random P, Q ,x and z random of rank weight r. He sends $c_1 = hash(P|Q|H(x + z)^t - i), c_3 = hash(Q * (x + z)P)$ and $c_2 = hash(Q * xP)$. Now for $b = 1$ the cheater reveals $x + z$ and P, Q, and for $b = 2$: $Q * xP$ and $Q * zP$. The verification follows.

Overall the probability of cheating is therefore exactly $\frac{2}{3}$.

Security. The security of the secret key is based on the Syndrome decoding problem for the rank distance in the same way than for the Chen protocol.

Zero-Knowledge. The aim of this proof is to construct a simulator of the scheme. If we can simulate the scheme, we can use this construction to attack the secret key without needing to actually run the scheme. This implies that a transcript of the actual scheme gives no usable information.

Let be B a verifier and S be the simulator that we construct. We construct a simulator which can answer any of the three challenges because the simulator always makes a good guess (in particular only 2 of the 3 commitments need to be verified). The protocol can be simulated as follows:

1. S chooses random $x \in V_n$ and $P \in \mathrm{GL}_n(\mathrm{GF}(q))$, $Q \in GL_m(q)$ and z with $rank(z) = r$.
   The simulator also chooses $j \in \{0, 1, 2\}$ corresponding to the challenge that he tries to guess in advance.
   If $j = 0$, it sends c_1, c_2 and c_3 such that :
   $$c_1 = hash(Q|P|Hx^t), c_2 = hash(Q * xP), c_3 = hash(x)$$
   If $j = 1$ it sends c_1, c_2 and c_3 such that:
   $$c_1 = hash(Q|P|Hx^t - i), c_2 = hash(Q * xP), c_3 = hash(Q * xP)$$

If $j = 2$ it sends c_1, c_2 and c_3 such that:

$$c_1 = hash(x), c_2 = hash(Q * xP), c_3 = hash(Q * (x + z)P)$$

2. B chooses $b \in \{0, 1, 2\}$.
3. If $b = 0$, S sends x, Q and P.
   If $b = 1$, S sends x, Q and P.
   If $b = 2$, S sends $Q * xP$ and $Q * zP$.

4. If $b = j$ the execution works (it was made for it) and the simulator saves the execution, otherwise the simulator restarts this execution.

The simulation succeeds if it can save l executions of this scheme with $b = j$. This can be done in about $3l$ executions since the probability that $b = j$ is $\frac{1}{3}$. The simulator creates a simulated execution of the protocol which is indistinguishable from a real execution because of the use of the hash function, it was not the case in the Chen protocol. The value x, Q and P are clearly constructed with a uniform distribution, which is the same as the distribution given by $Q * (x + s)P$ in the real scheme. The last value z has the same distribution as the value $Q * sP$ because of the use of Q and P. Indeed, if Q and P are randomly chosen, for s a vector of rank r, we saw in Section 2.2 that the vector $Q * sP$ could be *any vector* of rank r with a uniform probability. It is possible for the simulator to guess in advance all challenge values b with a uniform distribution. All distributions are equal to those in the real scheme, it makes this simulator indistinguishable from a real execution. With this simulator anyone can simulate an execution of the scheme without knowing the secret key, which proves the zero-knowledge property.

Remark: This proof is made possible because of the use of matrices P and Q on one hand and of the hash function on the other hand which assure the indistinguishability between a real execution of the protocol and a simulation. The Chen protocol did not use these features, for instance for l executions of the Chen protocol it is possible to distinguish between random words z of rank r (in the simulation) and vectors sP which by construction always belong to the same vector space. The same phenomenon occurs with the use of hash functions in the commitment step.

7 Parameters, Improvements and Comparison

7.1 Parameters

The soundness proof we proposed in the previous section enables us to take for the weight of the secret s a value just below the minimum distance of the code. If one considers random codes, it is known that with high probability they lie on the Gilbert-Varshamov bound (cf. [9] for the details on the exact formulae for these parameters). If we consider, $q = 2, n = 20, m = 20$ and $k = 11$ one obtains a minimal distance of 7, hence we can take $r = 6$ for the rank weight of the secret. In that case the known attacks lead to a complexity of at least 2^{83} operations.

The fact that one can take a rank weight r close to d enables one to greatly decrease the size of the parameters.

Parameters of the repaired Chen protocol with parameters $q = 2, n = 20, m = 20, k = 11, r = 6$

Public matrix H : $(n - k) \times k \times m = 1980$bits
Public key i : $(n - k) \times m = 180$ bits
Secret key s : $n \times m = 400$ bits
Average number of bits exchanged in one round: 2 hash + one word of $\mathrm{GF}(q^m) \sim 800$bits.

7.2 Possible Improvements

The protocol described in the previous section follows Stern's SD scheme, the only difference being that the notion of permutation for Hamming distance, which gives the indistinguishability is replaced by an equivalent notion for rank distance, the product $P*xQ$. Following this analogy other protocols like Veron's protocol ([15]) which improves on Stern's protocol can be adapted in this case.

It is also possible to improve the cheating probability to $1/2$ as in the Cayrel et al. protocol [3] by increasing the size of q. For the protocol we proposed we took $q = 2$, which gives a cheating probability of $2/3$, increasing q to large q permits to reach asymptotically a cheating probability of $1/2$ but it may also increase the cost of communication.

7.3 Comparison with Other Protocols

In terms of communication costs our protocol is 20% better than Stern's SD scheme and faster, however the syndrome decoding problem remains less studied for the rank distance and there is no underlying problem proven to be NP-hard (although the problem is considered as hard by the community). The new protocol is less interesting in terms of communication by a factor 20% compared to Shamir's PKP protocol, however the security of PKP has been even less studied than the syndrome decoding problem for the rank distance. Another advantage of the new protocol is that it benefits from a small public key compared to SD and PKP. Overall these three protocols are all interesting for different reasons for low-cost cryptography.

8 Conclusion

In this paper we presented two polynomial attacks on the Chen identification protocol, the first attack takes advantage of the structure of the masking of the secret and the second attack takes advantage of the fact that no hash function is used in the protocol. Overall our attacks completely break the system when previous cryptanalysis remained of exponential complexity. We propose a new repaired protocol but with a hash function: it comes with a rigorous zero-knowledge proof.

Besides the theoretical security of our new protocol, analysis of the proof of soundness considerably decreases the size of parameters compared to Chen's original protocol. This work shows that it is crucial to have correct zero-knowledge proof of protocols, otherwise the door is open to a total break. Besides a rigorous zero-knowledge proof the new protocol presents interesting features which can make it a good candidate for low-cost cryptography.

References

1. Berger, T.P.: Isometries for rank distance and permutation group of gabidulin codes. IEEE Transactions on Information Theory 49(11), 3016–3019 (2003)
2. Berlekamp, E., McEliece, R., van Tilborg, H.: On the inherent intractability of certain coding problems (Corresp.). IEEE Transactions on Information Theory 24(3), 384–386 (1978)
3. Cayrel, P.-L., Véron, P., El Yousfi Alaoui, S.M.: A Zero-Knowledge Identification Scheme Based on the Q-ary Syndrome Decoding Problem. In: Biryukov, A., Gong, G., Stinson, D.R. (eds.) SAC 2010. LNCS, vol. 6544, pp. 171–186. Springer, Heidelberg (2011)
4. Chabaud, F., Stern, J.: The Cryptographic Security of the Syndrome Decoding Problem for Rank Distance Codes. In: Kim, K.-c., Matsumoto, T. (eds.) ASIACRYPT 1996. LNCS, vol. 1163, pp. 368–381. Springer, Heidelberg (1996)
5. Chen, K.: A New Identification Algorithm. In: Dawson, E.P., Golić, J.D. (eds.) Cryptography: Policy and Algorithms 1995. LNCS, vol. 1029, pp. 244–249. Springer, Heidelberg (1996)
6. Finiasz, M., Sendrier, N.: Security Bounds for the Design of Code-based Cryptosystems. In: Matsui, M. (ed.) ASIACRYPT 2009. LNCS, vol. 5912, pp. 88–105. Springer, Heidelberg (2009)
7. Gabidulin, E.M.: Theory of Codes with Maximum Rank Distance. Probl. Peredachi Inf. 21(1), 3–16 (1985)
8. Gaborit, P., Girault, M.: Lightweight code-based authentification and signature. In: IEEE International Symposium on Information Theory, ISIT 2007, pp. 191–195 (2007)
9. Loidreau, P.: Properties of codes in rank metric. CoRR, abs/cs/0610057 (2006)
10. Ourivski, A.V., Johansson, T.: New technique for decoding codes in the rank metric and its cryptography applications. Probl. Inf. Transm. 38, 237–246 (2002)
11. Shamir, A.: An Efficient Identification Scheme Based on Permuted Kernels. In: Brassard, G. (ed.) CRYPTO 1989. LNCS, vol. 435, pp. 606–609. Springer, Heidelberg (1990)
12. Stein, W.A., et al.: Sage Mathematics Software (Version 3.3). The Sage Group (2009), http://www.sagemath.org
13. Stern, J.: A New Identification Scheme Based on Syndrome Decoding. In: Stinson, D.R. (ed.) CRYPTO 1993. LNCS, vol. 773, pp. 13–21. Springer, Heidelberg (1994)
14. Stern, J.: Designing Identification Schemes with Keys of Short Size. In: Desmedt, Y.G. (ed.) CRYPTO 1994. LNCS, vol. 839, pp. 164–173. Springer, Heidelberg (1994)
15. Véron, P.: Improved identification schemes based on error-correcting codes. Appl. Algebra Eng. Commun. Comput. 8(1), 57–69 (1996)

Decoding One Out of Many

Nicolas Sendrier

INRIA Paris-Rocquencourt, Project-Team SECRET
nicolas.sendrier@inria.fr

Abstract. Generic decoding of linear codes is the best known attack against most code-based cryptosystems. Understanding and measuring the complexity of the best decoding techniques is thus necessary to select secure parameters. We consider here the possibility that an attacker has access to many cryptograms and is satisfied by decrypting (i.e. decoding) only one of them. We show that, for the parameter range corresponding to the McEliece encryption scheme, a variant of Stern's collision decoding can be adapted to gain a factor almost $\sqrt{N}$ when N instances are given. If the attacker has access to an unlimited number of instances, we show that the attack complexity is significantly lower, in fact the number of security bits is divided by a number slightly smaller than 3/2 (but larger than 1). Finally we give indications on how to counter those attacks.

1 Introduction

Code-based cryptography has attracted a lot of interest in the past few years, accompanying the rise of post-quantum cryptography. It allows public-key encryption scheme [21,22], zero-knowledge protocols [28,29,16], digital signature [11], hash functions [1,7], stream ciphers [15,17] to mention only the most classical primitives. The common point of all code-based cryptographic primitives is the fact that they rely on the hardness of decoding a linear code with no apparent algebraic structure. This problem is NP-hard [4], and in fact, the parameter selection for those systems is based on the best knows decoding techniques, usually the collision decoding [27] and its variants, and sometimes the generalized birthday algorithm (GBA) [8,30].

In this work, we consider the case where the attacker is given many instances of the decoding problem for the same linear code and wishes to solve only one of them. Bleichenbacher's attack against [11] (unpublished but described for instance in [23]) is a variant of GBA which offers a theoretical speedup of $\sqrt{N}$ when the attacker tries to sign one out of N messages. The cost of the attack will drop from T initially to $\max(T/\sqrt{N}, N)$ whose minimal value is $T^{2/3}$ (when $N = T^{2/3}$). A variant of ISD for multiple instances has been proposed [18], but its cost analysis does not allow an easy measure of the gain.

We consider in this paper a modification of ISD (similar to [18]) with a complete cost analysis. We will show that, when the number of errors to decode is smaller than the Gilbert-Varshamov distance[1] (corresponding to McEliece's or

[1] The (binary) Gilbert-Varshamov distance is the largest integer d_0 such that $\binom{n}{d_0} \leq 2^r$.

B.-Y. Yang (Ed.): PQCrypto 2011, LNCS 7071, pp. 51–67, 2011.

Niederreiter's encryption schemes), collision decoding can be adapted to save a factor $N^{0.5-c}$ (for some small positive c) when decoding one out of N instances. Also, if the number of instances is unlimited, we show that the cost of the decoding is raised to the power $2/3 + c'$ (for some small positive c'). In other words, the number of security bits (i.e. the log in base 2 of the cost of the best attack) is divided by some number close (but smaller to) 1.5.

We will first analyze an abstract variant of ISD, similar to the one of [14]. We will then show how this algorithm and its analysis can be extended to the case of many instances and provide some estimates of what this modified algorithm can gain. This new attack constitutes a threat which must be considered. We briefly explain in the conclusion how to completely avoid it. The countermeasures are simple but it is a new feature to consider when implementing code-based cryptography.

Notation:

- $\mathcal{S}_n(\mathbf{0}, w)$ denotes the sphere of radius w centered in $\mathbf{0}$ in the Hamming space $\{0,1\}^n$, more generally $\mathcal{S}_n(x, w)$ denotes the same sphere centered in x.
- $|X|$ denotes the cardinality of the set X.

2 The Decoding Problem in Cryptology

The security of code-based cryptography heavily relies on the hardness of decoding in a random linear code. The computational syndrome decoding problem is NP-hard and is conjectured difficult in the average case.

Problem 1 (Computational Syndrome Decoding - CSD**).** *Given a matrix* $H \in \{0,1\}^{r\times n}$, *a word* $s \in \{0,1\}^r$, *and an integer* $w > 0$, *find* $e \in \{0,1\}^n$ *of Hamming weight* $\le w$ *such that* $eH^T = s$.

We will denote CSD(H, s, w) the above problem and the set of its solutions. Decoding is one of the prominent algorithmic problems in coding theory for more than fifty years. So far, no subexponential algorithm is known which correct a constant proportion of errors in a linear code. Code-based cryptography has been developed on that ground and for many code-based cryptosystems, public-key encryption [21,22] and digital signature [11], zero-knowledge protocols based on codes [28,29,16], hash-function [1], PRNG and stream ciphers [15,17] and many others, decoding is the most threatening attack and therefore is a key point in the parameter selection.

2.1 Generic Decoding Algorithms

The most ancient technique for addressing CSD in cryptology is Information Set Decoding (ISD). It can be traced back to Prange [25]. The variants useful today in cryptology all derive more or less from Stern's algorithm [27], which we will call collision decoding, following [6,24]. It was implemented (with various improvements) in [9] then in [5] which reports the first successful attack on the original parameter set. General lower bounds were proposed [14]. The last published variant is ball-collision decoding [6] which features a better decoding exponent than collision decoding.

The other main technique is the Generalized Birthday Algorithm (GBA) [30] (order 2 GBA was previously published in [8]). The first use of GBA for decoding was proposed in [10] for attacking an early version of FSB [2]. It is sometimes faster than ISD.

The security of the various code-based cryptographic primitives corresponds to a wide range of parameters for the CSD problem. To determine which attack is the most efficient, one should compare the error weight w with the Gilbert-Varshamov distance d_0 (which is a function of the code length and size). *For a single instance*, the situation is the following: (1) when $w < d_0$ (for encryption schemes) ISD is always better, (2) when $w \approx d_0$ (for ZK-protocols, digital signature, stream cipher), the best attack is also ISD, and (3) when $w > d_0$ (for hashing) the best attack is either ISD or GBA (with no easy rule to predict which is the best). Let us also mention that $w > r/4$ is insecure because Saarinen's attack [26].

For multiple instances the situation is not known precisely, but in one case at least (namely Bleichenbacher's attack against CFS signature scheme) GBA with multiple instances has become the most efficient attack. This was a motivation to consider whether a similar improvement was possible with ISD.

2.2 Decoding One Out of Many Instances

In this work we will consider the scenario where the attacker has many instances (H, s, w) at disposal where the parity check matrix H and the error weight w are identical, but the syndrome s runs over some large set.

Problem 2 (Computational Syndrome Decoding - Multi). *Given a matrix $H \in \{0,1\}^{r \times n}$, a set $\mathcal{S} \subset \{0,1\}^r$, and an integer $w > 0$, find a word $e \in \{0,1\}^n$ of Hamming weight $\leq w$ such that $eH^T \in \mathcal{S}$.*

For convenience, we will also denote CSD$(H, \mathcal{S}, w)$ this problem and the set of its solutions. It has been addressed already using GBA by Bleichenbacher (unpublished, reported in [23]) for attacking the digital signature CFS. In practice, the attacker builds a large number N of instances of a decoding problem (corresponding to N favorable messages) solves one of them with an order 2 GBA with a speedup of $\sqrt{N}$ compared with the decoding of a single instance with a birthday attack. This reduces the order of magnitude of the cost for forging a signature from $O(2^{r/2})$ to $O(2^{r/3})$. A variant of CFS resistant to this attack was recently published [13].

An attempt at using ISD with multiple instances was already made in [18]. We revisit here that work in a more general setting and with a more thorough complexity analysis.

3 A Generalized Information Set Decoding Algorithm

Following other works [19,20], J. Stern describes in [27] an algorithm to solve CSD. We present in Algorithm 1 a generalized version, similar to the one presented in [14], which acts on the parity check matrix H_0 of the code (instead of the generator matrix). The partial Gaussian elimination of H_0P consists in

Algorithm 1. Generalized ISD algorithm

For any fixed values of n, r and w, the following algorithm uses four parameters: two integers $p > 0$ and $\ell > 0$ and two sets $W_1 \subset \mathcal{S}_{k+\ell}(\mathbf{0}, p_1)$ and $W_2 \subset \mathcal{S}_{k+\ell}(\mathbf{0}, p_2)$ where p_1 and p_2 are positive integers such that $p_1 + p_2 = p$.

procedure main_isd
input: $H_0 \in \{0,1\}^{r\times n}$, $s_0 \in \{0,1\}^r$
  `repeat`
(ISD 0) {
    $P \leftarrow$ random $n \times n$ permutation matrix
    $(H', H'', U) \leftarrow \text{PartialGaussElim}(H_0P)$ // as in (1)
    $s \leftarrow s_0 U^T$
  }
    $e \leftarrow \text{isd\_loop}(H', H'', s)$
  `while` $e = \text{FAIL}$
  `return` (P, e)

procedure isd_loop
input: $H' \in \{0,1\}^{\ell\times(k+\ell)}$, $H'' \in \{0,1\}^{(r-\ell)\times(k+\ell)}$, $s \in \{0,1\}^r$
  `for all` $e_1 \in W_1$
(ISD 1) {
    $i \leftarrow e_1 H'^T$, $s_1'' \leftarrow e_1 H''^T$
    $\text{write}(e_1, s_1'', i)$ // stores (e_1, s_1'') at index i
  }
  `for all` $e_2 \in W_2$
(ISD 2) {
    $i \leftarrow s' + e_2 H'^T$, $s_2'' \leftarrow s'' + e_2 H''^T$
    $\text{Elts} \leftarrow \text{read}(i)$ // extracts the elements stored at index i
  }
    `for all` $(e_1, s_1'') \in \text{Elts}$
(ISD 3) {
      `if` $\text{wt}\,(s_1'' + s_2'') = w - p$
        `return` $e_1 + e_2$ (SUCCESS)
  }
  `return` FAIL (FAIL)

finding U and H (and H', H'') such that[2]

$$UH_0P = H = \begin{array}{c} \\ \\ \ell \end{array}\overset{\begin{array}{cc} r-\ell & \qquad k+\ell \end{array}}{\left[\begin{array}{c|c} \begin{matrix} 1 & & \\ & \ddots & \\ & & 1 \end{matrix} & H'' \\ \hline 0 & H' \end{array}\right]}, \quad s^T = U s_0^T = \left[\begin{array}{c} s''^T \\ \hline s'^T \end{array}\right] \qquad (1)$$

[2] If the first $r - \ell$ columns are dependent, we change P.

where U is a non-singular $r \times r$ matrix. We have $e \in \mathrm{CSD}(H, s, w)$ if and only if $eP^T \in \mathrm{CSD}(H_0, s_0, w)$. Let (P, e') be the output of the algorithm and $e'' = s'' + e'H''^T$ the word $e = (e'', e')$ is in $\mathrm{CSD}(H, s, w)$.

Definition 1. *For any fixed value of n, r and w, we denote $\mathrm{WF}_{\mathrm{ISD}}(n, r, w)$ the minimal work factor (average cost in elementary operations) of Algorithm 1 to produce a solution to* CSD *(provided there is a solution), for any choices of parameters ℓ, p, W_1 and W_2.*

In the literature, elementary operations are often binary instructions. Our purpose here is to obtain a quantity allowing us to compare algorithms and to measure the impact of decoding one out of many. Any reasonably fixed polynomial time (in n) "elementary operation" will serve that purpose.

3.1 A Preview of the Analysis

When there is a single solution to CSD ("small" w, corresponding to encryption) we can provide some intuition on the significance of the parameters.

Significance of W_1 and W_2. Given p and ℓ, we would like $W_1 + W_2 = \{e_1 + e_2 \mid (e_1, e_2) \in W_1 \times W_2\}$ to contain as many distinct elements of $\mathcal{S}_{k+\ell}(\mathbf{0}, p)$ as possible, but no (or not too many) duplicate sums[3]. Typically the elements of W_1 and those of W_2 are chosen with distinct supports, for instance in [12]

$$W_1 = \left\{(e, 0) \mid e \in \mathcal{S}_{\frac{k+\ell}{2}}(\mathbf{0}, \tfrac{p}{2})\right\} \text{ and } W_2 = \left\{(0, e) \mid e \in \mathcal{S}_{\frac{k+\ell}{2}}(\mathbf{0}, \tfrac{p}{2})\right\}$$

(assuming p and $k + \ell$ are even). A proper choice of W_1 and W_2 will allow us to find most solutions $e' \in \mathrm{CSD}(H', s', p)$ (see (1) for the notations) for a relatively moderate cost (exploring $W_1 \times W_2$ uses the birthday paradox and essentially consists in exploring W_1 then W_2).

Significance of p and ℓ. The optimal size of W_1 and W_2 depends on p and ℓ. Given p, the best value for ℓ keeps a balance between the costs of the various steps of the algorithm and it is best to choose $2^\ell \approx |W_1| = |W_2|$. There is no easy interpretation of the optimal p, but an easy analysis shows that the extremal cases $p = 0$ or w (either H' or H'' vanishes in (1)) are not optimal. So there has to be an optimal value for p between 0 and w.

3.2 Links With the Other Variants of Collision Decoding

Information set decoding is an old decoding technique [25], the variants of interest today for cryptanalysis derive from Stern's collision decoding [27]. The algorithm we present here is closer to the "Punctured Split Syndrome Decoding" of Dumer [12,3]. Depending on how the sets W_1 and W_2 are chosen, we

---

[3] That is $(e_1, e_2) \neq (e'_1, e'_2)$ in $W_1 \times W_2$ such that $e_1 + e_2 = e'_1 + e'_2$.

may obtain any known variant, including the recent ball-collision decoding [6]. Of course the Algorithm 1 is an abstraction. An effective algorithm, not to speak of its implementation must include a description of how the parameters p and ℓ are chosen (something we will do) and how the sets W_1 and W_2 are selected (something we will not do completely). Our main purpose in this work is to estimate the impact of having multiple instances. This requires some flexibility in the choice of the sizes of W_1 and W_2 which is relatively natural in our abstract model, but not straightforward, though probably possible, in the above mentioned variants. We believe that the evolution of the complexity given in (10) and (11) between the single and multiple instances scenarios can be obtained for most variants of collision decoding after proper adjustments.

4 Cost Estimation

We will neglect all control instructions and assume that counting only the instructions in blocks (ISD i) will give an accurate estimation of the algorithm cost. For $i = 0, 1, 2, 3$ we will denote K_i the average cost in elementary operations (whatever that means) for executing the block of instructions (ISD i).

We are given all the algorithm parameters n, r, w, p, ℓ, W_1, and W_2. For computing probabilities (and thus cost estimates) we will make the usual random coding assumption (pseudo-randomness of syndromes) and also assume that Algorithm 1 runs on instances which have a solution (this makes sense for a cryptanalysis). We also admit the following.

Assumptions and approximations:

1. K_0, K_1, K_2, and K_3 are independent of p, ℓ, W_1 and W_2.
2. All sums $e_1 + e_2$ for $(e_1, e_2) \in W_1 \times W_2$ are distinct and $|W_1||W_2| \leq \binom{k+\ell}{p}$.
3. Up to a (small) constant factor we have for any $x \ll 1$ and any integer N

$$1 - (1 - x)^N \approx \min(1, xN)$$

Those assumptions and approximations will not cost more than a small constant factor on the cost estimations we will compute later in this paper.

All the formulas we will give in the rest of the paper will depend of one fundamental quantity denoted $\varepsilon(p, \ell)$. It is equal to the probability for some $e' \in \mathcal{S}_{k+\ell}(\mathbf{0}, p)$ to be a valid output of a particular execution of isd_loop. The following estimates helps to understand how it varies with p and ℓ

$$\varepsilon(p, \ell) \approx \frac{\binom{r-\ell}{w-p}}{\min\left(2^r, \binom{n}{w}\right)}. \tag{2}$$

Proof. (of equation (2), sketch) We consider one particular execution of isd_loop and use all the notations of the algorithm. Given H' and H'', for any $e' \in \mathcal{S}_{k+\ell}(\mathbf{0}, p)$ we count how many $s = (s', s'')$ are such that e' is a valid output of isd_loop(H', H'', s). We must have $s' = e'H'^T$ and $s'' \in \mathcal{S}_{r-\ell}(e'H''^T, w - p)$, that is $\binom{r-\ell}{w-p}$ "good" values of s (1 for s' multiplied by $\binom{r-\ell}{w-p}$ for s''). Because Algorithm 1 is executed on an instance having solutions, we must have $s \in \mathcal{U} = \{eH^T \mid e \in \mathcal{S}_n(\mathbf{0}, w)\}$. It follows that $\varepsilon(p,\ell) = \binom{r-\ell}{w-p}/|\mathcal{U}|$. Within our assumptions, the set $\mathcal{U}$ can viewed as a set of $\binom{n}{w}$ randomly chosen elements of $\{0,1\}^r$ and thus *on average*

$$|\mathcal{U}| = 2^r\left(1-\left(1-\frac{1}{2^r}\right)^{\binom{n}{w}}\right) \approx \min\left(2^r, \binom{n}{w}\right)$$

from which we deduce the expression (2) of $\varepsilon(p,\ell)$. □

Let $W_1 + W_2 = \{e_1 + e_2 \mid (e_1, e_2) \in W_1 \times W_2\}$, we also introduce

$$\mathcal{P}(p,\ell) = 1 - (1-\varepsilon(p,\ell))^{|W_1+W_2|}, \tag{3}$$

the probability of one particular execution of isd_loop succeed. Note that within our assumptions $|W_1 + W_2| = |W_1 \times W_2| = |W_1||W_2|$.

Proposition 1. *For an input (H_0, s_0) such that $\mathrm{CSD}(H_o, s_0, w) \neq \emptyset$, the Algorithm 1 will stop after executing*

$$\approx \mathcal{T}(p,\ell) = \frac{K_0}{\mathcal{P}(p,\ell)} + \frac{K_1|W_1|}{\mathcal{P}(p,\ell)} + \frac{K_2}{|W_1|\varepsilon(p,\ell)} + \frac{K_3}{2^\ell\varepsilon(p,\ell)} \tag{4}$$

elementary operations on average.

Proof. The two leftmost terms are straightforward as the average number of calls to isd_loop is equal $1/\mathcal{P}(p,\ell)$. One particular execution of (ISD 2) will inspect $|W_1|$ different sums $e_1 + e_2$ and thus succeeds with probability $\pi_2 = 1 - (1-\varepsilon(p,\ell))^{|W_1|}$. When the parameters are optimal we have $\varepsilon(p,\ell)|W_1| \ll 1$ and thus $\pi_2 \approx \varepsilon(p,\ell)|W_1|$ which accounts for the third term in (4). Finally, if the call to isd_loop fails, the block (ISD 3) will be called on average $|W_1||W_2|/2^\ell$ times. Thus if π_3 is its probability of success, we have (remember $|W_1+W_2| = |W_1||W_2|$)

$$1 - \mathcal{P}(p,\ell) = (1-\pi_3)^{\frac{|W_1||W_2|}{2^\ell}} \text{ and thus } \pi_3 = 1 - (1-\varepsilon(p,\ell))^{2^\ell}.$$

As $\varepsilon(p,\ell)2^\ell \ll 1$, we have $\pi_3 = \varepsilon(p,\ell)2^\ell$ and thus the rightmost term of (4). □

A consequence of this proposition is that the minimal cost for Algorithm 1 is obtained when $|W_2|$ is maximal (everything else being fixed), that is when $|W_1||W_2| = \binom{k+\ell}{p}$. At this point, $\mathcal{P}(p,\ell)$ is independent of W_1 and the complexity is minimal when the two middle terms of (4) are equal, that is when

$$|W_1| = \mathcal{L}(p,\ell) = \sqrt{\frac{K_2\mathcal{P}(p,\ell)}{K_1\varepsilon(p,\ell)}} = \sqrt{\frac{K_2}{K_1}}\min\left(\sqrt{\frac{1}{\varepsilon(p,\ell)}}, \sqrt{\binom{k+\ell}{p}}\right) \tag{5}$$

which is consistent with the results of [14]. We have

$$\mathrm{WF}_{\mathrm{ISD}}(n,r,w) \approx \min_{p,\ell} \mathcal{T}(p,\ell)$$

where

$$\mathcal{T}(p,\ell) = \frac{K_0}{\mathcal{P}(p,\ell)} + \frac{2K_2}{\mathcal{L}(p,\ell)\varepsilon(p,\ell)} + \frac{K_3}{2^\ell \varepsilon(p,\ell)}. \tag{6}$$

Note that when $\varepsilon(p,\ell)\binom{k+\ell}{p} < 1$, the "min" in (5) is obtained for rightmost term and W_1 and W_2 have (approximatively) the same size. Else $\mathcal{P}(p,\ell) = 1$ (which happens only when w is large) and the optimal choice consists in choosing W_1 smaller than W_2.

4.1 Lower Bound

Assuming that $K_0 = 0$ (we neglect the cost for the Gaussian elimination step), the cost estimate becomes

$$\mathcal{T}(p,\ell) = \frac{2K_2}{\mathcal{L}(p,\ell)\varepsilon(p,\ell)} + \frac{K_3}{2^\ell \varepsilon(p,\ell)} \tag{7}$$

and because the first term is increasing and the second is decreasing with ℓ (for parameters of cryptologic interest), for all p we have $\mathcal{T}(p,\ell_1)/2 \leq \min_\ell \mathcal{T}(p,\ell) \leq \mathcal{T}(p,\ell_1)$ where $\ell_1(p)$, or ℓ_1 for short, is the unique integer in $[0,r[$ such that the two terms in $\mathcal{T}(p,\ell)$ are equal, that is

$$\ell_1 = \log_2\left(\frac{K_3}{2K_2}\mathcal{L}(p,\ell_1)\right) = \log_2\left(\frac{K_3}{2\sqrt{K_1K_2}}\sqrt{\frac{\mathcal{P}(p,\ell_1)}{\varepsilon(p,\ell_1)}}\right). \tag{8}$$

The lower bound is $\mathrm{WF}_{\mathrm{ISD}}(n,r,w) \geq \min_p \mathcal{T}(p,\ell_1)/2$ and the various forms of $\mathcal{T}(p,\ell_1)$ give various interpretations of the complexity

$$\mathcal{T}(p,\ell_1) = \frac{2K_1\mathcal{L}(p,\ell_1)}{\mathcal{P}(p,\ell_1)} = \frac{2K_3}{2^{\ell_1}\varepsilon(p,\ell_1)} = \frac{2K_2}{\mathcal{L}(p,\ell)\varepsilon(p,\ell_1)} = \frac{2\sqrt{K_1K_2}}{\sqrt{\mathcal{P}(p,\ell)\varepsilon(p,\ell_1)}}$$

This bound is very tight if the Gaussian elimination cost is negligible (which is often the case in practice, see Table 2). Numbers in Table 2 may seem different from other estimates [5,14]. This difference comes from the fact that we consider column operations rather than binary operations. In fact they are very close.

4.2 Some Numbers

For Table 3 we will assume that $K_0 = nr$, $K_1 = K_2 = 1$, and $K_3 = 2$. The elementary operation being a "column operation": a column addition or the computation of a Hamming weight, possibly accompanied by a memory access. The cost for (ISD 1) and (ISD 2) can be reduced to 1 by "reusing additions", as explained in [5]. The "column" has size r bits ($r - \ell$ for (ISD 3)), however we need in practice ℓ bits for computing the index in (ISD 1) and (ISD 2), and for (ISD 3) we only need on average $2(w-p)$ additional bits [5] for deciding whether or not we reach the target weight. This sets the "practical column size" to $\ell + 2(w-p)$ instead of r. We claim that up to a small constant factor, this measure will give a realistic account for the cost of a software implementation.

Table 2. Workfactor estimates and lower bounds for generalized ISD. The code parameters of the first block of numbers corresponds to encryption, the second to the CFS digital signature scheme and the third to collision search in the (non-regular) FSB hash function.

(n, r, w)	$\log_2(\mathrm{WF}_{\mathrm{ISD}})$	$\min_p \log_2 \frac{\mathcal{T}(p, \ell_1)}{2}$
$(2048, 352, 32)$	81.0	80.5
$(2048, 781, 71)$	100.7	100.1
$(4096, 252, 21)$	80.4	80.0
$(4096, 540, 45)$	128.3	127.9
$(8192, 416, 32)$	128.8	128.4
$(2^{16}, 144, 11)$	70.2	70.1
$(2^{16}, 160, 12)$	79.4	79.3
$(2^{18}, 162, 11)$	78.9	78.8
$(2^{20}, 180, 11)$	87.8	87.7
$(5 \cdot 2^{18}, 640, 160)$	91.8	90.9
$(7 \cdot 2^{18}, 896, 224)$	126.6	125.7
$(2^{21}, 1024, 256)$	144.0	143.1
$(23 \cdot 2^{16}, 1472, 368)$	205.9	205.0
$(31 \cdot 2^{16}, 1984, 496)$	275.4	274.6

4.3 Variations with the Parameter p

With (8), we have an expression for the optimal, or nearly optimal value $\ell_1(p)$ of ℓ for a given n, r, w, and p. Even though it defines $\ell_1(p)$ implicitly, it gives an intuition of the significance and variations of ℓ_1. Finding something similar for p given n, r, and w (with $\ell = \ell_1(p)$ of course) seems to be more challenging. However, we observe that, when w is much smaller than the Gilbert-Varshamov distance (typically for encryption), the value of $\mathcal{T}(p, \ell_1(p))$ varies relatively slowly with p when p is close to the optimal.

As an illustration, we give in Table 3 values of $\mathcal{T}(p, \ell)$ (computed with (6)) for various optimal pairs (p, ℓ) and code parameters.

5 Decoding One Out of Many

We assume now that we have to solve $\mathrm{CSD}(H_0, \mathcal{S}_0, w)$ for a set of $\mathcal{S}_0$ of N independent syndromes which all have a solution. We describe a procedure for that in Algorithm 4. This algorithm is very similar to Algorithm 1. The differences are related to the set of syndromes $\mathcal{S}_0$. In the block (DOOM 0) we compute $\mathcal{S} = \{s_0 U^T \mid s_0 \in \mathcal{S}_0\}$ instead of just $s = s_0 U^T$ and in the procedure doom_loop, the second loop we run through $W_2 \times \mathcal{S}$ instead of W_2. It is still optimal to have $W_1 + W_2$ close to $\mathcal{S}_{k+\ell}(\mathbf{0}, p)$, but instead of $|W_1| = |W_2|$ in Algorithm 1, it is better now to choose $|W_1| = |W_2 \times \mathcal{S}| = N|W_2|$.

Table 3. Cost estimate for various optimal (p, ℓ) the first (top) table corresponds to encryption, the second to digital signature and the third to hashing

$(n,r,w) = (4096, 540, 45)$												
p	6	7	8	9	10	11	**12**	13	14	15	16	17
ℓ	34	38	43	47	51	56	**60**	64	68	72	76	80
$\log_2 \mathcal{T}(p,\ell)$	129.4	129.0	128.7	128.5	128.4	128.3	**128.3**	128.4	128.6	128.9	129.2	129.6

$(n,r,w) = (2^{20}, 180, 11)$							
p	4	5	6	7	8	9	**10**
ℓ	41	50	59	68	77	86	**94**
$\log_2 \mathcal{T}(p,\ell)$	106.1	102.1	98.2	94.6	91.2	88.1	**87.7**

$(n,r,w) = (2^{21}, 1024, 256)$												
p	11	12	13	14	15	**16**	17	18	19	20	21	22
ℓ	103	112	121	129	138	**144**	145	146	147	148	148	149
$\log_2 \mathcal{T}(p,\ell)$	158.4	155.1	151.8	148.5	145.3	**144.0**	144.9	145.8	146.7	147.7	148.6	149.5

We keep the same notations and use the same assumptions and approximations as in §4. We denote

$$\mathcal{P}_N(p,\ell) = 1 - (1 - \varepsilon(p,\ell))^{N|W_1||W_2|} \approx \min\left(1, \varepsilon(p,\ell)N|W_1||W_2|\right)$$

the probability for one execution of doom_loop to succeed. We have a statement very similar to Proposition 1.

Proposition 2. *For an input $(H_0, \mathcal{S}_0)$ such that $\mathrm{CSD}(H_o, s_0, w) \neq \emptyset$ for all $s_0 \in \mathcal{S}_0$ the Algorithm 4 will stop after executing*

$$\approx \mathcal{T}_N(p,\ell) = \frac{K_0}{\mathcal{P}_N(p,\ell)} + \frac{K_1|W_1|}{\mathcal{P}_N(p,\ell)} + \frac{K_2}{|W_1|\varepsilon(p,\ell)} + \frac{K_3}{2^\ell \varepsilon(p,\ell)} \tag{9}$$

elementary operations on average.

We omit the proof which is similar to the proof of Proposition 1 with an identical expression for the complexity except for $\mathcal{P}_N(p,\ell)$ (which grows with N).

5.1 Cost of Linear Algebra

The constant K_0 will include, in addition to the Gaussian elimination, the computation of all the s_oU^T for $s_0 \in \mathcal{S}_0$. This multiplies the cost, at most, by a factor $N = |\mathcal{S}_0|$. On the other hand, as long as $N \leq 1/\varepsilon(p,\ell)\binom{k+\ell}{p}$ (with larger N just reading the instances would be the bottleneck, so we discard that possibility) the probability $\mathcal{P}_N(p,\ell)$ is N times larger than before and thus the ratio $K_0/\mathcal{P}_N(p,\ell)$ do not increase. The total cost $\mathcal{T}_N(p,\ell)$ is smaller than $\mathcal{T}(p,\ell)$, so the relative contribution of the linear algebra will increase, but the simplification $K_0 = 0$ remains reasonable as long as $\mathcal{P}_N(p,\ell) \ll 1$.

When N is close or equal to $1/\varepsilon(p,\ell)\binom{k+\ell}{p}$, as in §5.3, the situation is not so simple. With fast binary linear algebra computing all the s_oU^T will require

Algorithm 4. DOOM ISD algorithm

For any fixed values of n, r and w, the following algorithm uses four parameters: two integers $p > 0$ and $\ell > 0$ and two sets $W_1 \subset \mathcal{S}_{k+\ell}(\mathbf{0}, p_1)$ and $W_2 \subset \mathcal{S}_{k+\ell}(\mathbf{0}, p_2)$ where p_1 and p_2 are positive integers such that $p_1 + p_2 = p$.

procedure main_doom
input: $H_0 \in \{0,1\}^{r\times n}$, $\mathcal{S}_0 \subset \{0,1\}^r$
  `repeat`
(DOOM 0) $\left\{\begin{array}{l} P \leftarrow \text{random } n \times n \text{ permutation matrix} \\ (H', H'', U) \leftarrow \text{PartialGaussElim}(H_0 P) \\ \mathcal{S} \leftarrow \{s_0 U^T \mid s_0 \in \mathcal{S}_0\} \end{array}\right.$ // as in (1)
    $e \leftarrow \text{doom\_loop}(H', H'', \mathcal{S})$
  `while` $e =$ FAIL
  `return` (P, e)

procedure doom_loop
input: $H' \in \{0,1\}^{\ell\times(k+\ell)}$, $H'' \in \{0,1\}^{(r-\ell)\times(k+\ell)}$, $\mathcal{S} \subset \{0,1\}^r$
  `for all` $e_1 \in W_1$
(DOOM 1) $\left\{\begin{array}{l} i \leftarrow e_1 H'^T,\ s''_1 \leftarrow e_1 H''^T \\ \text{write}(e_1, s''_1, i) \end{array}\right.$ // stores (e_1, s''_1) at index i
  `for all` $e_2 \in W_2$
    `for all` $s = (s', s'') \in \mathcal{S}$
(DOOM 2) $\left\{\begin{array}{l} i \leftarrow s' + e_2 H'^T,\ s''_2 \leftarrow s'' + e_2 H''^T \\ \text{Elts} \leftarrow \text{read}(i) \end{array}\right.$ // extracts the elements stored at index i
      `for all` $(e_1, s''_1) \in$ Elts
(DOOM 3) $\left\{\begin{array}{l} \texttt{if } \text{wt}(s''_1 + s''_2) = w - p \\ \quad \texttt{return } e_1 + e_2 \end{array}\right.$ (SUCCESS)
  `return` FAIL (FAIL)

about $Nr/\log_2 N$ column operations. For the extremal values of N of §5.3 (the case most favorable to the attacker), assuming $K_1 = K_2 = K_3/2 = 1$, we have $\mathcal{P}_n(p, \ell) = 1$ and a complexity $\approx Nr/\log_2 N + 2^{\ell+2}$ with $N = 2^{2\ell}/\binom{k+\ell}{p} \leq 2^\ell$. Unless we precisely use the optimal value of p, for which $N \approx \binom{k+\ell}{p} \approx 2^\ell$, the ratio $N/2^\ell$ will be significantly smaller than 1 and $K_0 = 0$ provides an accurate estimate. Finally when p minimizes the formula for the cost (this value, by the way, is not necessarily an integer and does not correspond to a practical implementation) we have a complexity of the form $2^\ell(r/\ell + 4)$ and we cannot neglect r/ℓ compared with 4. For the sake of simplicity, we do it nevertheless.

5.2 Complexity Gain from Multiple Instances

We will denote

$$\text{WF}^{(N)}_{\text{ISD}}(n, r, w) = \min_{p,\ell} \mathcal{T}_N(p, \ell)$$

and the gain we wish to estimate is the ratio

$$\gamma = \log_N \frac{\mathrm{WF}_{\mathrm{ISD}}(n,r,w)}{\mathrm{WF}^{(N)}_{\mathrm{ISD}}(n,r,w)}$$

which we expect to be close to 1/2. First, we must have

$$N \leq \frac{1}{\varepsilon(p,\ell)\binom{k+\ell}{p}} = \frac{\min\left(2^r, \binom{n}{w}\right)}{\binom{r-\ell}{w-p}\binom{k+\ell}{p}}$$

else there is nothing to gain. Within this bound, we have

$$\mathcal{P}_N(p,\ell) = N\varepsilon(p,\ell)\binom{k+\ell}{p} \text{ and } \mathcal{L}_N(p,\ell) = \sqrt{\frac{K_2}{K_1}}\sqrt{N\binom{k+\ell}{p}}$$

and (assuming $K_0 = 0$)

$$\mathcal{T}_N(p,\ell) = \frac{2\sqrt{K_1K_2}}{\sqrt{N\binom{k+\ell}{p}}\varepsilon(p,\ell)} + \frac{K_3}{2^\ell\varepsilon(p,\ell)}.$$

The same analysis as in §4.1 will tell us that the above sum is minimal (up to a factor at most two) when its two terms are equal, that is when $\ell = \ell_N(p)$, or ℓ_N for short, where

$$\ell_N = \log_2\left(\frac{K_3\sqrt{N\binom{k+\ell_N}{p}}}{2\sqrt{K_1K_2}}\right).$$

Proposition 3. *For a given p, we have*

$$\log_N \frac{\mathcal{T}(p,\ell_1)}{\mathcal{T}_N(p,\ell_N)} = \frac{1}{2} - c(p) \text{ where } c(p) \approx \frac{1}{2\ln 2}\frac{w-p}{r-\ell_1-\frac{w-p-1}{2}}.$$

Proof. We have

$$\ell_N = \log_2\left(\frac{K_3\sqrt{N\binom{k+\ell_N}{p}}}{2\sqrt{K_1K_2}}\right) \text{ and } \ell_1 = \log_2\left(\frac{K_3\sqrt{\binom{k+\ell_1}{p}}}{2\sqrt{K_1K_2}}\right)$$

and if we consider only the first order variations, we have $\ell_N \approx \ell_1 + \frac{1}{2}\log_2 N$. Because we have

$$\frac{d}{da}\binom{a}{b} = \binom{a}{b}\Delta(a,b) \text{ where } \Delta(a,b) = \sum_{i=0}^{b-1}\frac{1}{a-i} \approx \frac{b}{a-\frac{b-1}{2}}$$

it follows that, keeping only the first order variations, we have

$$\varepsilon(p,\ell_N) = \varepsilon(p,\ell_1)\exp(-c(p)\log N)$$

where $c(p) \approx \Delta(r-\ell_1, w-p)/2\ln(2)$. Finally

$$\frac{\mathcal{T}(p,\ell_1)}{\mathcal{T}_N(p,\ell_N)} = \frac{2^{\ell_N}\varepsilon(p,\ell_N)}{2^{\ell_1}\varepsilon(p,\ell_1)} = \sqrt{N}\exp(-c(p)\log N).$$

□

Impact of the Variations of p. The optimal value of p for large N might not be the same as for $N = 1$. In practice when $\mathcal{T}(p, \ell_1)$ vary slowly with p (parameters corresponding to encryption) the behavior of Proposition 3 can be extended to the workfactor and, as long as N is not too large, we have

$$\mathrm{WF}_{\mathrm{ISD}}^{(N)}(n,r,w) = \frac{\mathrm{WF}_{\mathrm{ISD}}(n,r,w)}{N^{\gamma}} \text{ where } \gamma \approx \frac{1}{2} - 0.721 \frac{w-p}{r-\ell_1-\frac{w-p-1}{2}} \tag{10}$$

where p and ℓ_1 are the optimal parameters of the algorithm when $N = 1$. For parameters corresponding to digital signature and hash function, the algorithm does not seem to take full benefit of multiple instances.

Table 5. Decoding N instances

(n,r,w)	$\log_2 N$	p	ℓ	$\mathrm{WF}_{\mathrm{ISD}}^{(N)}$	observed γ	expected γ
$(4096, 540, 45)$	0	12	60	128.4	–	–
$(4096, 540, 45)$	40	12	80	110.5	0.4486	0.4487
$(4096, 540, 45)$	83.7	10	94	91.6	0.4398	0.4487
$(2048, 352, 32)$	0	6	30	81.0	–	–
$(2048, 352, 32)$	40	7	54	63.4	0.4403	0.4394
$(2048, 352, 32)$	51.4	7	60	58.8	0.4324	0.4394
$(2^{20}, 180, 11)$	0	10	94	87.8	–	–
$(2^{20}, 180, 11)$	40	6	79	79.6	0.2038	0.4856
$(2^{20}, 180, 11)$	70.3	4	76	74.6	0.1875	0.4856
$(2^{21}, 1024, 256)$	0	16	144	144.0	–	–
$(2^{21}, 1024, 256)$	40	6	79	141.5	0.0640	0.2724
$(2^{21}, 1024, 256)$	117.6	4	76	137.1	0.0597	0.2724

5.3 Unlimited Number of Instances

We assume that the attacker can let N grow indefinitely. Because any algorithm must at least read its input there is a limit to the growth of N. By “unlimited” we mean that the attacker has reached this limit (whatever it is). We will denote

$$\mathrm{WF}_{\mathrm{ISD}}^{(\infty)}(n,r,w) = \min_{N,p,\ell} \mathcal{T}_N(p,\ell)$$

and we wish to compare this cost with $\mathrm{WF}_{\mathrm{ISD}}(n,r,w)$. The best strategy for the attacker is to take a number of instances equal to

$$N = \frac{1}{\varepsilon(p,\ell)\binom{k+\ell}{p}} = \frac{\min\left(2^r, \binom{n}{w}\right)}{\binom{r-\ell}{w-p}\binom{k+\ell}{p}}$$

in which case (assuming $K_0 = 0$, see the discussion in §5.1) the complexity is

$$\mathcal{T}_\infty(p,\ell) = \frac{2\sqrt{K_1K_2}}{\sqrt{\varepsilon(p,\ell)}} + \frac{K_3}{2^\ell \varepsilon(p,\ell)}$$

The minimal value is reached, up to a constant factor, when $\ell = \ell_\infty(p)$ such that

$$\ell_\infty(p) = \log_2\left(\frac{K_3}{2\sqrt{K_1K_2\varepsilon(p,\ell_\infty(p))}}\right).$$

Interestingly $\ell_\infty(p)$ is increasing with p and so is the complexity $\mathcal{T}(p,\ell_\infty(p))$. We thus want to choose p as small as possible. On the other hand, we have $|W_1||W_2| = \binom{k+\ell}{p}$ and $|W_2|$ must be a positive integer which limits the decrease of p. We must have

$$|W_1| \leq \binom{k+\ell}{p} \Rightarrow \sqrt{\frac{K_2}{K_1\varepsilon(p,\ell)}} \leq \binom{k+\ell}{p},$$

with equality for the optimal p. Finally the optimal pair (p,ℓ) is the unique one such that we have simultaneously

$$\ell = \log_2\left(\frac{K_3}{2\sqrt{K_1K_2}}\sqrt{\frac{\min\left(2^r,\binom{n}{w}\right)}{\binom{r-\ell}{w-p}}}\right) = \log_2\left(\frac{K_3}{2K_2}\binom{k+\ell}{p}\right).$$

An Estimate of the Improvement. Let p is the optimal value obtained above with an unlimited number of instances. In that case (we take $K_0 = 0$, $K_1 = K_2 = 1$, $K_3 = 2$)

$$\ell_1 = \log_2\sqrt{\binom{k+\ell_1}{p}} \text{ and } \ell_\infty = \log_2\binom{k+\ell_\infty}{p}.$$

Keeping the first order variations we have $\ell_\infty \approx 2\ell_1$. From Proposition 3 we have

$$\log_N \frac{\mathcal{T}(p,\ell_1)}{\mathcal{T}_\infty(p,\ell_\infty)} = \frac{1}{2} - c(p) \text{ where } c(p) \approx 0.721\frac{w-p}{r-\ell_1}$$

where $N \approx \mathcal{T}_\infty(p,\ell_\infty) \approx 2^{\ell_\infty}$. Thus $\mathcal{T}(p,\ell_1) \approx \mathcal{T}_\infty(p,\ell_\infty)^{\frac{3}{2}-c(p)}$

Proposition 4. *For a given p, we have*

$$\frac{\log\mathcal{T}(p,\ell_1)}{\log\mathcal{T}_\infty(p,\ell_\infty)} = \frac{2}{3} + \frac{4}{9}c(p) \text{ where } c(p) \approx \frac{1}{2\ln 2}\frac{w-p}{r-\ell_1-\frac{w-p-1}{2}}.$$

Coming back to the single instance case, and assuming that $\mathcal{T}(p,\ell_1)$ varies very slowly with p, we may assume that $\mathrm{WF}_{\mathrm{ISD}}(n,r,w) \approx \mathcal{T}(p,\ell_1)$. This means that when an attacker has access to an unlimited number of instances and needs to decode one of them only, the decoding exponent is multiplied by a quantity, slightly larger than 2/3, close to the one given in the above proposition.

$$\mathrm{WF}_{\mathrm{ISD}}^{(\infty)}(n,r,w) = \mathrm{WF}_{\mathrm{ISD}}(n,r,w)^\beta \text{ where } \beta \approx \frac{2}{3} + 0.321\frac{w-p}{r-\ell_1-\frac{w-p-1}{2}} \quad (11)$$

where p and ℓ_1 are the optimal parameters of the algorithm when $N = 1$.

We can observe that in Table 6, as for formula (10) and Table 5, the behavior is close to what we expect when encryption is concerned (when w is significantly smaller than the Gilbert-Varshamov distance). For parameters for code-based signature schemes there is a gain but not as high as expected. For parameters for code-based hashing, multiple instances does not seem to provide a big advantage. The values of p and ℓ given in the fifth and sixth columns are real numbers which minimize the formula for $\log_2(\mathrm{WF}_{\mathrm{ISD}}^{(\infty)})$. In an implementation they must be integers and the real cost will be (marginally) different.

Table 6. Workfactor with unlimited number of instances with the same code parameters as in Table 2

	$\log_2(\mathrm{WF}_{\mathrm{ISD}})$			$\log_2(\mathrm{WF}_{\mathrm{ISD}}^{(\infty)})$		observed	expected
(n, r, w)	p	ℓ		p	$=\ell$	β	β
$(2048, 352, 32)$	6	30	81.0	6.01	55.2	.682	.694
$(2048, 781, 71)$	6	29	100.7	8.20	69.2	.688	.696
$(4096, 252, 21)$	10	52	80.4	5.27	55.3	.688	.685
$(4096, 540, 45)$	12	60	128.4	9.00	88.0	.685	.689
$(8192, 416, 32)$	15	81	128.8	8.10	89.2	.693	.683
$(2^{16}, 144, 11)$	10	75	70.2	3.69	55.1	.785	.671
$(2^{16}, 160, 12)$	11	81	79.4	4.16	61.7	.777	.671
$(2^{18}, 162, 11)$	10	85	78.9	3.77	63.7	.808	.671
$(2^{20}, 180, 11)$	10	94	87.8	3.83	72.3	.824	.670
$(5 \cdot 2^{18}, 640, 160)$	10	91	91.8	4.45	84.8	.924	.768
$(7 \cdot 2^{18}, 896, 224)$	14	126	126.6	6.12	117.6	.929	.768
$(2^{21}, 1024, 256)$	16	144	144.0	6.96	134.0	.930	.768
$(23 \cdot 2^{16}, 1472, 368)$	24	206	205.9	10.48	191.7	.931	.768
$(31 \cdot 2^{16}, 1984, 496)$	32	275	275.4	14.01	257.2	.934	.767

6 Conclusion

Decoding one out of many with collision decoding provides a significant advantage to an attacker. For the digital signature scheme, the threat is real because the attacker can create many syndromes by hashing many messages (favorable to him), however what we gain with ISD is less than what Bleichenbacher obtained with GBA. Anyway it is possible to completely avoid those attacks by signing several syndromes (see [13]).

For very large values of w (used for instance in hashing) we have seen that the attack is not so worrying, moreover the actual FSB [1] or RFSB [7] use regular words and using ISD threatens an idealized version used for the security proofs. Decoding regular words is harder, and the question of how to decode one out of many and how to use it for an attack is still open.

Finally, when w is significantly smaller than the Gilbert-Varshamov distance (for public-key encryption) there is a gain. If the attacker has access to many cryptograms and is satisfied by decoding only one of them, the present work must be taken into account. We consider two scenarios: (1) the encryption scheme is used to exchange session keys, and (2) the encryption scheme is used to encrypt a long stream of data. In the first scenario the number of session keys in a public key lifetime must be used to select the security parameters according to the result of the present study. The second scenario is plausible because code-based encryption is very fast, but in that case, it is enough to introduce some kind of chaining between encrypted blocks to counter the attack. Decrypting a single block will then be of no use to the attacker.

Acknowledgements. The author would like to thank the anonymous reviewers for their valuable comments and suggestions to improve the quality of the paper.

References

1. Augot, D., Finiasz, M., Gaborit, P., Manuel, S., Sendrier, N.: SHA-3 proposal: FSB. Submission to the SHA-3 NIST Competition (2008), `http://www-rocq.inria.fr/secret/CBCrypto/index.php?pg=fsb`
2. Augot, D., Finiasz, M., Sendrier, N.: A fast provably secure cryptographic hash function. Cryptology ePrint Archive, Report 2003/230 (2003), `http://eprint.iacr.org/`
3. Barg, A.: Complexity issues in coding theory. In: Pless, V., Huffman, W. (eds.) Handbook of Coding Theory, vol. I, ch. 7, pp. 649–754. North-Holland (1998)
4. Berlekamp, E., McEliece, R., van Tilborg, H.: On the inherent intractability of certain coding problems. IEEE Trans. on Information Theory 24(3) (May 1978)
5. Bernstein, D.J., Lange, T., Peters, C.: Attacking and Defending the Mceliece Cryptosystem. In: Buchmann, J., Ding, J. (eds.) PQCrypto 2008. LNCS, vol. 5299, pp. 31–46. Springer, Heidelberg (2008)
6. Bernstein, D.J., Lange, T., Peters, C.: Smaller Decoding Exponents: Ball-Collision Decoding. In: Rogaway, P. (ed.) CRYPTO 2011. LNCS, vol. 6841, pp. 743–760. Springer, Heidelberg (2011)
7. Bernstein, D.J., Lange, T., Peters, C., Schwabe, P.: Really Fast Syndrome-Based Hashing. In: Nitaj, A., Pointcheval, D. (eds.) AFRICACRYPT 2011. LNCS, vol. 6737, pp. 134–152. Springer, Heidelberg (2011)
8. Camion, P., Patarin, J.: The Knapsack Hash Function Proposed at Crypto'89 can be Broken. In: Davies, D.W. (ed.) EUROCRYPT 1991. LNCS, vol. 547, pp. 39–53. Springer, Heidelberg (1991)
9. Canteaut, A., Chabaud, F.: A new algorithm for finding minimum-weight words in a linear code: Application to McEliece's cryptosystem and to narrow-sense BCH codes of length 511. IEEE Trans. on Information Theory 44(1), 367–378 (1998)
10. Coron, J.S., Joux, A.: Cryptanalysis of a provably secure cryptographic hash function. Cryptology ePrint Archive, Report 2004/013 (2004), `http://eprint.iacr.org/`
11. Courtois, N.T., Finiasz, M., Sendrier, N.: How to Achieve a McEliece-Based Digital Signature Scheme. In: Boyd, C. (ed.) ASIACRYPT 2001. LNCS, vol. 2248, pp. 157–174. Springer, Heidelberg (2001)
12. Dumer, I.: On minimum distance decoding of linear codes. In: Proc. 5th Joint Soviet-Swedish Int. Workshop Inform. Theory, Moscow, pp. 50–52 (1991)
13. Finiasz, M.: Parallel-CFS: Strengthening the CFS McEliece-Based Signature Scheme. In: Biryukov, A., Gong, G., Stinson, D.R. (eds.) SAC 2010. LNCS, vol. 6544, pp. 159–170. Springer, Heidelberg (2011)

14. Finiasz, M., Sendrier, N.: Security Bounds for the Design of Code-Based Cryptosystems. In: Matsui, M. (ed.) ASIACRYPT 2009. LNCS, vol. 5912, pp. 88–105. Springer, Heidelberg (2009)
15. Fischer, J.B., Stern, J.: An Efficient Pseudo-Random Generator Provably as Secure as Syndrome Decoding. In: Maurer, U.M. (ed.) EUROCRYPT 1996. LNCS, vol. 1070, pp. 245–255. Springer, Heidelberg (1996)
16. Gaborit, P., Girault, M.: Lightweight code-based identification and signature. In: IEEE Conference, ISIT 2007, pp. 191–195. IEEE, Nice (2007)
17. Gaborit, P., Laudaroux, C., Sendrier, N.: Synd: a very fast code-based stream cipher with a security reduction. In: IEEE Conference, ISIT 2007, pp. 186–190. IEEE, Nice (2007)
18. Johansson, T., Jönsson, F.: On the complexity of some cryptographic problems based on the general decoding problem. IEEE-IT 48(10), 2669–2678 (2002)
19. Lee, P.J., Brickell, E.F.: An Observation on the Security of McEliece's Public-Key Cryptosystem. In: Günther, C.G. (ed.) EUROCRYPT 1988. LNCS, vol. 330, pp. 275–280. Springer, Heidelberg (1988)
20. Leon, J.: A probabilistic algorithm for computing minimum weights of large error-correcting codes. IEEE Trans. on Information Theory 34(5), 1354–1359 (1988)
21. McEliece, R.: A public-key cryptosystem based on algebraic coding theory. DSN Prog. Rep., Jet Prop. Lab., California Inst. Technol., Pasadena, CA pp. 114–116 (January 1978)
22. Niederreiter, H.: Knapsack-type cryptosystems and algebraic coding theory. Prob. Contr. Inform. Theory 15(2), 157–166 (1986)
23. Overbeck, R., Sendrier, N.: Code-based cryptography. In: Bernstein, D., Buchmann, J., Dahmen, E. (eds.) Post-Quantum Cryptography, pp. 95–145. Springer, Heidelberg (2009)
24. Peters, C.: Curves, Codes, and Cryptography. Ph.D. thesis, Technische Universiteit Eindhoven (2011)
25. Prange, E.: The use of information sets in decoding cyclic codes. IRE Transactions IT-8, S5–S9 (1962)
26. Saarinen, M.J.: Linearization Attacks against Syndrome Based Hashes. In: Srinathan, K., Rangan, C.P., Yung, M. (eds.) INDOCRYPT 2007. LNCS, vol. 4859, pp. 1–9. Springer, Heidelberg (2007)
27. Stern, J.: A Method for Finding Codewords of Small Weight. In: Wolfmann, J., Cohen, G. (eds.) Coding Theory 1988. LNCS, vol. 388, pp. 106–113. Springer, Heidelberg (1989)
28. Stern, J.: A New Identification Scheme Based on Syndrome Decoding. In: Stinson, D.R. (ed.) CRYPTO 1993. LNCS, vol. 773, pp. 13–21. Springer, Heidelberg (1994)
29. Véron, P.: Improved identification schemes based on error-correcting codes. AAECC 8(1), 57–69 (1997)
30. Wagner, D.: A Generalized Birthday Problem. In: Yung, M. (ed.) CRYPTO 2002. LNCS, vol. 2442, pp. 288–303. Springer, Heidelberg (2002)

On Provable Security of UOV and HFE Signature Schemes against Chosen-Message Attack

Koichi Sakumoto, Taizo Shirai, and Harunaga Hiwatari

Sony Corporation
5-1-12 Kitashinagawa Shinagawa-ku, Tokyo 141-0001, Japan
{Koichi.Sakumoto,Taizo.Shirai,Harunaga.Hiwatari}@jp.sony.com

Abstract. The multivariate public key cryptosystem (MPKC) is considered to be one of the candidates of post-quantum cryptography. Unbalanced Oil-Vinegar (UOV) scheme and Hidden Field Equation (HFE) scheme are well-known schemes in MPKC. However, little attention has been given to provable security for these schemes. In this paper, we study the provable security of the UOV and the HFE signature schemes in the sense of the existential unforgeability against adaptive chosen-message attack (EUF-CMA). Concretely, we suggest that a usual security proof for the Full-Domain Hash scheme cannot directly apply to that of the UOV and the HFE signature schemes. However, we show that the UOV and the HFE signature schemes can be modified into ones achieving the EUF-CMA in the random oracle model, without changing each underlying trapdoor function.

Keywords: signature scheme, MPKC, multivariate, distribution, UOV, HFE, provable security.

1 Introduction

One of the recent research challenges in public key cryptography is to find alternative public key cryptosystem which has resistance to a quantum computer [3]. The multivariate public key cryptosystem (MPKC) is one of the candidates for post-quantum cryptography. MPKC is based on a problem of solving a system of multivariate quadratic polynomials, which is called an MQ problem [9]. The Unbalanced Oil-Vinegar (UOV) scheme [16] and the Hidden Field Equation (HFE) scheme [22] are well-known and deeply studied schemes in MPKC. These schemes use a trapdoor one-way function whose security relies both on the MQ problem and on the Isomorphism of Polynomials (IP) problem. Compared with RSA, the computation in MPKC can be implemented efficiently [4,7], since the arithmetic operations are performed over a small finite field.

Many digital signature schemes based on integer factoring or discrete logarithm achieve the existential unforgeability against adaptive chosen-message attack (EUF-CMA) [14]. Recently, Sakumoto et al. proposed provably secure identification/signature schemes based on the MQ problem [25]. By contrast,

B.-Y. Yang (Ed.): PQCrypto 2011, LNCS 7071, pp. 68–82, 2011.

little attention has been given to provable security for the UOV and the HFE schemes. Although Courtois studied provable security against key-only attack on Quartz which is a variant of HFE [8], the security against chosen-message attack is unclear. In this paper, we study provable security of the UOV and the HFE signature schemes in the sense of EUF-CMA.

We note that it is not clear whether the UOV and the HFE signature schemes satisfy EUF-CMA or not, even if their underlying trapdoor functions are assumed to be one-way. The UOV and the HFE signature schemes employ the well-known "hash-and-sign" paradigm like the Full-Domain Hash (FDH) scheme, which is formalized by Bellare and Rogaway [1]. Let f be a trapdoor one-way permutation and f^{-1} its inverse. Specifically, they showed that if H is a hash function from $\{0,1\}^*$ to the domain of f^{-1}, then the FDH is provably secure against chosen-message attack in the random oracle model. However, unfortunately each underlying trapdoor function of the UOV and the HFE is not permutation.

Our Contribution. First, we suggest that a usual security proof for the Full-Domain Hash scheme cannot directly apply to that of the UOV and the HFE signature schemes. In the security proof for FDH-like schemes, a signing oracle has to sample signatures from actual distribution. However, the simulation of the signing oracle for the UOV and the HFE schemes might not be done by the usual manner, since their signatures are not uniformly distributed.

The UOV function consists of a secret non-bijective function $F_{\mathtt{UOV}}(x_1,\ldots,x_n, x_{n+1},\ldots,x_{n+v})$. We call $x_1,\ldots,x_n$ oil variables and $x_{n+1},\ldots,x_{n+v}$ vinegar variables. Once a set of randomly chosen vinegar variables is fixed as $x'_{n+1},\ldots,x'_{n+v}$, the number of elements included in the range of the map $(x_1,\ldots,x_n) \mapsto F_{\mathtt{UOV}}(x_1,\ldots,x_n,x'_{n+1},\ldots,x'_{n+v})$ is determined by the choice of $(x'_{n+1},\ldots,x'_{n+v})$. In the signing algorithm, a signer repeatedly chooses sets of vinegar variables until the range covers the chosen hash value. This makes the vinegar variables non-uniform.

On the other hand, the HFE function consists of a secret non-bijective function $F_{\mathtt{HFE}}$. Since $F_{\mathtt{HFE}}$ is a univariate mapping on a big field with degree which is less than some parameter d, it is known that each element in target space of $F_{\mathtt{HFE}}$ has 0 to d preimages. Note that, for a map $f : A \to B$, we call B target and $\{f(a)|a \in A\}$ range of f, respectively. In the signing algorithm, a signer randomly chooses one of preimages of a hashed message. Thus the difference of the number of preimages causes that signatures are not uniformly distributed.

Then, we show that the UOV and the HFE signature schemes can be modified into ones achieving the EUF-CMA without changing each underlying trapdoor function. The modifications require some additional cost, but are simple and straightforward. In particular, the modified signature-generation process makes the distribution of signatures uniform. The security of the modified schemes is proved in the random oracle model under an assumption that the underlying trapdoor function is one-way.

Related Work. Gentry et al. introduced a new notion of a trapdoor function with preimage sampling and gave a lattice-based concrete example which is surjective and many-to-one [13]. Their trapdoor function f satisfies two crucial properties which are used in the proof of the security for FDH-like signature schemes. First, $f(x)$ is uniformly distributed when x is sampled from a certain distribution D. Second, the inversion algorithm of the trapdoor function does not just find an arbitrary preimage of the input but samples from its preimages under the appropriate conditional distribution related to D. On the other hand, we just modify the signing algorithm to provide uniform distribution of the signatures, keeping the underlying trapdoor function untouched.

Organization. In Section 2 we briefly review some notations and definitions used throughout this paper. In Section 3 we review the UOV and the HFE signature schemes and analyze their signature distribution. In Section 4, we describe the modifications of these signature schemes and prove their EUF-CMA. Section 5 and Section 6 give some extensions and conclusion, respectively.

2 Preliminaries

We start by recalling the definition of a signature scheme.

Definition 1. *A signature scheme* $(\texttt{Gen}, \texttt{Sig}, \texttt{Ver})$ *is defined as follows:*

The key-generation algorithm $\texttt{Gen}$ *is a probabilistic algorithm which given* 1^λ*, outputs a pair of matching public and private keys,* (pk, sk)*.*

The signing algorithm $\texttt{Sig}$ *is a probabilistic algorithm which takes the message* M *to be signed and a secret key* sk*, and returns a signature* $\sigma = \texttt{Sig}_{sk}(M)$*.*

The verification algorithm $\texttt{Ver}$ *takes a message* M*, a candidate signature* σ *and* pk*, and returns a bit* $\texttt{Ver}_{pk}(M, \sigma)$*. The signature is accepted, only if the bit is equal to one. Otherwise, it is rejected. If* $\sigma \leftarrow \texttt{Sig}_{sk}(M)$*, then* $\texttt{Ver}_{pk}(M, \sigma) = 1$*.*

In the existential unforgeability under the adaptive chosen message attack scenario [14], the forger can obtain signatures on messages of his adaptive choices and attempt to output a valid forgery. A valid forgery is a message/signature pair (M, σ) such that $\texttt{Ver}_{pk}(M, \sigma) = 1$ whereas the forger never requests the signature on M. The security against chosen-message attack, in the random oracle model, is defined as follows.

Definition 2. *We say that a signature scheme* $(\texttt{Gen}, \texttt{Sig}, \texttt{Ver})$ *is* $(t(\lambda), q_s(\lambda), q_H(\lambda), \epsilon(\lambda))$*-secure if there is no forger* $\texttt{A}$ *who takes a public key* pk *generated via* $(pk, \cdot) \leftarrow \texttt{Gen}(1^\lambda)$*, after at most* $q_H(\lambda)$ *queries to the random oracle,* $q_s(\lambda)$ *signature queries, and* $t(\lambda)$ *processing time, then outputs a valid signature with probability at least* $\epsilon(\lambda)$*.*

Rank of a Random Matrix. Especially, we refer the equation (3) in the paper [19]. This formula gives the probability $p(q, m, n, i)$ that a random $m \times n$ matrix A on a field k with cardinality q has the rank i $(i > 0)$ where $m \leq n$. Then $p(q, m, n, i)$ is equal to

$$\frac{\prod_{j=m-i+1}^{m}(1-q^{-j})\prod_{j=n-i+1}^{n}(1-q^{-j})}{\prod_{j=1}^{i}(1-q^{-j})}. \tag{1}$$

We use this formula for analyzing the distribution and the efficiency on the UOV signature schemes.

The FDH and the PFDH Schemes. In the FDH scheme, a signature on a message M is generated via $\sigma \leftarrow f^{-1}(H(M))$ and verified via $f(\sigma) = H(M)$. Moreover, in the probabilistic FDH (PFDH) scheme which is parameterized by the length parameter l of random salt, a signature $\sigma = (x, r)$ on a message M is generated via $r \in_R \{0,1\}^l$ and $x \leftarrow f^{-1}(H(M||r))$ and verified via $f(x) = H(M||r)$.

3 Existing Schemes and Their Analyses

There are many FDH-like schemes in MPKC, for instance, the Matsumoto-Imai (MI) [20], the HFE [22], and the UOV [16] signature schemes. In this paper, we study the UOV and the HFE signature schemes, each of which employs a non-bijective trapdoor one-way function. It is well-known that FDH-like signature schemes using a trapdoor permutation achieve the EUF-CMA. However, it is unclear in the case that a non-bijective trapdoor function is used. The Rabin signature scheme in [2], which is a PFDH-like signature scheme using non-bijective trapdoor function, has the provable EUF-CMA, since the Rabin function has the useful property where every element in the range always has four preimages and the signature is uniformly distributed. On the other hand, in the UOV and the HFE functions, every element in the range has non-constant number of preimages. Therefore it seems to be difficult to sample from distribution of signatures without knowledge of trapdoor, because the distribution is non-uniform. In this section, we review the UOV and the HFE signature schemes, and analyze distribution of their signatures.

3.1 UOV Signature Scheme

In [23], Patarin designed the Oil and Vinegar (OV) signature scheme. The original OV signature scheme was broken by Kipnis and Shamir [17]. However, Kipnis et al. suggested that the attack does not work if we use more vinegar variables than oil variables [16]. They also presented such a scheme called the Unbalanced Oil-Vinegar (UOV) signature scheme. Cao et al. revisited the Kipnis-Shamir attack on the UOV scheme [6], and Braeken et al. studied its security in several aspects [5]. However, no general efficient attack for one-wayness of the UOV function has been reported so far. Even though Faugère and Perret reported that the one-wayness of the UOV function with only 64-bit output is broken in a complexity bounded by $2^{40.3}$ [12], it is applicable to limited sizes.

The UOV function and its security notion are defined as follows.

Definition 3. *Let $S : k^{n+v} \to k^{n+v}$ be an invertible affine transformation, n and v positive integers, k a finite field with cardinality q, and $F_{\mathtt{UOV}} : k^{n+v}$ a map $\to k^n, (x_1, \ldots, x_{n+v}) \mapsto (f_1, \ldots, f_n)$. f_ξ is defined by*

$$f_\xi = \sum_{1 \le i \le n < j \le n+v} \alpha_{\xi ij} x_i x_j + \sum_{n < i,j \le n+v} \alpha_{\xi ij} x_i x_j + \sum_{1 \le i \le n+v} \beta_{\xi i} x_i + \gamma_\xi,$$

for $\xi = 1, \ldots, n$. In this paper, we denote $\boldsymbol{x}_n = (x_1, \ldots, x_n)$, $\boldsymbol{x}_v = (x_{n+1}, \ldots, x_{n+v})$, and the map $F_{\mathtt{UOV}}$ as

$$F_{\mathtt{UOV}}(\boldsymbol{x}_n, \boldsymbol{x}_v) = A(\boldsymbol{x}_v)\boldsymbol{x}_n^T + (g_1(\boldsymbol{x}_v), \ldots, g_n(\boldsymbol{x}_v))^T,$$

where $A(\boldsymbol{x}_v) = [a_{i,j}(\boldsymbol{x}_v)]$ is an $n \times n$ matrix whose entries $a_{i,j}$ are polynomials of the first degree on $x_{n+1}, \ldots, x_{n+v}$ and $g_i(\boldsymbol{x}_v)$ is a quadratic polynomial on $x_{n+1}, \ldots, x_{n+v}$. $x_1, \ldots, x_n$ are called oil variables and $x_{n+1}, \ldots, x_{n+v}$ are called vinegar variables. The UOV function is defined as $P_{\mathtt{UOV}} = F_{\mathtt{UOV}} \circ S$ and its generation algorithm `GenUOVfunc` *is a probabilistic algorithm which takes a security parameter 1^λ and output $(P_{\mathtt{UOV}}, (S, F_{\mathtt{UOV}}))$, where q, n, and v are bounded by polynomial on λ.*

Definition 4. *We say that the UOV function generator* `GenUOVfunc` *is $(t(\lambda), \epsilon(\lambda))$-secure if there is no inverting algorithm that takes as input $P_{\mathtt{UOV}}$ generated via $(P_{\mathtt{UOV}}, \cdot) \leftarrow$* `GenUOVfunc`$(1^\lambda)$ *and a challenge $y \in_R k^n$, then finds a preimage x such that $P_{\mathtt{UOV}}(x) = y$ at $t(\lambda)$ processing time with probability at least $\epsilon(\lambda)$.*

The UOV signature scheme (`GenUOV`, `SigUOV`, `VerUOV`) is an FDH-like scheme using the UOV function [16]. If the vinegar variable are fixed to $\boldsymbol{x}'_v$, the function $F_{\mathtt{UOV}}(\boldsymbol{x}_n, \boldsymbol{x}'_v)$ is defined by a set of linear functions on oil variables $\boldsymbol{x}_n$. The signing algorithm, for a hash value y, first randomly chooses the vinegar variables $\boldsymbol{x}'_v$ and then computes unknown oil variables $\boldsymbol{x}_n$ by solving the linear equation system $F_{\mathtt{UOV}}(\boldsymbol{x}_n, \boldsymbol{x}'_v) = y$. If there is no solution, then we simply retry by choosing new random vinegar variables $\boldsymbol{x}'_v$.

The details of the scheme are the follows. The key-generation algorithm `GenUOV`(1^λ), on input 1^λ, runs $(P_{\mathtt{UOV}}, (S, F_{\mathtt{UOV}})) \leftarrow$ `GenUOVfunc`(1^λ). It outputs (pk, sk), where $pk = P_{\mathtt{UOV}}$ and $sk = (S, F_{\mathtt{UOV}})$. The signing and the verification algorithms use a hash function $H : \{0,1\}^* \to k^n$ which maps a bit string of arbitrary length to an element in k^n. The signing algorithm `SigUOV`$_{sk}(M)$ is as follows.

Signing algorithm `SigUOV`$_{sk}(M)$

1: $y \leftarrow H(M)$;
2: **repeat**
3: $\quad \boldsymbol{x}'_v \in_R k^v$;
4: **until** $\{\boldsymbol{z}_n | F_{\mathtt{UOV}}(\boldsymbol{z}_n, \boldsymbol{x}'_v) = y\} \neq \emptyset$
5: $\boldsymbol{x}'_n \in_R \{\boldsymbol{z}_n | F_{\mathtt{UOV}}(\boldsymbol{z}_n, \boldsymbol{x}'_v) = y\}$;
6: $x \leftarrow S^{-1}(\boldsymbol{x}'_n, \boldsymbol{x}'_v)$;
7: **return** $\sigma = x$

The verification algorithm $\texttt{VerUOV}_{pk}(\sigma, M)$, on a signature $\sigma = x$ and a message M, returns 1 if $P_{\text{UOV}}(x) = H(M)$. Otherwise, it returns 0. Note that the step 5 can be computed by using the Gaussian elimination.

Analysis of the scheme. First, we point out that a rank of the $n \times n$ matrix $A(\boldsymbol{x}'_v)$ defined in Definition 3 depends on a set of vinegar variables $\boldsymbol{x}'_v$. From the formula (1) in Section 2, $n \times n$ random matrix with element on k with cardinality q has rank $i \in \{1, \ldots, n\}$ with probability $p(q, n, n, i) = q^{-(n-i)^2}(\prod_{j=n-i+1}^{n}(1-q^{-j}))^2/\prod_{j=1}^{i}(1-q^{-j})$. For example, $p(2, 80, 80, 80) \approx 0.289$ and $p(2^8, 10, 10, 10) \approx 0.996$.

Then, we mention the non-uniformity of signatures on the UOV scheme. At the step 3, suppose that it chooses a set of vinegar variables $\boldsymbol{x}'_v$ such that the rank of $A(\boldsymbol{x}'_v)$ is equal to i. In this case, the map $k^n \to k^n, \boldsymbol{x}_n \mapsto F_{\text{UOV}}(\boldsymbol{x}_n, \boldsymbol{x}'_v)$ is a q^{n-i}-to-1 mapping. Therefore, the ratio of elements y in k^n such that $\{\boldsymbol{z}_n | F_{\text{UOV}}(\boldsymbol{z}_n, \boldsymbol{x}'_v) = y\}$ is not empty is $1/q^{n-i}$. The higher the rank of $A(\boldsymbol{x}'_v)$ is, the higher the probability of ending the loop is. As a result, such a set of vinegar variables tends to be output more frequently. Thus the signature distribution is non-uniform.

To simulate the signing oracle in the security proof, the simulator has to sample signatures from the real distribution of UOV. However, it is difficult for the UOV scheme, since the structure of the secret map F_{UOV} is hidden due to a secret mapping S. To our knowledge, no security proof against chosen-message attack has been presented on a signature scheme based on the UOV function.

3.2 HFE Signature Scheme

The HFE cryptosystem is proposed by Patarin [22], which is a generalized version of the Matsumoto-Imai cryptosystem [20]. The cryptanalyses on HFE for certain sets of parameters are presented in many papers [18,11,15]. In particular, Granboulan et al. estimated the complexity $O(n^{O(\log d)})$ of the attack using Gröbner basis algorithm where $q = 2$ and d is polynomial on n [15]. In the attack, field equations $x_i^2 - x_i = 0$, $i = 1, \ldots, n$, are used in the computations of the Gröbner basis. The field equations can be easily used in the case $q = 2$.

The HFE-minus (HFE$^-$) function is a variant of the HFE function, which is not considered to be broken yet. The HFE$^-$ function is defined as $P_{\text{HFE}^-}(x) = (p_1(x), \ldots, p_{n-m}(x))$ where the HFE function is $P_{\text{HFE}}(x) = (p_1(x), \ldots, p_n(x))$, that is, the last m components of the output vector of the HFE function is removed. However, we note that reasonable parameters of the HFE$^-$ function are still controversial, and the parameters should be appropriately chosen. For simplicity, we analyze the plain HFE signature scheme in this section, but its minus variant also has a similar problem.

The HFE function and its security notion are defined as follows.

Definition 5. *Let $S, T : k^n \to k^n$ be invertible affine transformations, $\phi : K \to k^n$ the standard linear isomorphism given by $\phi(a_0 + a_1 x + \ldots + a_{n-1}x^{n-1}) = (a_0, a_1, \ldots, a_{n-1})$, q, n, and d positive integers, k a finite field with cardinality*

q, $g(x) \in k[x]$ an irreducible polynomial of degree n, $K = k[x]/g(x)$ a field with a degree n extension of k, and $F_{\mathtt{HFE}} : K \to K$ defined by

$$F_{\mathtt{HFE}}(X) = \sum_{i=0}^{n-1}\sum_{j=0}^{n-1} a_{ij} X^{q^i+q^j} + \sum_{i=0}^{n-1} b_i X^{q^i} + c,$$

where $a_{ij}, b_i, c \in K$ such that $\forall a_{ij}, 1 \leq i, j \leq n, q^i + q^j > d \Rightarrow a_{ij} = 0$ and $\forall b_i, 1 \leq i \leq n, q^i > d \Rightarrow b_i = 0$. That is, d is the maximum degree of $F_{\mathtt{HFE}}$. The HFE function is $P_{\mathtt{HFE}} = T \circ \phi \circ F_{\mathtt{HFE}} \circ \phi^{-1} \circ S$ and its generator `GenHFEfunc` *is a probabilistic algorithm which takes a security parameter 1^λ and output $(P_{\mathtt{HFE}}, (S, F_{\mathtt{HFE}}, T))$, where q, n, and d are bounded by polynomial on λ.*

Definition 6. *We say that the HFE function generator* `GenHFEfunc` *is $(t(\lambda), \epsilon(\lambda))$-secure if there is no inverting algorithm that takes $P_{\mathtt{HFE}}$ generated via $(P_{\mathtt{HFE}}, \cdot) \leftarrow \mathtt{GenHFEfunc}(1^\lambda)$ and a challenge $y \in_R k^n$, then finds a preimage x such that $P_{\mathtt{HFE}}(x) = y$ at $t(\lambda)$ processing time with probability at least $\epsilon(\lambda)$.*

The HFE signature scheme (GenHFE, SigHFE, VerHFE) is a PFDH-like scheme using a neither surjective nor injective function, the HFE function $P_{\mathtt{HFE}}$. The key-generation algorithm $\mathtt{GenHFE}(1^\lambda)$, on input 1^λ, runs $(P_{\mathtt{HFE}}, (S, F_{\mathtt{HFE}}, T)) \leftarrow \mathtt{GenHFEfunc}(1^\lambda)$. It outputs (pk, sk), where $pk = P_{\mathtt{HFE}}$ and $sk = (S, F_{\mathtt{HFE}}, T)$. The signing and the verification algorithms use a hash function $H : \{0,1\}^* \to k^n$ which maps a bit string of arbitrary length to an element in k^n. The signing algorithm $\mathtt{SigHFE}_{sk}(M)$ is as follows.

Signing algorithm $\mathtt{SigHFE}_{sk}(M)$

1: $r \leftarrow 0$;
2: **repeat**
3: $\quad y \leftarrow \phi^{-1}(T^{-1}(H(r, M)))$; $r \leftarrow r + 1$;
4: **until** $\{z | F_{\mathtt{HFE}}(z) = y\} \neq \emptyset$
5: $x' \in_R \{z | F_{\mathtt{HFE}}(z) = y\}$; $x \leftarrow S^{-1}(\phi(x'))$;
6: **return** $\sigma = (x, r)$

The verification algorithm $\mathtt{VerHFE}_{pk}(\sigma, M)$, on a signature $\sigma = (x, r)$ and a message M, returns 1 if $P_{\mathtt{HFE}}(x) = H(r, M)$. Otherwise, it returns 0. Note that, $x' \in_R \{z | F_{\mathtt{HFE}}(z) = y\}$ at the step 5 can be computed by using the Berlekamp algorithm.

In the signing algorithm, it repeats the loop until $\{z | F_{\mathtt{HFE}}(z) = y\} \neq \emptyset$. In Quartz [24] which uses a variant of the HFE function, it loops until $\#\{z | F_{\mathtt{HFE}}(z) = y\} = 1$. The algorithm outputs only x such that x always has no 2nd preimage.

Analysis of the scheme. It is known that each element in target space of the HFE function has various number of preimage [9]. This means that the output of the signing algorithm is non-uniformly distributed even though $H(r, M)$ distributes uniformly. Suppose that x_0 is an element in domain such that $P_{\mathtt{HFE}}(x_0)$ has i preimages. Then, the signing algorithm returns x_0 without repeating loops with probability $1/q^n \cdot 1/i$. In particular, the probability that x_1 is chosen is i_2/i_1 times higher than x_2 is chosen, where the number of preimage of $P_{\mathtt{HFE}}(x_1)$ and $P_{\mathtt{HFE}}(x_2)$ be i_1 and i_2, respectively.

To simulate the signing oracle in the security proof, it is required to sample signatures from the real distribution. However, it seems to be difficult for the HFE scheme, since the structure of $F_{\mathtt{HFE}}$ is hidden due to the pair of secret mappings S and T. The same thing may be said of the Quartz-like scheme.

4 Slightly Modified Schemes

In the UOV-based and the HFE-based schemes, it might be difficult to sample from the actual distribution of signatures without knowledge of the secret key. This is an obstacle for the provable security against chosen-message attack. To solve this problem, we slightly modify the signing algorithm whose output is uniformly distributed. Note that we do not change the underlying trapdoor function.

In this section, for each of the UOV and the HFE functions, we present basic idea for the simple modification of the signing algorithm, a proof of the security against chosen-message attack, and analysis of its efficiency.

4.1 Modified UOV Signature Scheme

A basic idea for the modification of the UOV signing algorithm is to use a random salt hashed with a message, and to re-choose a random salt instead of vinegar variables. In the original signing algorithm, the higher the rank of $A(\boldsymbol{x}'_v)$ is, the higher the probability of ending the loop is. This means that such a set of vinegar variables causing higher rank tends to be output more frequently by signing algorithm. In our modified UOV signature scheme, hash values y are generated via $y \leftarrow H(M||r)$ where r is a random salt. If the linear equation system represented as $F_{\mathtt{UOV}}(\boldsymbol{x}_n, \boldsymbol{x}'_v) = H(M||r)$ has no solution, then the signer tries again with new random salt r and generates hash value $H(M||r)$ instead of choosing new vinegar variables. As a result, the output $\boldsymbol{x}'_v$ is uniformly distributed. The output $\boldsymbol{x}'_n$ is also uniformly distributed, since if $A(\boldsymbol{x}'_v)$ has rank i, a map $\boldsymbol{x}_n \mapsto F_{\mathtt{UOV}}(\boldsymbol{x}_n, \boldsymbol{x}'_v)$ is q^{n-i}-to-1 mapping for each element in the range.

The modified UOV signature scheme (`GenUOV*`, `SigUOV*`, `VerUOV*`) is defined as follows. The key-generation algorithm `GenUOV*`, on input 1^λ, runs $(P_{\mathtt{UOV}}, (S, F_{\mathtt{UOV}})) \leftarrow \mathtt{GenUOVfunc}(1^\lambda)$. It outputs (pk, sk), where $pk = (P_{\mathtt{UOV}}, l)$, $sk = (S, F_{\mathtt{UOV}}, l)$, and l is a length of the random salt bounded by polynomial on λ. `GenUOV*` is the same as the original `GenUOV` except that public and secret keys contain l. The signing algorithm `SigUOV*` is the follows.

Signing algorithm $\mathtt{SigUOV}^*_{sk}(M)$

1: $\boldsymbol{x}'_v \in_R k^v$;
2: **repeat**
3: $\quad r \in_R \{0,1\}^l$; $y \leftarrow H(M||r)$;

4: **until** $\{\boldsymbol{z}_n | F_{\mathtt{UOV}}(\boldsymbol{z}_n, \boldsymbol{x}'_v) = y\} \neq \emptyset$
5: $\boldsymbol{x}'_n \in_R \{\boldsymbol{z}_n | F_{\mathtt{UOV}}(\boldsymbol{z}_n, \boldsymbol{x}'_v) = y\}$;
6: $x \leftarrow S^{-1}(\boldsymbol{x}'_n, \boldsymbol{x}'_v)$;
7: **return** $\sigma = (x, r)$

The verification algorithm $\mathtt{VerUOV}^*_{pk}(\sigma, M)$ returns 1 if $P_{\mathtt{UOV}}(x) = H(M||r)$, otherwise returns 0. We treat the hash function H as the random oracle in our analysis. If l is large enough, then a random salt r is fresh every time with overwhelming probability. Therefore, each y is independently and uniformly distributed over k^n, and the output x and r of the above signing algorithm are also uniformly distributed over k^{n+v} and over $\{0,1\}^l$, respectively.

Next, we show the security of the slightly modified UOV scheme against chosen-message attack.

Theorem 1. *If the UOV function is (ϵ', t')-secure, the modified UOV scheme is (ϵ, t, q_H, q_s)-secure, where $\epsilon = \epsilon'(q_H + q_s + 1)/(1 - (q_H + q_s)q_s 2^{-l})$, $t = t' - (q_H + q_s + 1)(t_{\mathtt{UOV}} + O(1))$, and $t_{\mathtt{UOV}}$ is running time to compute the UOV function $P_{\mathtt{UOV}}$.*

Proof sketch. This proof is similar to a usual proof for FDH-like signature schemes. Here we briefly describe the simulation of the random oracle and the signing oracle. In the simulation of the random oracle, the simulator answers just a random value $h \in_R k^n$, but returns the given challenge $y \in k^n$ only once. In the simulation of the signing oracle, the simulator answers a signature $\sigma = (x, r)$ and sets $H(M||r) \leftarrow P_{\mathtt{UOV}}(x)$ where $x \in_R k^{n+v}$ and $r \in_R \{0,1\}^l$. Note that, on condition where random variables (x, r) are output of the signing algorithm taking input M, the conditional distribution of hash values $H(M||r)$ is identical to the distribution of $P_{\mathtt{UOV}}(x)$ where $x \in_R k^{n+v}$. Because the value x output by the signing algorithm is uniformly distributed over k^{n+v} and satisfies $P_{\mathtt{UOV}}(x) = H(M||r)$. The details of this proof are given in Appendix A. □

Efficiency of the modified UOV. Let p_i be the probability that the rank of $A(x'_{n+1}, \ldots, x'_{n+v})$ is equal to i where $(x'_{n+1}, \ldots, x'_{n+v}) \in_R k^v$. We assume that the probability p_i follows the formula (1). On condition that the rank of $A(x'_{n+1}, \ldots, x'_{n+v})$ is equal to i at the step 1, the conditional probability that it does not repeat the loop again is $1/q^{n-i}$ and the conditional expectation of the number of loops is q^{n-i}. For example, in the case of $(q, n) = (2^8, 10)$, the probabilities p_i that the expected number of loop q^{n-i} is 1, 2^8, 2^{16}, and 2^{24} are about 0.996, 2^{-8}, 2^{-32}, and 2^{-72}, respectively. The expected number of loops $\sum_{i=0}^{n} p_i q^{n-i}$ is 2.0.

On the other hand, in the original signing algorithm, the probability to escape the loop with a set of vinegar variables $\boldsymbol{x}'_v$ such that the rank of $A(\boldsymbol{x}'_v)$ is equal to i is p_i/q^{n-i}. Thus, it returns without repeating loops with probability $\sum_{i=0}^{n} p_i/q^{n-i}$. Accordingly, the expected number of loops is $1/(\sum_{i=0}^{n} p_i/q^{n-i})$. Using the formula (1), $1/(\sum_{i=0}^{n} p_i/q^{n-i}) = 1.004$ if $(q, n) = (2^8, 10)$. Therefore, the impact on efficiency on the modified UOV is limited.

Then, we mention the size of the signature. Since the modification requires an additional random salt, the length of the signature increases by at least $\log(q_s(q_H + q_s))$-bits, This is about 90-bits.

Another Approach for UOV Scheme. For a vinegar variable $\boldsymbol{x}'_v$, we consider the map $\boldsymbol{x}_n \mapsto F_{\mathtt{UOV}}(\boldsymbol{x}_n, \boldsymbol{x}'_v)$. In a typical parameter-setting of UOV, it uses $n \times n$ square matrix $A(\boldsymbol{x}'_v)$. Unfortunately, from the formula (1), a rank of random $n \times n$ square matrix is less than n with probability $\delta = 1 - \prod_{j=1}^{n}(1 - q^{-j})$. This means that the map $\boldsymbol{x}_n \mapsto F_{\mathtt{UOV}}(\boldsymbol{x}_n, \boldsymbol{x}'_v)$ is not 1-to-1 mapping with non-negligible probability. For example, $\delta \approx 0.004$ if $q = 2^8$ and $n = 10$.

We mention another approach which reduces δ to a negligible probability. Let w be a positive integer. In this approach, the $m \times n$ matrix $A(\boldsymbol{x}'_v)$ where $n = m + w$ is used instead of $n \times n$ matrix. Consequently, the UOV function becomes $P_{\mathtt{UOV}} = F_{\mathtt{UOV}} \circ S$ where $F_{\mathtt{UOV}} : k^{n+v} \to k^m$ is defined by

$$F_{\mathtt{UOV}}(\boldsymbol{x}_n, \boldsymbol{x}_v) = A(\boldsymbol{x}_v)\boldsymbol{x}_n^T + (g_1(\boldsymbol{x}_v), \ldots, g_m(\boldsymbol{x}_v))^T,$$

where $A(\boldsymbol{x}_v) = [a_{i,j}(\boldsymbol{x}_v)]$ is a $m \times n$ matrix and $a_{i,j}$ and g_i are the same in Definition 3. From the formula (1), a rank of random $m \times n$ matrix is less than m with probability $\delta = 1 - \prod_{j=w+1}^{n}(1 - q^{-j}) \leq q^{-w}/(q-1)$. For example, if $q = 2^8$ and $w = 9$ then δ is less than about 2^{-80}. This means that, the map $\boldsymbol{x}_n \mapsto F_{\mathtt{UOV}}(\boldsymbol{x}_n, \boldsymbol{x}'_v)$ is surjective with probability $1 - \delta$. That is, the UOV function $P_{\mathtt{UOV}}$ is always q^v-to-1 mapping except negligible ratio of domain. As a result, we can easily prove that the UOV signature scheme is secure against chosen-message attack. This approach can be also applied to the multi-layer OV schemes [10].

4.2 Modified HFE Signature Scheme

Basic idea for the simple modification of the HFE signing algorithm is to sample no element with a certain probability which depends on a hash value. In the original signing algorithm of HFE, as explained in Section 3.2, the signatures are not uniformly distributed. To make the distribution uniform, we adjust probability for repeating the loop.

Here we introduce a positive integer N which is used as a threshold for a number of preimages. For simplicity, we suppose $N = d$. For an element y in target space of $F_{\mathtt{HFE}}$ such that y has i preimages $\{x_1, ..., x_i\}$, our modified signing algorithm randomly chooses an element from $\{x_1, ..., x_i\}$ with probability i/N. Accordingly, it chooses no element from $\{x_1, ..., x_i\}$ and repeats the loop again with probability $1 - i/N$. As a result, the signing algorithm returns x without repeating the loop with probability $1/q^n \cdot \omega(x)/N \cdot 1/\omega(x) = 1/q^n \cdot 1/N$ where $\omega(x)$ is the number of preimages of $P_{\mathtt{HFE}}(x)$. The probability apparently does not depend on x. We note that, for any element y in target space of $F_{\mathtt{HFE}}$, the probability i/N is at most 1 where $N = d$, since y has at most d preimages in the HFE function, where d is the maximum degree of $F_{\mathtt{HFE}}$. In a practical setting, the threshold N can be set less than d, e.g., $N = 30$, because an element in the target space of $F_{\mathtt{HFE}}$ has a few preimages with overwhelming probability.

The modified HFE$^-$ signature scheme ($\mathtt{GenHFE}^{-*}, \mathtt{SigHFE}^{-*}, \mathtt{VerHFE}^{-*}$) is described as follows. Note that we employ the minus version of the HFE for security concern of the underlying trapdoor function, but the plain HFE scheme can also be modified similarly. Let the HFE$^-$ function generator $\mathtt{GenHFEfunc}^-$ be a probabilistic algorithm which takes a security parameter 1^λ and outputs $(P_{\mathsf{HFE}^-}, (S, F_{\mathsf{HFE}}, T, m))$, where m is the number of neglected polynomials which is less than n. The key-generation algorithm $\mathtt{GenHFE}^{-*}$, on input 1^λ, runs $(P_{\mathsf{HFE}^-}, (S, F_{\mathsf{HFE}}, T, m)) \leftarrow \mathtt{GenHFEfunc}^-(1^\lambda)$. It outputs (pk, sk), where $pk = (P_{\mathsf{HFE}^-}, l)$, $sk = (S, F_{\mathsf{HFE}}, T, l, m, N)$, l is the length of a random salt bounded by polynomial on λ, and N is a threshold which is d or less. In the signing and the verification algorithms, a random salt r of l bits is concatenated to the message M before hashing, and a hash function $H : \{0,1\}^* \rightarrow k^{n-m}$ is used. The signing algorithm $\mathtt{SigHFE}^{-*}$ is the follows.

Signing algorithm $\mathtt{SigHFE}_{sk}^{-*}(M)$

1: **repeat**
2: $\quad r \in_R \{0,1\}^l$; $(h_1, \ldots, h_{n-m}) \leftarrow H(M||r)$; $(h_{n-m+1}, \ldots, h_n) \in_R k^m$;
3: $\quad y \leftarrow \phi^{-1}(T^{-1}(h_1, \ldots, h_n))$; $u \in_R \{1, \ldots, N\}$;
4: **until** $1 \leq u \leq \#\{z | F_{\mathsf{HFE}}(z) = y\}$
5: $x' \in_R \{z | F_{\mathsf{HFE}}(z) = y\}$; $x \leftarrow S^{-1}(\phi(x'))$;
6: **return** $\sigma = (x, r)$

The verification algorithm $\mathtt{VerHFE}_{pk}^{-*}(\sigma, M)$, on input a signature $\sigma = (x, r)$ and a message M, returns 1 if $P_{\mathsf{HFE}^-}(x) = H(M||r)$. Otherwise, it returns 0. We treat the hash function as the random oracle in our analysis. If l is large enough, then a random salt r is fresh every time with overwhelming probability. Therefore, each $H(M||r)$ is independently and uniformly distributed over k^{n-m}, and the output x and r of the above signing algorithm are also uniformly distributed over k^n and over $\{0,1\}^l$, respectively.

Then, we show the security of the above modified HFE scheme against chosen-message attack.

Theorem 2. *When a threshold N is equal to the max degree d of F_{HFE}, if the HFE$^-$ function generator is (ϵ', t')-secure, then the modified HFE$^-$ scheme is (ϵ, t, q_H, q_s)-secure, where $\epsilon = \epsilon'(q_H + q_s + 1)/(1 - (q_H + q_s)q_s 2^{-l})$, $t = t' - (q_H + q_s + 1)(t_{\mathsf{HFE}^-} + O(1))$, and t_{HFE^-} is running time to compute the HFE$^-$ function P_{HFE^-}.*

Since the proof of Theorem 2 is almost the same proof to that of Theorem 1, it is omitted in this paper.

Efficiency of the modified scheme. In the case of $N = d$, for each element x in domain, the signing algorithm returns x without repeating loops with probability $1/q^n \cdot 1/N$. For the case of $N < d$, the probability is almost the same if N is large enough. Therefore, it returns without repeating loops with probability $1/q^n \cdot 1/N \cdot q^n = 1/N$. The expected number of loops is N. On the other hand, in the original signing algorithm, it returns without repeating loops with probability $1 - 1/e$. So the expected number of loops is $1/(1-1/e) \approx 1.58$. We can see that the expected number of loops in our modified scheme is $(1 - 1/e)N \approx 0.63N$ times more than in the original scheme. The cost of key generation and verification are identical to the original one.

Then, we mention the size of the signature. In the original scheme [22], r is a "small" number. However, at a viewpoint of provable security, we estimate that the length of the random salt r is required at least $\log(q_s(q_H + q_s))$-bits. This is about 90-bits.

5 Extension for HFEV Signature Scheme

The HFEV function is presented by Kipnis et al. [16] and is generated from the combination of HFE and UOV. The HFEV function uses a variant of the map F_{HFE} such that it has additional vinegar variables, and its coefficients b_i and c are a first degree function and a quadratic function on the vinegar variables, respectively. Signatures of this scheme are not uniformly distributed, because of the same reason both of the UOV and of the HFE signature schemes.

By combining the approaches for UOV in Section 4.1 and for HFE in Section 4.2, the HFEV signature scheme can be also modified. In short, the modified signing algorithm first fixes a set of vinegar variables and then computes a preimage by the same way to the modified HFE$^-$ scheme in Section 4.2. Assuming that the HFEV function generator is secure, we can also prove the EUF-CMA of the modified scheme as Theorem 2. Due to the combination of the approaches for HFE and UOV, the signatures are also uniformly distributed. The details of the modified HFEV scheme is described in Appendix B.

6 Conclusions

We analyzed distribution of signatures of the UOV and the HFE signature schemes, and suggested that it might be difficult to sample from the distribution without knowledge of trapdoor. It implies that a usual security proof of FDH-like schemes cannot directly apply to that of the UOV and the HFE schemes. Moreover, we showed that the UOV and the HFE signature schemes can be simply modified into ones achieving the EUF-CMA without changing the underlying trapdoor functions.

References

1. Bellare, M., Rogaway, P.: Random Oracles are Practical: A Paradigm for Designing Efficient Protocols. In: ACM Conference on Computer and Communications Security, pp. 62–73 (1993)
2. Bellare, M., Rogaway, P.: The Exact Security of Digital Signatures - How to Sign with RSA and Rabin. In: Maurer [21], pp. 399–416
3. Bernstein, D.J., Buchmann, J., Dahmen, E.: Post-Quantum Cryptography. Springer, Heidelberg (2009)
4. Bogdanov, A., Eisenbarth, T., Rupp, A., Wolf, C.: Time-Area Optimized Public-Key Engines: Cryptosystems as Replacement for Elliptic Curves? In: Oswald, E., Rohatgi, P. (eds.) CHES 2008. LNCS, vol. 5154, pp. 45–61. Springer, Heidelberg (2008)
5. Braeken, A., Wolf, C., Preneel, B.: A Study of the Security of Unbalanced Oil and Vinegar Signature Schemes. In: Menezes, A. (ed.) CT-RSA 2005. LNCS, vol. 3376, pp. 29–43. Springer, Heidelberg (2005)

6. Cao, W., Hu, L., Ding, J., Yin, Z.: Kipnis-Shamir Attack on Unbalanced Oil-Vinegar Scheme. In: Bao, F., Weng, J. (eds.) ISPEC 2011. LNCS, vol. 6672, pp. 168–180. Springer, Heidelberg (2011)
7. Chen, A.I.-T., Chen, M.-S., Chen, T.-R., Cheng, C.-M., Ding, J., Kuo, E.L.-H., Lee, F.Y.-S., Yang, B.-Y.: SSE Implementation of Multivariate PKCs on Modern X86 CPUs. In: Clavier, C., Gaj, K. (eds.) CHES 2009. LNCS, vol. 5747, pp. 33–48. Springer, Heidelberg (2009)
8. Courtois, N.: Generic Attacks and the Security of Quartz. In: Desmedt, Y.G. (ed.) PKC 2003. LNCS, vol. 2567, pp. 351–364. Springer, Heidelberg (2002)
9. Ding, J., Gower, J.E., Schmidt, D.S.: Multivariate Public Key Cryptosystems. Springer, Heidelberg (2006)
10. Ding, J., Schmidt, D.: Rainbow, a New Multivariable Polynomial Signature Scheme. In: Ioannidis, J., Keromytis, A.D., Yung, M. (eds.) ACNS 2005. LNCS, vol. 3531, pp. 164–175. Springer, Heidelberg (2005)
11. Faugère, J.-C., Joux, A.: Algebraic Cryptanalysis of Hidden Field Equation (HFE) Cryptosystems Using Gröbner Bases. In: Boneh, D. (ed.) CRYPTO 2003. LNCS, vol. 2729, pp. 44–60. Springer, Heidelberg (2003)
12. Faugère, J.-C., Perret, L.: On the Security of UOV. Cryptology ePrint Archive, Report 2009/483 (2009), `http://eprint.iacr.org/`
13. Gentry, C., Peikert, C., Vaikuntanathan, V.: Trapdoors for Hard Lattices and New Cryptographic Constructions. In: Ladnerand, R.E., Dwork, C. (eds.) STOC, pp. 197–206. ACM (2008)
14. Goldwasser, S., Micali, S., Rivest, R.L.: A Digital Signature Scheme Secure Against Adaptive Chosen-Message Attacks. SIAM J. Comput. 17(2), 281–308 (1988)
15. Granboulan, L., Joux, A., Stern, J.: Inverting HFE is Quasipolynomial. In: Dwork, C. (ed.) CRYPTO 2006. LNCS, vol. 4117, pp. 345–356. Springer, Heidelberg (2006)
16. Kipnis, A., Patarin, J., Goubin, L.: Unbalanced Oil and Vinegar Signature Schemes. In: Stern, J. (ed.) EUROCRYPT 1999. LNCS, vol. 1592, pp. 206–222. Springer, Heidelberg (1999)
17. Kipnis, A., Shamir, A.: Cryptanalysis of the Oil & Vinegar Signature Scheme. In: Krawczyk, H. (ed.) CRYPTO 1998. LNCS, vol. 1462, pp. 257–266. Springer, Heidelberg (1998)
18. Kipnis, A., Shamir, A.: Cryptanalysis of the HFE Public Key Cryptosystem by Relinearization. In: Wiener, M. (ed.) CRYPTO 1999. LNCS, vol. 1666, pp. 19–30. Springer, Heidelberg (1999)
19. Levitskaya, A.A.: Systems of Random Equations over Finite Algebraic Structures. Cybernetics and Sys. Anal. 41(1), 67–93 (2005)
20. Matsumoto, T., Imai, H.: Public Quadratic Polynomial-Tuples for Efficient Signature-Verification and Message-Encryption. In: Günther, C.G. (ed.) EUROCRYPT 1988. LNCS, vol. 330, pp. 419–453. Springer, Heidelberg (1988)
21. Maurer, U.M. (ed.): EUROCRYPT 1996. LNCS, vol. 1070. Springer, Heidelberg (1996)
22. Patarin, J.: Hidden Fields Equations (HFE) and Isomorphisms of Polynomials (IP): Two New Families of Asymmetric Algorithms. In: Maurer [21], pp. 33–48
23. Patarin, J.: The Oil and Vinegar Signature Scheme. Presented at the Dagstuhl Workshop on Cryptography (September 1997); transparencies
24. Patarin, J., Courtois, N.T., Goubin, L.: QUARTZ, 128-Bit Long Digital Signatures. In: Naccache, D. (ed.) CT-RSA 2001. LNCS, vol. 2020, pp. 282–297. Springer, Heidelberg (2001)

25. Sakumoto, K., Shirai, T., Hiwatari, H.: Public-Key Identification Schemes Based on Multivariate Quadratic Polynomials. In: Rogaway, P. (ed.) CRYPTO 2011. LNCS, vol. 6841, pp. 706–723. Springer, Heidelberg (2011)

A Proof of Theorem 1

Assume that, on the modified UOV signature scheme, there is an adversary A who takes a public key pk generated via $(pk, \cdot) \leftarrow \texttt{GenUOV}^*(1^\lambda)$, after at most $q_H(\lambda)$ queries to the random oracle, $q_s(\lambda)$ signature queries, and $t(\lambda)$ processing time, then outputs a valid signature with probability at least $\epsilon(\lambda)$. Then, we construct an inverting algorithm B that takes $P_{\texttt{UOV}}$ generated via $(P_{\texttt{UOV}}, \cdot) \leftarrow \texttt{GenUOVfunc}(1^\lambda)$ and a challenge $y \in_R k^n$, then finds a preimage x such that $P_{\texttt{UOV}}(x) = y$ at $t(\lambda)$ processing time with probability at least $\epsilon(\lambda)$. We also call B the simulator.

The simulator B takes as input $(P_{\texttt{UOV}}, y)$ generated via $(P_{\texttt{UOV}}, \cdot) \leftarrow \texttt{GenUOVfunc}(1^\lambda)$, and $y \in_R k^n$, sets a list $L \leftarrow \emptyset$ and $i \leftarrow 0$, randomly picks $\alpha \in \{1, \ldots, q_H + q_s + 1\}$, and selects an length of the random salt l which is large enough. The simulator B runs A on public key $pk = (P_{\texttt{UOV}}, l)$ and simulates the random oracle and the signing oracle as follows.

Answering random oracle queries. Suppose $(m_i||r_i)$ is a random oracle query. First, B increases i by 1. If $(m_i, r_i, \cdot) \in L$, then B answers h such that $(m_i, r_i, h) \in L$. Else if $i = \alpha$, then B sets $L \leftarrow L \cup \{(m_i, r_i, y)\}$ answers y. Else B chooses an element $h_i \in_R k^n$, sets $L \leftarrow L \cup \{(m_i, r_i, h_i)\}$, and answers h_i.

Answering signing oracle queries. Suppose m_i is a signing oracle query. First, the simulator B increases i by 1. The simulator B chooses $r_i \in_R \{0,1\}^l$ and $x_i \in_R k^{n+v}$ and computes $y_i \leftarrow P_{\texttt{UOV}}(x_i)$. If $(m_i, r_i, \cdot) \in L$ then B aborts. Else B sets $L \leftarrow L \cup \{(m_i, r_i, y_i)\}$ and answers (x_i, r_i).

Output. Eventually, A outputs a forgery (x, r) of some message m. Without loss of generality, we assume that A asked the hash query $m||r$ beforehand (if not, B can do it instead of A). If the answer was y, we get x such that $P_{\texttt{UOV}}(x) = y$, thus B outputs the preimage x. Otherwise, we do not learn anything, then B fails.

Analysis. The view of A in the successful simulation is properly distributed. In particular, we note that the value x output by the legitimate signing algorithm is uniformly distributed over k^{n+v}, because for any $(\boldsymbol{x}'_n, \boldsymbol{x}'_v) \in k^{n+v}$,

$$\sum_{i=1}^{\infty} \Pr\left[\begin{array}{l} \boldsymbol{x}_v \in_R k^v, H_1, \ldots, H_i \in_R k^n, \boldsymbol{x}_n \in_R \{\boldsymbol{z}_n | F_{\texttt{UOV}}(\boldsymbol{z}_n, \boldsymbol{x}_v) = H_i\}; \\ \{\boldsymbol{z}_n | F_{\texttt{UOV}}(\boldsymbol{z}_n, \boldsymbol{x}_v) = H_1\} = \emptyset, \ldots, \{\boldsymbol{z}_n | F_{\texttt{UOV}}(\boldsymbol{z}_n, \boldsymbol{x}_v) = H_{i-1}\} = \emptyset, \\ \{\boldsymbol{z}_n | F_{\texttt{UOV}}(\boldsymbol{z}_n, \boldsymbol{x}_v) = H_i\} \neq \emptyset, (\boldsymbol{x}_n, \boldsymbol{x}_v) = (\boldsymbol{x}'_n, \boldsymbol{x}'_v) \end{array}\right]$$

$$= \frac{1}{q^v} \frac{\Pr\left[\begin{array}{l} H \in_R k^n, \boldsymbol{x}_n \in_R \{\boldsymbol{z}_n | F_{\texttt{UOV}}(\boldsymbol{z}_n, \boldsymbol{x}_v) = H\}; \\ \{\boldsymbol{z}_n | F_{\texttt{UOV}}(\boldsymbol{z}_n, \boldsymbol{x}_v) = H\} \neq \emptyset, \boldsymbol{x}_n = \boldsymbol{x}'_n \end{array}\right]}{\Pr\left[H \in_R k^n; \{\boldsymbol{z}_n | F_{\texttt{UOV}}(\boldsymbol{z}_n, \boldsymbol{x}_v) = H\} \neq \emptyset\right]} = \frac{1}{q^{n+v}}.$$

The probability that B answers to all queries is at least $1-(q_H+q_s)q_s2^{-l}$. Therefore, A outputs a forgery of a certain message with probability at least $\epsilon(1-(q_H+q_s)q_s2^{-l})$. Since the simulation of the random oracle is perfect and reveals no information about α, $H(m||r)$ corresponds to the challenge y, rather than to another random value h_i, so that B does not fail with probability $1/(q_H+q_s+1)$. Therefore, the simulator B finds an inverse of y for $P_{\texttt{UOV}}$ with probability at least $\epsilon(1-(q_H+q_s)q_s2^{-l})/(q_H+q_s+1)$. Then, $\epsilon \le \epsilon'(q_H+q_s+1)/(1-(q_H+q_s)q_s2^{-l})$. The running time of B is at most $t+(q_H+q_s+1)(t_{\texttt{UOV}}+O(1))$. Then, $t \ge t'-(q_H+q_s+1)(t_{\texttt{UOV}}+O(1))$.

B HFEV Signature Schemes

Let T, ϕ, q, n, d, k, and K be the parameters defined in Definition 5. Let $S : k^{n+v} \to k^{n+v}$ be an invertible affine transformation v a positive integer, and a map $\hat{\phi}^{-1}$ defined by $\hat{\phi}^{-1} : k^{n+v} \to K \times k^v$, $(x_1,\dots,x_{n+v}) \mapsto (\phi^{-1}(x_1,\dots,x_n), x_{n+1},\dots,x_{n+v})$. The map $F_{\texttt{HFEV}}$ is defined by $F_{\texttt{HFEV}} : K \times k^v \to K, (X, x_{n+1},\dots,x_{n+v}) \mapsto$

$$\sum_{i=0}^{n-1}\sum_{j=0}^{n-1} a_{ij}X^{q^i+q^j} + \sum_{i=0}^{n-1} b_i(x_{n+1},\dots,x_{n+v})X^{q^i} + c(x_{n+1},\dots,x_{n+v}),$$

where $q^i+q^j > d \Rightarrow a_{ij}=0$ for $\forall a_{ij}$, $q^i > d \Rightarrow b_i \equiv 0$ for $\forall b_i$, a_{ij} are secret elements in K, and $b_i : k^n \to K$ and $c : k^n \to K$ are secret linear and quadratic maps, respectively. The HFEV function is $P_{\texttt{HFEV}} = T \circ \phi \circ F_{\texttt{HFEV}} \circ \hat{\phi}^{-1} \circ S$ and its generator `GenHFEVfunc` is a probabilistic algorithm which takes a security parameter 1^λ and output $(P_{\texttt{HFEV}}, (S, F_{\texttt{HFEV}}, T))$, where v is bounded by polynomial on λ.

Using the idea in Section 4.1 and 4.2, we define the modified HFEV signature scheme (`GenHFEV*`, `SigHFEV*`, `VerHFEV*`) as follows. The key-generation algorithm `GenHFEV*`(1^λ), on input 1^λ, runs $(P_{\texttt{HFEV}}, (S, F_{\texttt{HFEV}}, T)) \leftarrow$ `GenHFEVfunc`(1^λ). It outputs (pk, sk), where $pk = (P_{\texttt{HFEV}}, l)$, $sk = (S, F_{\texttt{HFEV}}, T, l, N)$, l is the length of a random salt, and N is a threshold. l and N are bounded by polynomial on λ. The signing and verification algorithms use a hash function $H : \{0,1\}^* \to k^n$ which maps a bit string of arbitrary length to an element in k^n. The signing algorithm $\texttt{SigHFEV}^*_{sk}(M)$ is as follows:

Signing algorithm $\texttt{SigHFEV}^*_{sk}(M)$

1: $(x'_{n+1},\dots,x'_{n+v}) \in_R k^v$;
2: **repeat**
3: $\quad r \in_R \{0,1\}^l$; $y \leftarrow \phi^{-1}(T^{-1}(H(M||r)))$; $u \in_R \{1,\dots,N\}$;
4: **until** $1 \le u \le \#\{z \in K | F_{\texttt{HFEV}}(z, x'_{n+1},\dots,x'_{n+v}) = y\}$
5: $x' \in_R \{z \in K | F_{\texttt{HFEV}}(z, x'_{n+1},\dots,x'_{n+v}) = y\}$;
6: $(x'_1,\dots,x'_n) \leftarrow \phi(x')$; $x \leftarrow S^{-1}(x'_1,\dots,x'_{n+v})$;
7: **return** $\sigma = (x, r)$

The verification algorithm $\texttt{VerHFEV}^*_{pk}(\sigma, M)$, on input a signature $\sigma = (x, r)$ and a message M, returns 1 if $P_{\texttt{HFEV}}(x) = H(M||r)$. Otherwise, it returns 0.

Roots of Square: Cryptanalysis of Double-Layer Square and Square+

Enrico Thomae and Christopher Wolf

Horst Görtz Institute for IT-security
Faculty of Mathematics
Ruhr-University of Bochum, 44780 Bochum, Germany
http://www.cits.rub.de/
{enrico.thomae,christopher.wolf}@rub.de, chris@christopher-wolf.de

Abstract. Square is a multivariate quadratic encryption scheme proposed in 2009. It is a specialization of Hidden Field Equations by using only odd characteristic fields and also X^2 as its central map. In addition, it uses embedding to reduce the number of variables in the public key. However, the system was broken at Asiacrypt 2009 using a differential attack. At PQCrypto 2010 Clough and Ding proposed two new variants named *Double-Layer Square* and *Square+*. We show how to break Double-Layer Square using a refined *MinRank* attack in 2^{45} field operations. A similar fate awaits Square+ as it will be broken in 2^{32} field operations using a mixed MinRank attack over both the extension and the ground field. Both attacks recover the private key, given access to the public key. We also outline how possible variants such as *Square–* or *multi-Square* can be attacked.

Keywords: Multivariate Cryptography, Algebraic Cryptanalysis, Square, Double-Layer Square, Square+, MinRank, Key Recovery.

1 Introduction

In the world of Post-Quantum cryptography, $\mathcal{M}$ultivariate $\mathcal{Q}$uadratic public key schemes have an important place. They were investigated as early as 1985 [14, 16] and have branched out into several systems.

In this article, we deal with the so-called *Square* system, which works both over a ground field $\mathbb{F}_q$ with q elements, as over an extension field $\mathbb{F}_{q^{n+\ell}}$. Its main feature is the operation X^2 over $\mathbb{F}_{q^{n+\ell}}$. Obviously, this is very simple to compute and invert—in particular when compared to the similar system Hidden Field Equations [17]. Inversion of X^2 utilizes the equation $X = \pm Y^{\frac{q^{n+\ell}+1}{4}}$. Hence, we need $q^{n+\ell} \equiv 3 \pmod 4$ and inverting $Y \in \mathbb{F}_q$ requires only one exponentiation in $\mathbb{F}_{q^{n+\ell}}$. Depending on the choice of q, n, the inversion is as efficient as for Sflash [1, 10].

Square itself was proposed 2009 in [7]. It was broken in the same year [4] using a differential attack. At PQCrypto 2010 Clough and Ding [9] proposed two new variants of Square, called *Double-Layer Square* and *Square+* which are claimed

B.-Y. Yang (Ed.): PQCrypto 2011, LNCS 7071, pp. 83–97, 2011.

to be secure against all known attacks. We will outline below how they differ from the original Square scheme—but can be broken nevertheless.

One thing which has also developed with $\mathcal{MQ}$ schemes is their cryptanalysis. In this article, we will concentrate on attacks from the so-called MinRank family. Idea is to find a linear combination of some matrices, such that the new matrix has a special (minimal) rank. Or more formally: Given k matrices $M_1, \ldots, M_k \in \mathbb{F}_q^{n\times n}$ and a scalar $r \in \mathbb{N}$, find a vector $\lambda \in \mathbb{F}_q^k$ such that

$$\text{Rank}\left(\sum_{i=1}^{k} \lambda_i M_i\right) \leq r.$$

We call this an *MinRank(q, k, r)-problem*. Note that the *general* MinRank problem is NP-complete [5]. We will see later how $\mathcal{M}$ultivariate $\mathcal{Q}$uadratic schemes relate to matrices in general and to MinRank in particular.

A first MinRank attack in the $\mathcal{M}$ultivariate $\mathcal{Q}$uadratic setting was launched against TTM [13]. Informally speaking, the authors exploited the existence of a so-called step-structure in the private key to reveal linear relations between the private and the public key. When enough of these relations were found, the whole private key could be unravelled. A similar approach was followed in [19]. Here, the step-width was made wider: Instead of allowing only rank differences of 1, rank differences up to r were allowed. Finally, [21] gave further ideas on discovering rank structure, in particular "crawling" attacks that exploit that areas of low rank might be close-by. A cryptanalysis of the Rainbow Signature Scheme using MinRank can be found in [3]. Our attack on Double-Layer Square (see sect. 3) will strongly refer to this paper.

Another algorithm to break MinRank-instances in practice is [12]. Here, Gröbner bases are used to actually calculate elements of the kernel and thus derive possible choices of $\lambda \in \mathbb{F}_q^k$. For some parameters this algorithm is much faster than sampling and therefore we use it in sect. 4 to break Square+.

1.1 Achievement and Organisation

In this paper, we describe an efficient cryptanalysis of the two public key schemes Double-Layer Square and Square+. We show how to break Double-Layer Square by a refined MinRank attack that is an extension of Billet and Gilbert [3] attack against Rainbow. The overall attack complexity is 2^{45}. Furthermore we break Square+ using methods from the cryptanalysis of odd characteristic HFE [2] and a MinRank attack [12]. In both cases, the attack is in polynomial time of (nearly) all parameters. In particular, the schemes are completely broken for all possible, practical choices of parameters.

In sect. 2, we introduce the Square cryptosystem and fix some notation. Double-Layer Square and its attack is discussed in sect. 3. We deal with Square+ and the corresponding MinRank problem in sect. 4. This paper concludes with sect. 5. There, we also outline possible extensions to Square– or multi-Square. Due to space limitations, we had to remove most of the experiments. A full version is available at `http://eprint.iacr.org/2011/431`.

2 Notation

In this section we shortly recap the Square encryption scheme [7]. We start by giving some general outline on $\mathcal{M}$ultivariate $\mathcal{Q}$uadratic public key systems and some notation.

Each $\mathcal{MQ}$-scheme uses a public $\mathcal{M}$ultivariate $\mathcal{Q}$uadratic map $\mathcal{P} : \mathbb{F}_q^n \to \mathbb{F}_q^m$ with

$$\mathcal{P} := \begin{pmatrix} p^{(1)}(x_1, \ldots, x_n) \\ \vdots \\ p^{(m)}(x_1, \ldots, x_n) \end{pmatrix}$$

for $1 \le k \le m$ and

$$p^{(k)}(x_1, \ldots, x_n) := \sum_{1 \le i \le j \le n} \gamma_{ij}^{(k)} x_i x_j$$

as public key. The trapdoor is given by a structured central map $\mathcal{F} : \mathbb{F}_q^n \to \mathbb{F}_q^m$ with

$$\mathcal{F} := \begin{pmatrix} f^{(1)}(x_1, \ldots, x_n) \\ \vdots \\ f^{(m)}(x_1, \ldots, x_n) \end{pmatrix}$$

for $1 \le k \le m$ and

$$f^{(k)}(x_1, \ldots, x_n) := \sum_{1 \le i \le j \le n} \widetilde{\gamma}_{ij}^{(k)} x_i x_j.$$

In order to hide this trapdoor we choose two secret linear transformations $S \in \mathbb{F}_q^{n \times n}, T \in \mathbb{F}_q^{m \times m}$ and define $\mathcal{P} := T \circ \mathcal{F} \circ S$. Note that some proposals also use a linear and constant part of $p^{(k)}$ and $f^{(k)}$. However, as it is well known that quadratic terms only depend on quadratic terms from the secret map $\mathcal{F}$ and on linear terms from S, T, we can safely ignore the linear and constant parts in our cryptanalysis to ease explanation [3, 15, 18]. Where necessary, the affine case can be added easily.

Sometimes, as for Square, the trapdoor does not reveal itself over $\mathbb{F}_q^n$ but over the extension field $\mathbb{F}_{q^{n+\ell}}$. Let $\varphi : \mathbb{F}_q^{n+\ell} \to \mathbb{F}_{q^{n+\ell}}$ be the standard isomorphism between the vector space and the extension field and $\mathcal{F}' = \varphi \circ \mathcal{F} \circ \varphi^{-1}$. As outlined above, Square is defined for $q^{n+\ell} \equiv 3 \pmod 4$ and uses $\mathcal{F}' = X^2$ over $\mathbb{F}_{q^{n+\ell}}$. This can be easily inverted by the square root formula

$$X = \pm Y^{\frac{q^{n+\ell}+1}{4}}. \tag{1}$$

To make their scheme more resistant, the authors of Square have chosen S as a $(n+\ell) \times n$ matrix of rank n. This is equivalent to deleting ℓ variables from the secret map $\mathcal{F}$ in the public map $\mathcal{P}$. See figure 1 for an overall illustration of Square. The original parameters of the scheme are $n = 34, q = 31$ and $\ell = 3$ [7].

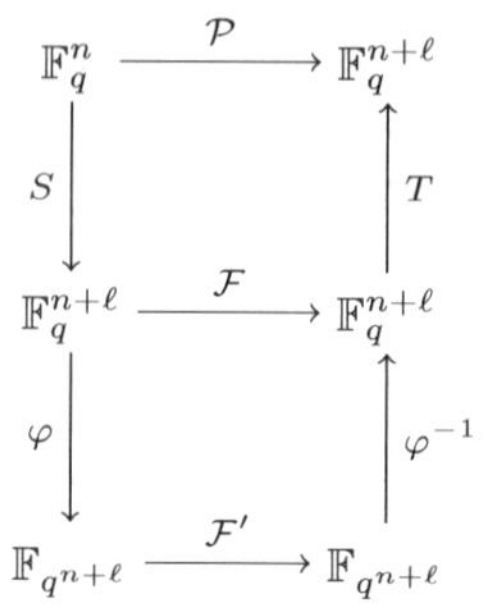

Fig. 1. The Square Scheme

In the sequel, we will make heavy use of the matrix representation of $\mathcal{M}$ultivariate $\mathcal{Q}$uadratic polynomials. As described above, we assume all polynomials $p^{(k)}$ and $f^{(k)}$ for $1 \leq k \leq n + \ell$ to be homogenized. As explained, we can do so as the linear and constant parts of the $p^{(k)}$ and $f^{(k)}$ do not carry any cryptographically relevant information. Let $\underline{x} = (x_1, \ldots, x_n)^\intercal$ respectively $\underline{\tilde{x}} = (\tilde{x}_1, \ldots, \tilde{x}_{n+\ell})^\intercal$ be a column vector and $\mathfrak{P}^{(k)} \in \mathbb{F}^{n \times n}$ respectively $\mathfrak{F}^{(k)} \in \mathbb{F}^{n+\ell \times n+\ell}$ the matrix describing the quadratic form of $p^{(k)} = \underline{x}^\intercal \mathfrak{P}^{(k)} \underline{x}$ respectively $f^{(k)} = \underline{\tilde{x}}^\intercal \mathfrak{F}^{(k)} \underline{\tilde{x}}$. We restrict to symmetric matrices (see figure 2). Using a minor twist, we can also represent univariate polynomials over the extension field $\mathbb{F}_{q^n}$ this way. By a slight abuse of notation, we obtain the same figure 2 for the univariate polynomial $P^{(k)}(X) = \sum_{0 \leq i \leq j < n} \gamma_{i,j}^{(k)} X^{q^i + q^j}$ over the extension field $\mathbb{F}_{q^n}$ for $\underline{x} = (X, X^q, \ldots, X^{n-1})^\intercal$.

$$\mathfrak{P}^{(k)} = \begin{pmatrix} \gamma_{1,1}^{(k)} & \gamma_{1,2}^{(k)}/2 & \cdots & \cdots & \gamma_{1,n}^{(k)}/2 \\ \gamma_{1,2}^{(k)}/2 & \gamma_{2,2}^{(k)} & & & \gamma_{2,n}^{(k)}/2 \\ \vdots & \vdots & \ddots & & \vdots \\ \gamma_{1,n-1}^{(k)}/2 & \gamma_{2,n-1}^{(k)}/2 & & \gamma_{n-1,n-1}^{(k)} & \gamma_{n-1,n}^{(k)}/2 \\ \gamma_{1,n}^{(k)}/2 & \gamma_{2,n}^{(k)}/2 & \cdots & \gamma_{n-1,n}^{(k)}/2 & \gamma_{n,n}^{(k)} \end{pmatrix}$$

Fig. 2. Matrix representation $\mathfrak{P}^{(k)}$ of the public key polynomial $p^{(k)}$

3 Double-Layer Square

Double-Layer Square as proposed in [9] uses the idea of Rainbow [11] to split the central map into two layers and thus destroy the differential properties in the public map that where used to break Square. The first layer is just the same mapping $\mathcal{F}$ as for Square. The second layer is defined by $\mathcal{G} : \mathbb{F}_q^{2n+\ell} \to \mathbb{F}_q^n$ with $\mathcal{G} = \varphi'^{-1} \circ G \circ (id \times \varphi')$ and $\varphi' : \mathbb{F}_q^n \to \mathbb{F}_{q^n}$ the standard isomorphism. It is explicitly given by

$$G((x_1,\ldots,x_{n+\ell}),X) = \alpha X^2 + \beta(x_1,\ldots,x_{n+\ell})X + \gamma(x_1,\ldots,x_{n+\ell}) \qquad (2)$$

where $\alpha \in \mathbb{F}_{q^n}$, β is affine and γ is quadratic over $\mathbb{F}_{q^n}$. The whole central map over the vector space is thus given by

$$\mathcal{F}||\mathcal{G} = \begin{pmatrix} f^{(1)}(x_1,\ldots,x_{n+\ell}) \\ \vdots \\ f^{(n+\ell)}(x_1,\ldots,x_{n+\ell}) \\ g^{(1)}(x_1,\ldots,x_{2n+\ell}) \\ \vdots \\ g^{(n)}(x_1,\ldots,x_{2n+\ell}) \end{pmatrix}.$$

With $||$ we denote concatenation of two vectors and $g^{(i)} = x^{\intercal}\mathfrak{G}^{(i)}x$ with $\mathfrak{G}^{(i)} \in \mathbb{F}_q^{2n+\ell\times 2n+\ell}$. By construction, we have $\mathrm{rank}(f^{(i)}) \leq n+\ell$ and $\mathrm{rank}(g^{(i)}) \leq 2n+\ell$, cf. fig. 3 for the overall structure of the two layers. In order to invert the central map we first use the square root formula (1) to determine $x_1,\ldots,x_{n+\ell}$. This solution is plugged into (2) which is then solved, *e.g.* by the school book root finding for quadratic equations or by Berlekamp's algorithm.

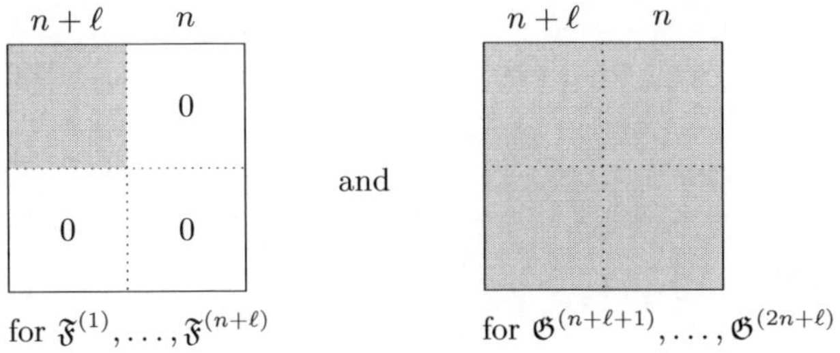

Fig. 3. Central maps of Double-Layer Square

3.1 MinRank Attack against Double-Layer Square

In this section we adapt the MinRank attack of Billet and Gilbert [3] to Double-Layer Square. In order to reconstruct T we have to solve the problem of finding a linear combination $\sum_{i=1}^{2n+\ell} \lambda_i \mathfrak{P}^{(i)}$ for $\lambda_i \in \mathbb{F}_q$ with minimal rank. In general this is a difficult problem, as Buss *et al.* [5] showed that the decisional version of MinRank over $\mathbb{F}_q$ is NP-complete.

The idea of [3] to calculate a solution of the MinRank problem is to sample a vector $\omega \in_R \mathbb{F}_q^{2n}$ and hope that it lies in the kernel of a linear combination of low-rank matrices. If this is the case solving the linear system of equations

$$\sum_{i=1}^{2n+\ell} \lambda_i \mathfrak{P}^{(i)}\omega = 0 \text{ for } \omega \in_R \mathbb{F}_q^{2n}, \lambda_i \in \mathbb{F}_q, \mathfrak{P}^{(i)} \in \mathbb{F}_q^{2n\times 2n} \qquad (3)$$

reveals a part of the secret transformation T. The crucial point is to calculate the probability over all ω that there exist values $\lambda_1, \ldots, \lambda_{n+\ell} \in \mathbb{F}_q$ such that

$$\omega \in \ker \left(\sum_{i=1}^{n+\ell} \lambda_i S^\intercal \mathfrak{F}^{(i)} S \right) . \tag{4}$$

For S being an $(2n+\ell) \times (2n+\ell)$ matrix of full rank and $\omega \in_R \mathbb{F}_q^{2n+\ell}$ this probability equals the likelihood of

$$S\omega \in \ker \left(\sum_{i=1}^{n+\ell} \lambda_i \mathfrak{F}^{(i)} \right) . \tag{5}$$

While the general idea is the same for Double-Layer Square, we need to be careful as S is a $(2n+\ell) \times 2n$ matrix of rank $2n$. We will tackle this problem after having calculated the probability that there exists $\lambda_i \in \mathbb{F}_q$ fulfilling (5).

It is well known that the probability of a random $(m \times n)$ matrix over $\mathbb{F}_q$ being regular is given by

$$\prod_{i=0}^{m-1} \left(1 - \frac{q^i}{q^n} \right) < 1 - \frac{1}{q^{n-m+1}} . \tag{6}$$

This implies that the probability of a random $(m \times n)$ matrix over $\mathbb{F}_q$ to be singular is bounded below by $1/(q^{n-m+1})$. In Fig. 4 we illustrate the coefficient matrix of the linear system

$$\left(\sum_{i=1}^{n+\ell} \lambda_i \mathfrak{F}^{(i)} + \sum_{i=n+\ell+1}^{2n+\ell} \lambda_i \mathfrak{G}^{(i)} \right) S\omega = 0 \tag{7}$$

for a random but fixed $\omega \in \mathbb{F}^{2n+\ell}$ and for $g^{(i)} = x^\intercal \mathfrak{G}^{(i)} x$ the associated matrix for a given secret polynomial $g^{(i)}$.

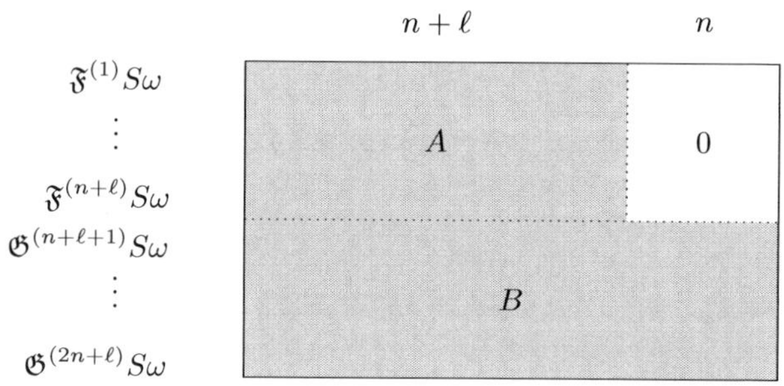

Fig. 4. Coefficient matrix of linear system (7)

The probability that there exist $\lambda_1, \ldots, \lambda_{n+\ell} \in \mathbb{F}_q$ such that (5) holds is the probability of matrix A in figure 4 to be singular, *i.e.* $1/q$. Note that it is not enough for our attack that such a linear combination exists. In order to *efficiently* obtain this solution using (3) we also need the rank of the whole matrix from fig. 4 to be rank$(A) + n$. This is true with overwhelming probability in our case. Otherwise we would obtain parasitic solutions by (3).

Up to this point, the overall complexity of the MinRank attack is

$$q(n+\ell)(2n+\ell)^3$$

as we expect to sample q vectors $\omega \in \mathbb{F}_q^{2n}$ until A becomes singular. We need to repeat this sampling until we have recovered $(n+\ell)$ linearly independent equations of small rank. Solving (3) requires Gaussian elimination in $(2n+\ell)$ variables.

Now we have to deal with the problem that S is not a $(2n+\ell) \times (2n+\ell)$ square matrix but a rectangular $(2n+\ell) \times 2n$ matrix. Obviously equation (4) and (5) are not equivalent any longer, but it holds

$$\sum_{i=1}^{n+\ell} \lambda_i \mathfrak{F}^{(i)} S\omega + \sum_{i=n+\ell+1}^{2n+\ell} \lambda_i \mathfrak{G}^{(i)} S\omega = 0 \Rightarrow \sum_{i=1}^{n+\ell} \lambda_i S^\intercal \mathfrak{F}^{(i)} S\omega + \sum_{i=n+\ell+1}^{2n+\ell} \lambda_i S^\intercal \mathfrak{G}^{(i)} S\omega = 0.$$

I.e. the probability of choosing ω in the kernel of low-rank matrices is still $1/q$. This argument hides a slight heuristic. If we choose $\omega \in_R \mathbb{F}_q^{2n}$, $S\omega$ is not a random element in $\mathbb{F}_q^{2n+\ell}$ any longer and thus the rows of the matrix in fig. 4 are not randomly chosen. Nevertheless they are independent and thus formula (6) should be a good approximation. Our experiments in table 1 confirm this. The backward direction is not true, as $2n+\ell$ vectors of lenght $2n$ are always linearly dependent and thus we obtain q^ℓ parasitic solutions. The overall attack cost is therefore

$$(n+\ell)q^{\ell+1}(2n+\ell)^3 .$$

Unfortunately, the authors of [9] did not provide concrete security parameters. However, using their security analysis, we derived $q = 31, n = 17, \ell = 4$ for a claimed security level of 2^{80}. Using our attack, this reduces to 2^{45} to separate the upper from the lower level. We have broken this set of parameters in about 1 day, see Table 1. Moreover, we can ignore the embedding modifier, as explained in [2, Sect. 5]. In a nutshell, we work on the maximal rank of the corresponding matrices. However, the embedding modifier will only *decrease* the rank and hence not increase its maximum. Hence, the difference from fig. 3 still holds. Once we have separated these layers, the rest of the attack is equal to Billet/Macario-Rat [4], although we have to take the Double-Layer structure into account. First, we separate out the two layers $\mathfrak{F}$ and $\mathfrak{G}$. Using the algorithm of Billet/Macario-Rat, we can separate the variables of the $\mathfrak{F}$-layer into $x_1, \ldots, x_{n+\ell}$ (output of Billet/Macario-Rat) and $x_{n+\ell+1}, \ldots, x_{2n+\ell}$ (others). Using these, we have the variable mixing S, the equation mixing T, and the inner layer X^2 for the *first layer* $\mathfrak{F}$. For the second, *i.e.* the $\mathfrak{G}$-layer, it is a bit more complicated as we are dealing with

$$\alpha X^2 + \beta(x_1, \ldots, x_{n+\ell})X + \gamma(x_1, \ldots, x_{n+\ell}) .$$

Table 1. Time to recover the hidden vector space T for fixed field size $q = 31$, field extension $n = 17$ and variable embedding degree ℓ. The number of samples from the vector space ($\#\omega$) is independent from ℓ, but close to qn. Each line is based on 11 independent experiments. The line with previously secure parameters is **highlighted in bold**.

q	n	ℓ	$\#\omega$	$\#\omega/n$	time [sec] min	avg	max
31	17	1	500	29.41	129	170	219
		2	460	27.06	210	268	375
		3	568	33.41	2416	3069	3903
		4	**651**	**38.30**	**87556**	**97911**	**117534**

here, so Billet/Macario-Rat does not apply directly. However, we see by inspection that all monomials depending on $x_1, \ldots, x_{n+\ell}$ come from the term γ, all monomials depending both on $x_1, \ldots, x_{n+\ell}$ and $x_{n+\ell+1}, \ldots, x_{2n+\ell}$ come from βX; and the rest comes from αX^2. Applying Billet/Macario-Rat to these gives us the complete variable change S and equation change T (up to equivalences, [20]). Hence, we have reconstructed the private key and are therefore in the same position as the legitimate user when computing $y = \mathcal{P}(x)$ for given $y \in \mathbb{F}_q^{2n+\ell}$.

4 Square+

Another version of the Square cryptosystem is called Square+. It was also suggested in the very same paper as Double-Layer Square by Clough and Ding [9]. As Square, it uses X^2 over the extension field $\mathbb{F}_{q^{n+\ell}}$ as its central monomial. In addition, we have $p \in \mathbb{N}$ random equations that blind the differential structure of X^2 in the public key. In total, we obtain $m := n + \ell + p$ equations for Square+. Obviously, Square+ is overdetermined—both due to the embedding of ℓ variables and the p extra polynomials. In order to prevent Gröbner based attacks, $(\ell + p)$ has to be chosen relatively small compared to n. In the original Square+ paper, proposed parameters are $q = 31, n = 48, \ell = 3, p = 5$ [9].

Let $\varphi : \mathbb{F}_q^{n+\ell} \to \mathbb{F}_{q^{n+\ell}}$ be the standard isomorphism between the vector space $\mathbb{F}_q^{n+\ell}$ and the finite field $\mathbb{F}_{q^{n+\ell}}$. Denote with $a_1, \ldots, a_p$ a total of p random, quadratic polynomials over $\mathbb{F}_q$, the so-called *plus-polynomials*. The mixing of the equations is realized by a full-rank matrix $T \in \mathbb{F}_q^{(n+\ell+p)\times(n+\ell+p)}$. The embedding modifier is realized via a matrix $S \in \mathbb{F}_q^{(n+\ell)\times n}$ with $\text{rank}(S) = n$. The Square part is expressed over the ground field as $\mathcal{C}$, the plus polynomials are given in $\mathcal{A}$, see

$$\begin{aligned}
\mathcal{C} : \mathbb{F}^{n+\ell} \to \mathbb{F}^{n+\ell} : (u_1, \ldots, u_{n+\ell}) &\to \varphi^{-1} \circ X^2 \circ \varphi(u_1, \ldots, u_{n+\ell}) \\
&= (v_1, \ldots, v_{n+\ell}), \\
\mathcal{A} : \mathbb{F}^{n+\ell} \to \mathbb{F}^{p} : (u_1, \ldots, u_{n+\ell}) &\to (a_1(u_1, \ldots, u_{n+\ell}), \ldots, a_p(u_1, \ldots, u_{n+\ell})) \\
&= (v_{n+\ell+1}, \ldots, v_{n+\ell+p})
\end{aligned}$$

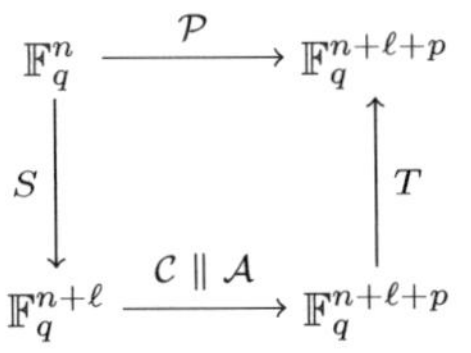

Fig. 5. The Double-Layer Square Scheme

Now we can write the public key $\mathcal{P}$ of Square+ as $\mathcal{P} := T \circ (\mathcal{C} \circ S, \mathcal{A} \circ S)$. See figure 5 for a graphical representation. Note that all intermediate operations are quadratic over $\mathbb{F}_q$, as is $\mathcal{P}$. If we leave out the embedding modifier for a moment (transformation S), there are two parts of Square+, namely the invertible, but "soft" part X^2, represented by transformation $\mathcal{C}$, and the not-invertible "hard" part $a_1, \ldots, a_p$, represented by transformation $\mathcal{A}$. If we manage to separate them, we are done as there is an efficient attack against Square [2].

4.1 Odd-Characteristic HFE Attack against Square+

To this aim, we have a closer look at "odd Characteristic HFE" (or odd-HFE) and its cryptanalysis [2, 6]. In particular we notice that the central map of odd-HFE is

$$\sum_{(i,j)\in\Delta(D)} \gamma_{i,j} X^{q^i+q^j}$$

for a set of admissible degrees $\Delta(D) := \{(i,j) \in \mathbb{N}^2 : i \le j, q^i + q^j \le D\}$, $\mathbb{N}$ the set of non-negative integers and $\gamma_{i,j} \in \mathbb{F}_{q^n}$ the coefficients of the corresponding private key. Setting $D = 2$ and $\gamma_{(0,0)} = 1$, we obtain $\Delta(2) = (0,0)$ and the central map of odd-HFE coincides with the one of Square+. As a result, we can apply the cryptanalysis of Bettale *et al.* [2] against odd-HFE also against Square+. Alas, this cryptanalysis does not include the case odd-HFE+, so we need to investigate this question closely to determine if we can break Square+ within this framework. As we will see below, it works but there are subtle changes to be made.

As for the original attack against odd-HFE, the key point is the observation that we can write X^2 as a matrix of small rank over the *extension* field. More to the point, we have $X^2 = \underline{x}^\intercal \mathfrak{F} \underline{x}$ over $\mathbb{F}_{q^{n+\ell}}$ for $\underline{x} = (X, X^q, X^{q^2}, \ldots, X^{q^{n-1}})^\intercal$ with $\mathfrak{F}_{1,1} = 1$ but $\mathfrak{F}_{i,j} = 0$ otherwise. As only $\mathfrak{F}_{1,1}$ is non-zero, we obviously have $\mathrm{rank}(\mathfrak{F}) = 1$. A similar observation for a MinRank attack against HFE was already used by Kipnis-Shamir [15]. Note that expressing X^2 over the ground field yields a much higher rank, in practice close to $(n + \ell)$.

To ease notation and to mount the attack, we follow the approach of [2] and start with the vector $(\theta_1, \ldots, \theta_{n+\ell}) \in \mathbb{F}_{q^{n+\ell}}^{n+\ell}$. Note that this vector has a double function: First, it fixes a basis of the vector space $\mathbb{F}_q^{n+\ell}$, *i.e.* over the ground field, and second, the elements $\theta_1, \ldots, \theta_{n+\ell}$ are simultaneously interpreted over the extension field $\mathbb{F}_{q^{n+\ell}}$. This way we can apply the homomorphism $\mathbb{F}_{q^{n+\ell}} \to \mathbb{F}_{q^{n+\ell}} : x \mapsto x^{q^k}$ for $k = 0, \ldots, (n + \ell - 1)$ within the extension field. Finally, this is used to construct a matrix $M_{n+\ell}$.

$$M_{n+\ell} := \begin{pmatrix} \theta_1 & \theta_1^q & \ldots & \theta_1^{q^{n+\ell-1}} \\ \theta_2 & \theta_2^q & \ldots & \theta_2^{q^{n+\ell-1}} \\ \vdots & & \ddots & \vdots \\ \theta_{n+\ell} & \theta_{n+\ell}^q & \ldots & \theta_{n+\ell}^{q^{n+\ell-1}} \end{pmatrix}$$

More precisely, for a vector $v := (v_1, \ldots, v_{n+l}) \in \mathbb{F}_q^{n+\ell}$ we have the mapping $\phi : \mathbb{F}_q^{n+\ell} \mapsto \mathbb{F}_{q^{n+\ell}}$ with

$$\phi(v) \mapsto V_1 : (v_1, \ldots, v_{n+\ell})M_{n+\ell} =: (V_1, \ldots, V_{n+\ell}).$$

Note that this mapping only uses the first component of the vector $(V_1, \ldots, V_{n+\ell})$. Moreover, the first column of $M_{n+\ell}$ consists only of base elements of $\mathbb{F}_q^{n+\ell}$. Hence, two values $V_1, \tilde{V_1} \in \mathbb{F}_{q^n}$ will only be equal if the corresponding vectors $v, \tilde{v} \in \mathbb{F}_q^n$ are the same. The inverse mapping needs to make use of the special structure of the matrix $M_{n+\ell}$ to map elements back into the ground field. We have $\phi^{-1} : \mathbb{F}_{q^{n+\ell}} \mapsto \mathbb{F}_q^{n+\ell}$ for

$$\phi^{-1}(V) \mapsto (v_1, \ldots, v_{n+\ell}) : (V, V^q, \ldots, V^{q^{n+\ell-1}})M_{n+\ell}^{-1} =: (v_1, \ldots, v_{n+\ell}).$$

Using the matrix $M_{n+\ell}$, we can now go back and forth between the two vector spaces $\mathbb{F}_q^{n+\ell}$ (ground field) and $\mathbb{F}_{q^{n+\ell}}^{n+\ell}$ (extension field). The latter is a very redundant version of the former as we could use any component of the vector $\underline{V} = (V, V^q, \ldots, V^{q^{n+\ell-1}})$ to reconstruct all other $(n + \ell - 1)$ elements. However, we will see below how it will help us to express the rank condition on $\mathfrak{F}$ using only publicly available information.

There are two minor ingredients missing before we can formulate the full attack. The first is the quadratic form of the plus polynomials $a_1, \ldots, a_p$. As for Double-Layer Square, we write them as symmetric matrices $A^{(i)} \in \mathbb{F}_q^{(n+\ell)\times(n+\ell)}$ with $x = (x_1, \ldots, x_{n+\ell})$ and $a_i = \underline{x}A^{(i)}\underline{x}^\intercal$ for $1 \le i \le p$. Hence, we work over the ground field here. Second, we define matrices $\mathfrak{F}^{(i)} \in \mathbb{F}_{q^{n+\ell}}^{(n+\ell)\times(n+\ell)}$ similar to $\mathfrak{F}$ from above as $\mathfrak{F}_{k,k}^{(i)} := 1$ but $\mathfrak{F}_{a,b}^{(i)} = 0$ for $k := (1 - i) \pmod{n + \ell} + 1$, $1 \le a, b \le k$. Or to rephrase this, we have the all-zero matrix with the a single 1, the matrix $\mathfrak{F}^{(1)}$ coincides with the originally defined matrix $\mathcal{F}$, and the 1 is traveling backwards on the main diagonal for each consecutive matrix $\mathfrak{F}^{(i)}$. Note that evaluating $M_{n+\ell}\mathfrak{F}^{(k)}M_{n+\ell}^\intercal$ yields exactly X^2 for each matrix $\mathfrak{F}^{(k)}$.

We now express the private key in terms of $S, T, A, \mathfrak{F}$ and study their corresponding ranks

$$\begin{aligned} P &= T \circ F \circ S \\ &= (\mathcal{C} \circ S, \mathcal{A} \circ S)T \end{aligned}$$

Replacing P on the left hand side with the public key matrices $\mathfrak{P}^{(k)}$ for $1 \leq k \leq (n+\ell+p)$, plugging in the definitions of $\mathcal{C}, \mathcal{A}$, and bringing the matrix T to the left we obtain

$$(\mathfrak{P}^{(1)}, \ldots, \mathfrak{P}^{(n+\ell+p)})T^{-1} = [(SM_{n+\ell}\mathfrak{F}^{(1)}M_{n+\ell}^{\intercal}S^{\intercal}, \ldots, SM_{n+\ell}\mathfrak{F}^{(n+\ell)}M_{n+\ell}^{\intercal}S^{\intercal})M_{n+\ell}^{-1} \\ \|(SA^{(1)}S^{\intercal}, \ldots, SA^{(p)}S^{\intercal})]$$

Again, "$\|$" denotes the concatenation of vectors. Note that the *overall* equation is over the ground field $\mathbb{F}_q$, while the matrices $\mathfrak{F}^{(i)}$ are over the extension field $\mathbb{F}_q^{n+\ell}$. There are two important remarks to be made: First, the matrices $A^{(i)}$ are with overwhelming probability of high rank, both over the ground field and the extension field $\mathbb{F}_{q^{n+\ell}}$. In contrast, each column $SM_{n+\ell}\mathfrak{F}^{(1)}M_{n+\ell}^{\intercal}S^{\intercal}$ has at most rank 1 over the extension field $\mathbb{F}_{q^{n+\ell}}$. Note that the embedding modifier does *not* change the latter rank property as the rank will only decrease, not increase by the embedding modifier, cf. [2, Sect. 5] for a more detailed explanation of this fact. Second, we are only interested in separating out the first $(n+\ell)$ columns of the right hand side from the last p ones. So we do not look for the full matrix T^{-1}, but only its first $(n+\ell)$ columns. We denote them by $\tilde{T} \in \mathbb{F}_q^{(n+\ell+p)\times(n+\ell)}$ and have rank $n+\ell$. Combining these two observations, our equation simplifies to

$$(\mathfrak{P}^{(1)}, \ldots, \mathfrak{P}^{(n+\ell+p)})\tilde{T}M_{n+\ell} = (SM_{n+\ell}\mathfrak{F}^{(1)}M_{n+\ell}^{\intercal}S^{\intercal}, \ldots, SM_{n+\ell}\mathfrak{F}^{(n+\ell)}M_{n+\ell}^{\intercal}S^{\intercal})$$

Note that the *whole* equation is now over the extension field while the coefficients of the matrices $\mathfrak{P}^{(i)}$ come from the ground field. For simplicity, write $U := \tilde{T}M_{n+\ell}$. By construction of $M_{n+\ell}$ we have $u_{i,j} = u_{i,j-1}^q$ and $u_{i,1} = u_{i,n+\ell}^q$ for $1 \leq i \leq n+\ell, 1 < j \leq n+\ell$, so the knowledge of *one* column of U is enough to determine the whole matrix. Hence we only concentrate on the first column of U and obtain

$$\sum_{i=1}^{n+\ell+p} \mathfrak{P}^{(i)}u_{i,1} = SM_{n+\ell}\mathfrak{F}^{(1)}M_{n+\ell}^{\intercal}S^{\intercal} =: H \text{ with } H \in \mathbb{F}_{q^{n+\ell}}^{n\times n}$$

for unknown S. As our final equation is over $\mathbb{F}_q^{n+\ell}$ we clearly have $\operatorname{rank}(H) \leq 1$ and can thus use a similar technique as in section 3 to determine values $\lambda_i \in \mathbb{F}_{q^{n+\ell}}$ such that

$$\operatorname{rank}(\sum_{i=1}^{n+\ell+p} \lambda_i \mathfrak{P}^{(i)}) \leq 1$$

by solving the corresponding MinRank$(q, n+\ell+p, 1)$ problem, *i.e.* for rank $r = 1$.

4.2 Solving MinRank for Square+

All in all, there are two methods available. The first is credited to Schnorr and works on determinants for $(r+1) \times (r+1)$ submatrices while the other was developed by Levy-dit-Vehel *et al.* [12] and uses Gröbner bases.

Table 2. Time to solve the MinRank problem for Square+ for varying embedding degree ℓ, but fixed field size $q = 31$, field extension $n = 17$, and plus equations $p = 5$. Each line is based on 11 independent experiments. We see that the running time does not depend on the embedding degree ℓ.

q	n	p	ℓ	time [sec] min	avg	max
31	17	5	0	1590.59	1610.13	1630.91
			1	1580.85	1605.42	1624.17
			2	1563.80	1600.54	1616.89
			3	1587.97	1603.67	1628.78
			4	1557.96	1604.47	1626.03
			5	1567.56	1610.80	1636.44
			6	1584.20	1606.61	1622.34
			7	1573.56	1604.07	1621.94
			8	1583.91	1609.04	1629.97
			9	1575.46	1603.57	1624.08
			10	1565.71	1597.58	1618.23

We start with Schorr's method. It uses the following observation: For given rank r, each sub-matrix of size $(r+1) \times (r+1)$ must have determinant zero. Hence, each such determinant gives rise to one equation of degree $(r+1)$. For a $(\tau \times \tau)$-matrix, we can form $\binom{\tau}{r+1}^2$ sub-matrices (selecting $r+1$ rows and columns, respectively) and hence equations. Assuming that a sufficiently high proportion of them is linearly independent, we are able to solve the corresponding system of equations by linearization. In our case, we have $r = 1$ and $\tau := (n + \ell + p)$ free variables, leading to a total of $\binom{n+\ell+p}{2}$ degree 2 monomials. For $\ell + p < n$, this allows to compute a solution in $\binom{n+\ell+p}{2}^3 \in O(n^6)$ computations over $\mathbb{F}_{q^{n+\ell}}$ and is hence polynomial in all security parameters. For the proposed parameters $n = 48, \ell = 3, p = 5$ we obtain a total workload of $\approx 2^{31.77}$ and have hence broken the scheme.

For the second method, we inspect the kernel of the matrix H. Remember that each kernel element $\omega \in \mathbb{F}^n_{q^{n+\ell}}$ has the form $\omega := SM_{n+\ell}(0, \tilde{\omega_2}, \ldots, \tilde{\omega_{n+\ell}})$ for $\tilde{\omega_i} \in \mathbb{F}_{q^{n+\ell}}$ and $2 \le i \le (n+\ell)$. So randomly sampling vectors $\omega \in_R \mathbb{F}^n_{q^{n+\ell}}$ needs $q^{n+\ell}$ trials on average to find a kernel element of H and is hence exponential in the security parameters n, ℓ. It is also impractical for the proposed parameters. Thus we use the more refined technique from Levy-dit-Vehel *et al.* [12] to solve instances of the MinRank problem. In a nutshell, they do not sample vectors ω but *calculate* them. This is done by generating an overdetermined $\mathcal{MQ}$-system and then solving it with Gröbner base techniques. Note that the attack complexity grows exponentially with the rank of the target matrix. However, as this rank is fixed to 1 in our case, we are not concerned by this.

The dimension of the kernel of H is $(n-1)$ in the extension field and thus we can fix all but one coefficient of ω at random and still expect a solution. The corresponding vector therefore becomes $(\omega_1, \ldots, \omega_{n-1}, x)$ with $\omega_i \in \mathbb{F}_{q^{n+\ell}}$

fixed values and x a free variable living over the extension field $\mathbb{F}_{q^{n+\ell}}$. Using this notation, we can formulate the following system of quadratic equations over the vector space $\mathbb{F}^n_{q^{n+\ell}}$:

$$\left(\sum_{i=1}^{n+\ell+p} \lambda_i \mathfrak{P}^{(i)}\right) \omega = 0^n$$

For rank 1, we can sample a total of $(n-1)$ linearly independent values $\omega^{(1)}, \ldots, \omega^{(n-1)}$ from the kernel and hence obtain $(n-1)n$ linearly independent equations in a total of $(n-1) + (n+\ell+p) = 2n+\ell+p-1$ unknowns. According to [12], we expect an overall complexity of $\binom{N+r+1}{r+2}^3$ for N the number of unknowns. For the proposed parameters $n = 48, \ell = 3, p = 5$ we obtain a workload of $\binom{2n+\ell+p+1}{3}^3 \approx 2^{52.55}$. This is clearly worse than Schnorr's method. However, the Gröbner method can exploit computing all intermediate steps in the ground field, so the authors of [2] report a substantial speed-up here. Moreover, for variations of Square+, we might be able to formulate side-conditions easier than for Schnorr's method.

Executing either the algorithm of Schnorr or of Levi-dit-Vehel *et al.*, we can reconstruct the initial Square system and are in the same position as a legitimate user.

We have implemented the attack of Schnorr and found the theory in line with the practical experiments. In particular, the matrices $\mathfrak{F}^{(i)}$ for $1 \leq i \leq (n+\ell)$ have rank 1 over the extension field and we can reconstruct the matrix H for public key matrices $\mathfrak{P}^{(k)}$ for $1 \leq k \leq (n+\ell+p)$ alone.

5 Conclusion

In this paper we have presented the first cryptanalysis of the two twin schemes Double-Layer Square and Square+. Both attacks relied heavily on the rank properties of the public key equations over the ground field (Double-Layer) or the extension field (Square+). In either case, each scheme is fully broken for any reasonable choice of parameters: For Double-Layer Square, the attack is exponential in the security parameter ℓ. However, as $\ell = 4$ and cannot be increased too much due to generic attacks against $\mathcal{M}$ultivariate $\mathcal{Q}$uadratic schemes, it is efficient in practice. For Square+, the attack is fully polynomial in all security parameters q, ℓ, p.

As we have established a strong link between odd characteristic Hidden Field Equations and Square, we know that any cryptanalytic result for the former can be exploited for the latter. So the relation between Square and odd-HFE is the same as for MIA/C* and HFE. All attacks to the latter (odd-HFE, HFE) will inevitable apply to the former (Square, MIA/C*). Hence, any strategy to repair Square will need to take these similarities into account. In addition, we have to remember that Square will always be much weaker than odd-HFE—for reasons similar to the pair MIA/C* and HFE. Moreover, we expect that any successful cryptanalysis of odd-HFE can be turned easily in a cryptanalysis of Square—maybe even *without* any further modification. For example, transferring Square

Table 3. Summary of the complexity of the attacks given in this paper. In both cases, we measure the number of computations over the corresponding field with q being the size of the ground field, n an intermediate extension degree, and ℓ the embedding degree.

Algorithm	Attack	Complexity	over
Double-Layer Square	Key Recovery	$(n+\ell)q^{\ell+1}(2n+\ell)^3$	$\mathbb{F}_q$
Square+	Key Recovery	$\binom{n+\ell+p}{2}^3$	$\mathbb{F}_{q^{n+\ell}}$

to the equivalent of "multi-HFE" [6] does not seem to be a good idea. It was already established that this variant actually leads to a *weaker* version of the original odd-HFE. Similarly, we can conclude that Square- is broken, as is MIA-. Both variations were suggested in [8], the first as "bivariate Square", the other as Square-. On the other hand, a secure version of Square will most certainly give rise to a secure version of MIA.

In particular, Square has exactly the same big advantage over odd-HFE that MIA/C* has over HFE: *Speed.* When it comes to signing/decrypting, both will outperform the more secure variants by orders of magnitudes. Hence, it seems to be too early to call the overall game "Square" being over but it seems a fair guess that some further modifications will be tried. If they will stand the test of time is a different question altogether.

Acknowledgments. We want to thank Gottfried Herold (Bochum) for fruitful discussions and helpful remarks. Furthermore we thank the reviewers for helpful comments.

The authors were supported by the German Science Foundation (DFG) through an Emmy Noether grant where the second author is principal investigator. All authors were in part supported by the European Commission through the IST Programme under contract *ICT-2007-216676 Ecrypt II.*

References

[1] Akkar, M.-L., Courtois, N.T., Duteuil, R., Goubin, L.: A Fast and Secure Implementation of Sflash. In: Desmedt, Y.G. (ed.) PKC 2003. LNCS, vol. 2567, pp. 267–278. Springer, Heidelberg (2002)

[2] Bettale, L., Faugère, J.-C., Perret, L.: Cryptanalysis of Multivariate and Odd-Characteristic HFE Variants. In: Catalano, D., Fazio, N., Gennaro, R., Nicolosi, A. (eds.) PKC 2011. LNCS, vol. 6571, pp. 441–458. Springer, Heidelberg (2011)

[3] Billet, O., Gilbert, H.: Cryptanalysis of Rainbow. In: De Prisco, R., Yung, M. (eds.) SCN 2006. LNCS, vol. 4116, pp. 336–347. Springer, Heidelberg (2006)

[4] Billet, O., Macario-Rat, G.: Cryptanalysis of the Square Cryptosystems. In: Matsui, M. (ed.) ASIACRYPT 2009. LNCS, vol. 5912, pp. 451–468. Springer, Heidelberg (2009)

[5] Buss, J.F., Frandsen, G.S., Shallit, J.O.: The computational complexity of some problems of linear algebra. Research Series RS-96-33, BRICS, Department of Computer Science, University of Aarhus, pages 39 (September 1996), `http://www.brics.dk/RS/96/33/`

[6] Chen, C.-H.O., Chen, M.-S., Ding, J., Werner, F., Yang, B.-Y.: Odd-char multivariate hidden field equations. Cryptology ePrint Archive, Report 2008/543 (2008), http://eprint.iacr.org/

[7] Clough, C., Baena, J., Ding, J., Yang, B.-Y., Chen, M.-S.: Square, a New Multivariate Encryption Scheme. In: Fischlin, M. (ed.) CT-RSA 2009. LNCS, vol. 5473, pp. 252–264. Springer, Heidelberg (2009)

[8] Clough, C.L.: Square: A New Family of Multivariate Encryption Schemes. PhD thesis, University of Cincinnati (2009)

[9] Clough, C.L., Ding, J.: Secure Variants of the Square Encryption Scheme. In: Sendrier, N. (ed.) PQCrypto 2010. LNCS, vol. 6061, pp. 153–164. Springer, Heidelberg (2010)

[10] Courtois, N., Goubin, L., Patarin, J.: Sflash: Primitive specification (second revised version) Submissions, Sflash, 11 pages (2002), https://www.cosic.esat.kuleuven.be/nessie

[11] Ding, J., Schmidt, D.: Rainbow, a New Multivariable Polynomial Signature Scheme. In: Ioannidis, J., Keromytis, A.D., Yung, M. (eds.) ACNS 2005. LNCS, vol. 3531, pp. 164–175. Springer, Heidelberg (2005)

[12] Faugère, J.-C., dit Vehel, F.L., Perret, L.: Cryptanalysis of MinRank. In: Wagner, D. (ed.) CRYPTO 2008. LNCS, vol. 5157, pp. 280–296. Springer, Heidelberg (2008)

[13] Goubin, L., Courtois, N.T.: Cryptanalysis of the TTM Cryptosystem. In: Okamoto, T. (ed.) ASIACRYPT 2000. LNCS, vol. 1976, pp. 44–57. Springer, Heidelberg (2000)

[14] Imai, H., Matsumoto, T.: Algebraic methods for constructing asymmetric cryptosystems. In: Calmet, J. (ed.) AAECC 1985. LNCS, vol. 229, pp. 108–119. Springer, Heidelberg (1986)

[15] Kipnis, A., Shamir, A.: Cryptanalysis of the HFE Public Key Cryptosystem by Relinearization. In: Wiener, M. (ed.) CRYPTO 1999. LNCS, vol. 1666, pp. 19–30. Springer, Heidelberg (1999), http://www.minrank.org/hfesubreg.ps, http://citeseer.nj.nec.com/kipnis99cryptanalysis.html

[16] Matsumoto, T., Imai, H., Harashima, H., Miyakawa, H.: A cryptographically useful theorem on the connection between uni and multivariate polynomials. Transactions of the IECE of Japan 68(3), 139–146 (1985)

[17] Patarin, J.: Hidden Fields Equations (HFE) and Isomorphisms of Polynomials (IP): Two New Families of Asymmetric Algorithms. In: Maurer, U.M. (ed.) EUROCRYPT 1996. LNCS, vol. 1070, pp. 33–48. Springer, Heidelberg (1996), http://www.minrank.org/hfe.pdf

[18] Wolf, C., Braeken, A., Preneel, B.: Efficient Cryptanalysis of RSE(2)PKC and RSSE(2)PKC. In: Blundo, C., Cimato, S. (eds.) SCN 2004. LNCS, vol. 3352, pp. 294–309. Springer, Heidelberg (2005), http://eprint.iacr.org/2004/237

[19] Wolf, C., Braeken, A., Preneel, B.: On the security of stepwise triangular systems. Designes, Codes and Cryptography 40(3), 285–302 (2006)

[20] Wolf, C., Preneel, B.: Equivalent keys in multivariate quadratic public key systems. Journal of Mathematical Cryptology 4(4), 375–415 (2011)

[21] Yang, B.-Y., Chen, J.-M.: Rank attacks and defence in Tame-like multivariate PKC's. Cryptology ePrint Archive, Report 2004/061, September 29, pages 21 (2004), http://eprint.iacr.org/

An Efficient Attack on All Concrete KKS Proposals

Ayoub Otmani[1,2] and Jean-Pierre Tillich[1]

[1] SECRET Project - INRIA Rocquencourt
Domaine de Voluceau, B.P. 105 78153 Le Chesnay Cedex - France
{ayoub.otmani,jean-pierre.tillich}@inria.fr
[2] GREYC - Université de Caen - Ensicaen
Boulevard Maréchal Juin, 14050 Caen Cedex, France

Abstract. Kabastianskii, Krouk and Smeets proposed in 1997 a digital signature scheme based on a couple of random error-correcting codes. A variation of this scheme was proposed recently and was proved to be EUF-1CMA secure in the random oracle model. In this paper we investigate the security of these schemes and suggest a simple attack based on (essentially) Stern's algorithm for finding low weight codewords. It efficiently recovers the private key of all schemes of this type existing in the literature. This is basically due to the fact that we can define a code from the available public data with unusual properties: it has many codewords whose support is concentrated in a rather small subset. In such a case, Stern's algorithm performs much better and we provide a theoretical analysis substantiating this claim. Our analysis actually shows that the insecurity of the proposed parameters is related to the fact that the rates of the couple of random codes used in the scheme were chosen to be too close. This does not compromise the security of the whole KKS scheme. It just points out that the region of weak parameters is really much larger than previously thought.

Keywords: Code-based cryptography, digital signature, random error-correcting codes, cryptanalysis.

1 Introduction

Digital signature schemes are probably among the most useful cryptographic algorithms. If quantum computers were to become reality, it would be useful to devise such schemes which would resist to it. A possible approach to meet this goal could be to build such schemes whose security relies on the difficulty of decoding linear codes. Two code based schemes of this kind have been proposed, namely the Courtois-Finiasz-Sendrier signature scheme [CFS01] and the Kabatianskii, Krouk and Smeets (KKS) scheme [KKS97, KKS05].

The Courtois-Finiasz-Sendrier (CFS) scheme presents the advantage of having an extremely short signature and its security has been proven to rely on the well-known syndrome decoding problem and the distinguishability of binary Goppa codes from a random code. However, it has been proved in [FGO$^+$10]

B.-Y. Yang (Ed.): PQCrypto 2011, LNCS 7071, pp. 98–116, 2011.

that the latter problem can be solved in the range of parameters used in the CFS signature algorithm. This does not prove that their proposal is insecure. However, it invalidates the hypotheses of their security proof. The main difficulty in suggesting a CFS type scheme is to come up with a family of very high rate codes with an efficient decoding algorithm and whose structure can be hidden in the same way as in the McEliece scheme. This narrows down quite a bit the families of codes which can be used in this setting and up to now only Goppa codes are known to meet this goal. It should be emphasized that it is precisely their rich algebraic structure which makes it possible to distinguish them from random codes.

On the other hand, the KKS proposal does not rely on Goppa codes and can be instantiated with random codes. Moreover, unlike in the CFS signature scheme, it does not compute a signature by using a decoding algorithm for the code and thus completely avoids the necessity of having to use restricted families of codes with a "hidden" trapdoor. Moreover, a variation of it has been proposed in [BMJ11] and has been proved to be EUF-1CMA secure in the random oracle model. The security of the KKS scheme has been investigated in [COV07]. It was shown that a passive attacker who may intercept just a few signatures can recover the private key. All the schemes proposed in [KKS97] can be broken in this way with the help of at most 20 signatures. Basically it uses the fact that a valid message-signature pair reveals on average half of the secret support J (see Section 3 where this set is defined precisely). Therefore with $O(\log |J|)$ message-signature pairs it is expected to recover the whole set J. The security of the scheme is not compromised by this attack however if only one signature is computed, and this especially in the variant proposed in [BMJ11] where some random noise is added on top of the signature.

The purpose of this article is to present a completely new security analysis of the KKS scheme and its variant proposed in [BMJ11]. Our approach for breaking the scheme is to define a certain error correcting code from the couple of public matrices used in the scheme and to notice that certain rather low weight codewords give actually valid signatures. It is therefore natural to use standard algorithms for finding low-weight codewords in this setting, such as Stern's algorithm [Ste88] or its Dumer variant [Dum96, FS09] (see also [BLP11]). It turns out that such algorithms are unusually successful in this setting due to the conjunction of three factors: (i) there are many low-weight codewords, (ii) they are localized on a rather small support, (iii) some part of this support is known to the attacker. It appears that all parameters suggested in [KKS97, KKS05, BMJ11] are easily broken by this approach and this without even knowing a single signature pair. Moreover, this approach can exploit the knowledge of a message-signature pair which speeds up the attack.

We provide an analysis of this attack which explains what makes it feasible for the parameters proposed in [KKS97, KKS05, BMJ11]. The KKS scheme relies on a couple of matrices which can be viewed as parity-check matrices of two linear codes. We show that when the first code has a rate which is smaller than the rate of the second one (or has approximately the same rate), then our attack is quite successful. This was exactly the case for all the parameters suggested in the past. In other words, our attack does not compromise the security of the whole KKS scheme. It just points out that the region of weak parameters is really much larger than previously thought.

2 Terminology and Notation

In the whole paper q denotes some prime power and we denote by $\mathbb{F}_q$ the finite field with q elements. Let n be a non-negative integer. The set of integers i such that $1 \leqslant i \leqslant n$ is denoted by $[1 \cdots n]$. The cardinality of a set A is denoted by $|A|$. The concatenation of the vectors $\boldsymbol{x} = (x_1, \dots, x_n)$ and $\boldsymbol{y} = (y_1, \dots, y_m)$ is denoted by $(\boldsymbol{x}||\boldsymbol{y}) \stackrel{\text{def}}{=} (x_1, \dots, x_n, y_1, \dots, y_m)$. The support $\mathsf{supp}(\boldsymbol{x})$ of $\boldsymbol{x} \in \mathbb{F}_q^n$ is the set of i's such that $x_i \neq 0$. The *(Hamming) weight* $|\boldsymbol{x}|$ is the cardinality of $\mathsf{supp}(\boldsymbol{x})$. For a vector $\boldsymbol{x} = (x_i)$ and a subset I of indices of $\boldsymbol{x}$, we denote by $\boldsymbol{x}_I$ its restriction to the indices of I, that is:

$$\boldsymbol{x}_I \stackrel{\text{def}}{=} (x_i)_{i \in I}.$$

We will also use this notation for matrices, in this case it stands for the submatrix formed by the columns in the index set, i.e. for any $k \times n$ matrix $\boldsymbol{H}$

$$\boldsymbol{H}_J \stackrel{\text{def}}{=} (h_{ij})_{\substack{1 \leqslant i \leqslant k \\ j \in J}}.$$

A linear code $\mathscr{C}$ of type $[n, k, d]$ over $\mathbb{F}_q$ is a linear subspace of $\mathbb{F}_q^n$ of dimension k and minimum distance d where by definition $d \stackrel{\text{def}}{=} \min\{|\boldsymbol{x}| \ : \ \boldsymbol{x} \in \mathscr{C} \text{ and } \boldsymbol{x} \neq \boldsymbol{0}\}$. The elements of $\mathscr{C}$ are *codewords.* A linear code can be defined either by a parity check matrix or a generator matrix. A *parity check matrix* $\boldsymbol{H}$ for $\mathscr{C}$ is an $(n-k) \times n$ matrix such that $\mathscr{C}$ is the right kernel of $\boldsymbol{H}$:

$$\mathscr{C} = \{\boldsymbol{c} \in \mathbb{F}_q^n \ : \ \boldsymbol{H}\boldsymbol{c}^T = 0\}$$

where $\boldsymbol{x}^T$ denotes the *transpose* of $\boldsymbol{x}$. A generator matrix $\boldsymbol{G}$ is a $k \times n$ matrix formed by a basis of $\mathscr{C}$. We say that $\boldsymbol{G}$ is in *systematic* form if there exists a set J such that $\boldsymbol{G}_J = \boldsymbol{I}_k$. The *syndrome* $\boldsymbol{s}$ by $\boldsymbol{H}$ of $\boldsymbol{x} \in \mathbb{F}_q^n$ is defined as $\boldsymbol{s}^T \stackrel{\text{def}}{=} \boldsymbol{H}\boldsymbol{x}^T$. A *decoding* algorithm for $\boldsymbol{H}$ is an algorithm such that, given $\boldsymbol{s}$ in $\mathbb{F}_q^r$, finds a vector $\boldsymbol{e}$ of *minimum* weight whose syndrome is $\boldsymbol{s}$.

3 The Kabatianskii-Krouk-Smeets Signature Scheme and Its Variant

This section is devoted to the description of two code-based signature schemes proposed in [KKS97] and more recently in [BMJ11], where the latter can be viewed as a "noisy" version of the former [KKS97]. Our presentation presents the main ideas without giving all the details which can be found in the original papers. We first focus on the scheme of [KKS97] whose construction relies on the following ingredients:

1. a full rank binary matrix $\boldsymbol{H}$ of size $(N-K) \times N$ with entries in a finite field $\mathbb{F}_q$.
2. a subset J of $\{1, \ldots, N\}$ of cardinality n,
3. a linear code $\mathscr{C}_{\text{hidden}}$ over $\mathbb{F}_q$ of length $n \leqslant N$ and dimension k defined by a generator matrix $\boldsymbol{G}$ of size $k \times n$. Let t_1 and t_2 be two integers such that with very high probability, we have that $t_1 \leqslant |\boldsymbol{u}| \leqslant t_2$ for any non-zero codeword $\boldsymbol{u} \in \mathscr{C}_{\text{hidden}}$.

The matrix $\boldsymbol{H}$ is chosen such that the best decoding algorithms cannot solve the following search problem.

Problem 1. Given the knowledge of $\boldsymbol{s} \in \mathbb{F}_q^{N-K}$ which is the syndrome by $\boldsymbol{H}$ of some $\boldsymbol{e} \in \mathbb{F}_q^N$ whose weight lies in $[t_1 \cdots t_2]$, find explicitly $\boldsymbol{e}$, or eventually $\boldsymbol{x}$ in $\mathbb{F}_q^N$ different from $\boldsymbol{e}$ sharing the same properties as $\boldsymbol{e}$.

Finally let $\boldsymbol{F}$ be the $(N-K)\times k$ matrix defined by $\boldsymbol{F} \stackrel{\text{def}}{=} \boldsymbol{H}_J \boldsymbol{G}^T$. The Kabatianskii-Krouk-Smeets (KKS) signature scheme is then described in Figure 1.

- Setup.
  1. The signer S chooses N, K n, k, t_1 and t_2 according to the required security level.
  2. S draws a random $(N-K) \times N$ matrix $\boldsymbol{H}$.
  3. S randomly picks a subset J of $\{1, \ldots, N\}$ of cardinality n.
  4. S randomly picks a random $k \times n$ generator matrix $\boldsymbol{G}$ that defines a code $\mathscr{C}_{\text{hidden}}$ such that with high probability $t_1 \leqslant |\boldsymbol{u}| \leqslant t_2$ for any non-zero codeword $\boldsymbol{u} \in \mathscr{C}_{\text{hidden}}$.
  5. $\boldsymbol{F} \stackrel{\text{def}}{=} \boldsymbol{H}_J \boldsymbol{G}^T$ where $\boldsymbol{H}_J$ is the restriction of $\boldsymbol{H}$ to the columns in J.
- Keys.
  - Private key. J and $\boldsymbol{G}$
  - Public key. $\boldsymbol{F}$ and $\boldsymbol{H}$
- Signature. The signature σ of a message $\boldsymbol{x} \in \mathbb{F}_q^k$ is defined as the unique vector σ of $\mathbb{F}_q^N$ such that $\sigma_i = 0$ for any $i \notin J$ and $\sigma_J = \boldsymbol{x}\boldsymbol{G}$.
- Verification. Given $(\boldsymbol{x}, \sigma) \in \mathbb{F}_q^k \times \mathbb{F}_q^N$, the verifier checks that $t_1 \leqslant |\sigma| \leqslant t_2$ and $\boldsymbol{H}\sigma^T = \boldsymbol{F}\boldsymbol{x}^T$.

Fig. 1. Description of the KKS scheme given in [KKS97]

The scheme was modified in [BMJ11] to propose a one-time signature scheme by introducing two new ingredients, namely a hash function f and adding an error vector $\boldsymbol{e}$ to the signature. It was proved that such a scheme is EUF-1CMA secure in the random oracle model. The description is given in Figure 2.

4 Description of the Attack

The purpose of this section is to explain the idea underlying our attack which aims at recovering the private key. The attack is divided in two main steps.

– Setup.
  1. The signer S chooses N, K n, k, t_1 and t_2 according to the required security level.
  2. S chooses a hash function $f : \{0,1\}^* \times \mathbb{F}_2^{N-K} \longrightarrow \mathbb{F}_2^k$.
  3. S draws a random binary $(N-K) \times N$ matrix $\boldsymbol{H}$.
  4. S randomly picks a subset J of $\{1, \ldots, N\}$ of cardinality n.
  5. S randomly picks a $k \times n$ generator matrix $\boldsymbol{G}$ that defines a binary code $\mathscr{C}_{\text{hidden}}$ such that with high probability $t_1 \leqslant |\boldsymbol{u}| \leqslant t_2$ for any non-zero codeword $\boldsymbol{u} \in \mathscr{C}_{\text{hidden}}$.
  6. $\boldsymbol{F} \stackrel{\text{def}}{=} \boldsymbol{H}_J \boldsymbol{G}^T$ where $\boldsymbol{H}_J$ is the restriction of $\boldsymbol{H}$ to the columns in J.
– Keys.
  • Private key. J and $\boldsymbol{G}$
  • Public key. $\boldsymbol{F}$ and $\boldsymbol{H}$
– Signature. The signature of a message $\boldsymbol{x} \in \{0,1\}^*$ is (h, σ) defined as follows:
  • S picks a random $\boldsymbol{e} \in \mathbb{F}_2^N$ such that $|\boldsymbol{e}| = n$.
  • Let $\boldsymbol{h} \stackrel{\text{def}}{=} f(\boldsymbol{x}, \boldsymbol{H}\boldsymbol{e}^T)$ and $\boldsymbol{y}$ be the unique vector of $\mathbb{F}_2^N$ such that (i) $\mathsf{supp}(\boldsymbol{y}) \subset J$, (ii) $\boldsymbol{y}_J = \boldsymbol{h}\boldsymbol{G}$. The second part of the signature σ is then given by $\sigma \stackrel{\text{def}}{=} \boldsymbol{y}+\boldsymbol{e}$.
– Verification. Given a signature $(\boldsymbol{h}, \sigma) \in \mathbb{F}_2^k \times \mathbb{F}_2^N$ for $\boldsymbol{x} \in \{0,1\}^*$, the verifier checks that $|\sigma| \leqslant 2n$ and $\boldsymbol{h} = f(\boldsymbol{x}, \boldsymbol{H}\sigma^T + \boldsymbol{F}\boldsymbol{h}^T)$.

Fig. 2. Description of the scheme of [BMJ11]

First, we produce a valid signature for some message using only the public key. To do so, we define a certain code from matrices $\boldsymbol{H}$ and $\boldsymbol{F}$. It turns out that low weight codewords of this code give valid message-signature pairs. Then we just apply Dumer's algorithm [Dum91] in order to find these low weight codewords. This attack can even be refined in the following way. Whenever we are able to produce one valid message-signature pair, and since each signature reveals partial information about the private key (especially about J as explained further in this section), we can use it to get another valid message-signature pair revealing more information about J. We repeat this process a few times until we totally recover the whole private key. More details will be given in the following sections.

In what follows, we make the assumption that all the codes are binary because all the concrete proposals are of this kind. The non-binary case will be discussed in the conclusion.

4.1 An Auxiliary Code

We give here the first ingredient we use to forge a valid message/signature pair for the KKS scheme just from the knowledge of the public pair $\boldsymbol{H}, \boldsymbol{F}$. This attack can also be used for the second scheme given by Figure 2. In the last case, it is not a valid message/signature pair anymore but an auxiliary quantity which helps in revealing J. This ingredient consists in a linear code $\mathscr{C}_{\text{pub}}$ of length $N + k$ defined as the kernel of $\hat{\boldsymbol{H}}$ which is obtained by the juxtaposition of the two public matrices $\boldsymbol{H}$ and $\boldsymbol{F}$ as given in Figure 3. The reason behind this definition lies in the following Fact 1.

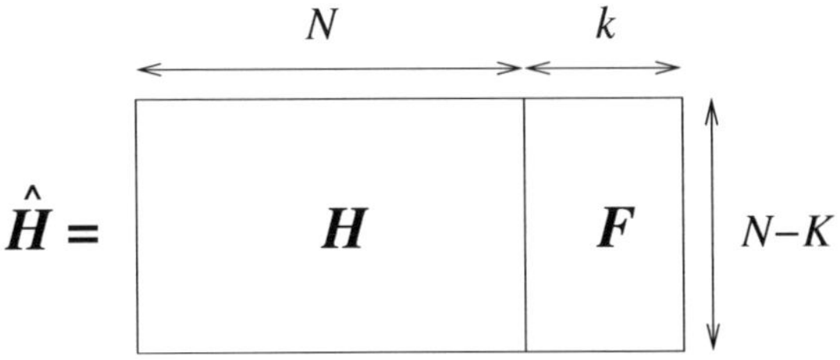

Fig. 3. Parity-check matrix $\hat{\boldsymbol{H}}$ of the code $\mathscr{C}_{\text{pub}}$

Fact 1. *Let $\boldsymbol{x}'$ be in $\mathbb{F}_2^{N+k}$ and set $(\sigma||\boldsymbol{x}) \stackrel{def}{=} \boldsymbol{x}'$ with σ in $\mathbb{F}_2^N$ and $\boldsymbol{x}$ in $\mathbb{F}_2^k$. Then σ is a signature of $\boldsymbol{x}$ if and only if:*

1. $\hat{\boldsymbol{H}}\boldsymbol{x}'^T = 0$
2. $t_1 \leqslant |\sigma| \leqslant t_2$.

The code $\mathscr{C}_{\text{pub}}$ is of dimension $k + K$, and of particular interest is the linear space $\mathscr{C}_{\text{sec}} \subset \mathscr{C}_{\text{pub}}$ that consists in words that satisfy both conditions of Fact 1 and that are obtained by all pairs $(\sigma, \boldsymbol{x})$ of valid message-signature pairs which are obtained by the secret signature algorithm, that is to say:

$$\mathscr{C}_{\text{sec}} \stackrel{\text{def}}{=} \left\{(\sigma||\boldsymbol{x}) \in \mathbb{F}_2^{N+k} \ : \ \boldsymbol{x} \in \mathbb{F}_2^k, \ \sigma \in \mathbb{F}_2^N, \ \sigma_J = \boldsymbol{x}\boldsymbol{G}, \ \sigma_{[1\cdots N]\setminus J} = 0\right\}. \quad (1)$$

Clearly, the dimension of $\mathscr{C}_{\text{sec}}$ is k. Additionally, we expect that the weight of σ is of order $n/2$ for any $(\sigma, \boldsymbol{x})$ in $\mathscr{C}_{\text{sec}}$, which is much smaller than the total length N. This strongly suggests to use well-known algorithms for finding low weight codewords to reveal codewords in $\mathscr{C}_{\text{sec}}$ and therefore message-signature pairs. The algorithm we used for that purpose is specified in the following subsection.

4.2 Finding Low-Weight Codewords

We propose to use the following variation on Stern's algorithm due to [Dum91] (See also [FS09]). The description of the algorithm is given in Algorithm 1. It consists in searching for low-weight codewords among the candidates that are of very low-weight $2p$ (where p is typically in the range $1 \leqslant p \leqslant 4$) when restricted to a set I of size slightly larger than the dimension $k + K$ of the code $\mathscr{C}_{\text{pub}}$, say $|I| = k + K + l$ for some small integer l. The key point in this approach is to choose I among a set S of test positions. The set S will be appropriately chosen according to the considered context. If no signature pair is known, then a good choice for S is to take:

$$S = [1 \cdots N]. \quad (2)$$

This means that we always choose the test positions among the N first positions of the code $\mathscr{C}_{\text{pub}}$ and never among the k last positions. The reason for this choice will be explained in the following subsection.

Algorithm 1. KKSforge: algorithm that forges a valid KKS signature.

PARAMETERS:

r : number of iterations,
l : small integer ($l \leqslant 40$),
p : very small integer ($1 \leqslant p \leqslant 4$).
S : a subset of $[1 \cdots N]$ from which in each iteration a subset of cardinality $K+k+l$ will be randomly chosen.

INPUT: $\hat{\boldsymbol{H}}$
OUTPUT: a list $\mathcal{L}$ containing valid signature/message pairs $(\sigma, \boldsymbol{x}) \in \mathbb{F}_2^N \times \mathbb{F}_2^k$.

```
ℒ ← ∅.
for 1 ⩽ t ⩽ r do
    Step 1: Randomly pick K + k + l positions among S to form the set I. This set
    is partitioned into I = I_1 ∪ I_2 such that ||I_1| − |I_2|| ⩽ 1.
    Step 2: Perform Gaussian elimination over the complementary set {1, 2, ..., N+
    k} \ I to put Ĥ in quasi-systematic form (as shown in Figure 4).
    Step 3:
    Generate all binary vectors x_1 of length ⌊(K + k + l)/2⌋ and weight p and store
    them in a table at the address H_1 x_1^T
    for all binary vectors x_2 of length ⌈(K + k + l)/2⌉ and weight p do
        for all x_1 stored at the address H_2 x_2^T do
            Compute x_3 =def (x_1||x_2)H_3^T and form the codeword x =def (x_1||x_2||x_3) of
            𝒞_pub
            if t_1 ⩽ |x_[1···N]| ⩽ t_2 then
                ℒ ← ℒ ∪ {x}
            end if
        end for
    end for
end for
return ℒ
```

4.3 Explaining the Success of the Attack

It turns out that this attack works extremely well on all the parameter choices made in the literature, and this even without knowing a single message-signature pair which would make life much easier for the attacker as demonstrated in [COV07]. In a first pass, the attack recovers easily message-signature pairs for all the parameters suggested in [BMJ11, KKS97, KKS05]. Once a signature-message pair is obtained, it can be exploited to bootstrap an attack that recovers the private key as we will explain later.

The reason why the attack works much better here than for general linear codes comes from the fact that $\hat{\boldsymbol{H}}$ *does not behave like a random matrix at all even if the two chosen matrices for the scheme, namely* $\boldsymbol{H}$ *and* $\boldsymbol{G}$ *are chosen at random.* The left part and the right part $\boldsymbol{H}$ and $\boldsymbol{F}$ are namely related by the equation:

$$\boldsymbol{F} = \boldsymbol{H}_J \boldsymbol{G}^T.$$

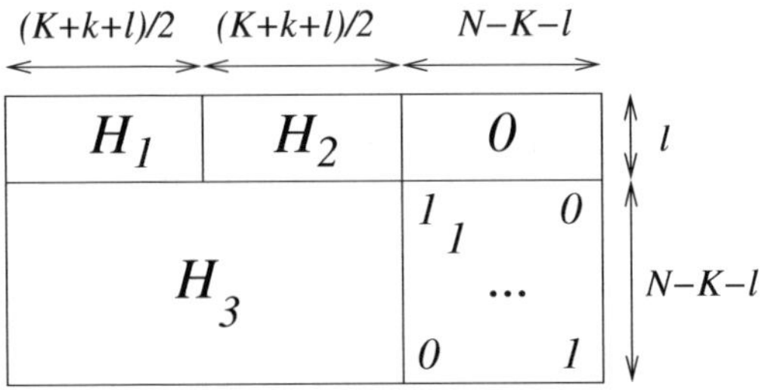

Fig. 4. A parity-check matrix for $\mathscr{C}_{\text{pub}}$ in quasi-systematic form

Indeed, the parity-check matrix $\hat{\boldsymbol{H}}$ displays peculiar properties: $\mathscr{C}_{\text{pub}}$ contains $\mathscr{C}_{\text{sec}}$ as a subcode and its codewords represent valid message-signature pairs. This subcode has actually a very specific structure that helps greatly the attacker:

1. There are many codewords in $\mathscr{C}_{\text{sec}}$, namely 2^k.
2. The support of these codewords is included in a fixed (and rather small) set of size $k + n$.
3. k positions of this set are known to the attacker.
4. These codewords form a linear code (of dimension k).

Because of all these properties, the aforementioned attack will work much better than should be expected from a random code. More precisely, let us bring in:

$$I' \stackrel{\text{def}}{=} I \cap J.$$

Notice that the expectation $\mathbb{E}\{|I'|\}$ of the cardinality of the set I' is equal to:

$$\mathbb{E}\{|I'|\} = \frac{n}{N}(k + K + l) = (R + \alpha\rho + \lambda)n \tag{3}$$

where we introduced the following notation:

$$R \stackrel{\text{def}}{=} \frac{K}{N}, \quad \rho \stackrel{\text{def}}{=} \frac{k}{n}, \quad \alpha \stackrel{\text{def}}{=} \frac{n}{N} \quad \text{and} \quad \lambda \stackrel{\text{def}}{=} \frac{l}{N}.$$

The point is that whenever there is a codeword $\boldsymbol{c}$ in $\mathscr{C}_{\text{sec}}$ which is such that $|\boldsymbol{c}_{I'}| = 2p$ we have a non-negligible chance to find it with Algorithm 1. This does not hold with certainty because the algorithm does not examine all codewords $\boldsymbol{x}$ such that $|\boldsymbol{x}_I| = 2p$, but rather it consists in splitting I in I_1 and I_2 of the same size and looking for codewords $\boldsymbol{x}$ such that $|\boldsymbol{x}_{I_1}| = |\boldsymbol{x}_{I_2}| = p$. In other words, we consider only a fraction δ of such codewords where:

$$\delta = \frac{\binom{(K+k+l)/2}{p}\binom{(K+k+l)/2}{p}}{\binom{K+k+l}{2p}} \approx \sqrt{\frac{(K+k+l)}{\pi p(K+k+l-2p)}}.$$

We will therefore obtain all codewords $\boldsymbol{c}$ in $\mathscr{C}_{\rm sec}$ which are such that $|\boldsymbol{c}_{I_1}| = |\boldsymbol{c}_{I_2}| = p$. Consider now the restriction $\mathscr{C}'_{\rm sec}$ of $\mathscr{C}_{\rm sec}$ to the positions belonging to I', that is:

$$\mathscr{C}'_{\rm sec} = \left\{ (\boldsymbol{x}_i)_{i\in I'} \ : \ \boldsymbol{x} = (\boldsymbol{x}_i)_{i\in[1\cdots N+k]} \in \mathscr{C}_{\rm sec} \right\}. \quad (4)$$

The crucial issue is now the following question:

Does there exist in $\mathscr{C}'_{sec}$ a codeword of weight $2p$?

The reason for this is explained by the following proposition.

Proposition 1. *Let $I'_s \stackrel{def}{=} I_s \cap J$ for $s \in \{1,2\}$. If there exists a codeword $\boldsymbol{x}'$ in $\mathscr{C}'_{sec}$ such that $|\boldsymbol{x}'_{I'_1}| = |\boldsymbol{x}'_{I'_2}| = p$, then it will be the restriction of a codeword $\boldsymbol{x}$ in $\mathscr{C}_{sec}$ which will belong to the list $\mathcal{L}$ output by Algorithm 1.*

Proof. Consider a codeword $\boldsymbol{x}'$ in $\mathscr{C}'_{\rm sec}$ such that $|\boldsymbol{x}'_{I'_1}| = |\boldsymbol{x}'_{I'_2}| = p$. For $s \in \{1,2\}$, extend $\boldsymbol{x}_{I'_s}$ with zeros on the other positions of I_s and let $\boldsymbol{x}_s$ be the corresponding word. Notice that $\boldsymbol{x}_1$ and $\boldsymbol{x}_2$ will be considered by Algorithm 1 and $\boldsymbol{x}_1$ will be stored at the address $\boldsymbol{H}_1\boldsymbol{x}_1^T$. By definition of $\boldsymbol{x}'$, $(\boldsymbol{x}_1||\boldsymbol{x}_2)$ is the restriction of a codeword $\boldsymbol{x}$ of $\mathscr{C}_{\rm sec}$ to I, say $\boldsymbol{x} = (\boldsymbol{x}_1||\boldsymbol{x}_2||\boldsymbol{y})$ with $\boldsymbol{y} \in \mathbb{F}_2^{N-K-l}$. Since $\mathscr{C}_{\rm sec} \subset \mathscr{C}_{\rm pub}$ we have $\hat{\boldsymbol{H}}\boldsymbol{x}^T = 0$. Let $\hat{\boldsymbol{H}}'$ be the matrix obtained from $\hat{\boldsymbol{H}}$ put in quasi-systematic form through a Gaussian elimination as given in Figure 4. We also have $\hat{\boldsymbol{H}}'\boldsymbol{x}^T = 0$ and hence:

$$\boldsymbol{H}_1\boldsymbol{x}_1^T + \boldsymbol{H}_2\boldsymbol{x}_2^T = 0 \quad (5)$$

and

$$\boldsymbol{H}_3(\boldsymbol{x}_1||\boldsymbol{x}_2)^T + \boldsymbol{y}^T = 0. \quad (6)$$

Equation (5) shows that $\boldsymbol{x}_1$ is stored at address $\boldsymbol{H}_2\boldsymbol{x}_2^T$ and will be considered at Step 8 of the algorithm. In this case, $\boldsymbol{x}$ will be stored in $\mathcal{L}$. □

We expect that the dimension of $\mathscr{C}'_{\rm sec}$ is still k and that this code behaves like a random code of the same length and dimension. Ignoring the unessential issue whether or not $\boldsymbol{x}'$ satisfies $|\boldsymbol{x}'_{I'_1}| = |\boldsymbol{x}'_{I'_2}| = p$, let us just assume that there exists $\boldsymbol{x}'$ in $\mathscr{C}'_{\rm sec}$ such that $|\boldsymbol{x}'| = 2p$. There is a non negligible chance that we have $|\boldsymbol{x}'_{I'_1}| = |\boldsymbol{x}'_{I'_2}| = p$ and that this codeword will be found by our algorithm. The issue is therefore whether or not there is a codeword of weight $2p$ in a random code of dimension k and length $|I'|$. This holds with a good chance (see [BF02] for instance) as soon as:

$$2p \geqslant d_{\rm GV}(|I'|, k) \quad (7)$$

where $d_{\rm GV}(|I'|, k)$ denotes the Gilbert-Varshamov distance of a code of length $|I'|$ and dimension k. Recall that [MS86]:

$$d_{\rm GV}(|I'|, k) \approx h^{-1}\left(1 - k/|I'|\right)|I'|$$

where $h^{-1}(x)$ is the inverse function defined over $[0, \frac{1}{2}]$ of the binary entropy function $h(x) \stackrel{\text{def}}{=} -x \log_2 x - (1-x)\log_2(1-x)$. Recall that we expect to have:

$$|I'| \approx (R + \alpha\rho + \lambda)n,$$

which implies

$$\frac{k}{|I'|} \approx \frac{\rho}{R + \alpha\rho + \lambda} \approx \frac{\rho}{R}$$

when α and λ are small. Roughly speaking, to avoid such an attack, several conditions have to be met:

1. ρ has to be significantly smaller than R,
2. n has to be large enough.

This phenomenon was clearly not taken into account in the parameters suggested in [KKS97, KKS05, BMJ11] as shown in Table 1. The values of $d_{\text{GV}}(|I'|, k)$ are extremely low (in the range $1-6$). In other words, taking $p = 1$ is already quite threatening for all these schemes. For the first parameter set, namely $(k, n, K, N) = (60, 1023, 192, 3000)$, this suggests to take $p = 3$. Actually taking $p = 1$ is already enough to break the scheme. The problem with these low values of p comes from the dependency of the complexity in p as detailed in the following section. For instance as long as p is smaller than 3 the complexity of one iteration is dominated by the Gaussian elimination Step 2.

Finally, let us observe that when this attack gives a message/signature pair, it can be used as a bootstrap for an attack that recovers the whole private key as will be explained in the following subsection.

Table 1. KKS Parameters with the corresponding value of $d_{\text{GV}}(n', k)$

Article	ρ	n	l	$n' \stackrel{\text{def}}{=} \mathbb{E}\{\lvert I'\rvert\}$	R	N	$d_{\text{GV}}(n', k)$
[KKS97]	$\frac{60}{1023} \approx 0.059$	1,023	8	89	$\frac{192}{3000} \approx 0.064$	3,000	6
[KKS05]	$\frac{48}{255} \approx 0.188$	255	8	65	$\frac{273}{1200} \approx 0.228$	1,200	5
[KKS97]	$\frac{48}{180} \approx 0.267$	180	8	64	$\frac{335}{1100} \approx 0.305$	1,100	4
[BMJ11]	1/2	320	12	165	1/2	11,626	1
[BMJ11]	1/2	448	13	230	1/2	16,294	1
[BMJ11]	1/2	512	13	264	1/2	18,586	1
[BMJ11]	1/2	768	13	395	1/2	27,994	2
[BMJ11]	1/2	1,024	14	527	1/2	37,274	2

4.4 Exploiting a Signature for Extracting the Private Key

If a signature σ of a message $\boldsymbol{x}$ is known, then $\boldsymbol{y} \stackrel{\text{def}}{=} (\sigma, \boldsymbol{x})$ is a codeword of $\mathscr{C}_{\text{sec}}$ which has weight about $n/2$ when restricted to its N first positions. This yields almost half of the positions of J. This can be exploited as follows. We perform the same attack as in the previous subsection, but we avoid choosing positions

i for which $\sigma_i = 1$. More precisely, if we let $J_\sigma \stackrel{\text{def}}{=} \mathsf{supp}(\sigma) = \{i : \sigma_i = 1\}$, then we choose $K + k + l$ positions among $[1 \cdots N] \setminus J_\sigma$ to form I. The point of this choice is that we have more chances to have a smaller size for $I' = I \cap J$. Let $n' \stackrel{\text{def}}{=} |I'|$, we have now:

$$\mathbb{E}\{n' \,|J_\sigma\} = \frac{n - |J_\sigma|}{N - |J_\sigma|}(k + K + l) \tag{8}$$

$$\mathbb{E}\{|I'|\} = \mathbb{E}\{\mathbb{E}\{n' \,|J_\sigma\}\} \approx \frac{n/2}{(N - n/2)}(k + K + l). \tag{9}$$

The last approximation follows from the fact that the weight $|\sigma|$ is quite concentrated around $n/2$. The same reasoning can be made as before, but the odds that the algorithm finds other valid signatures are much higher. This comes from the fact that the expectation $|I'|$ is half the expected size of I' in the previous case as given in Equation (3). Previously we had $\mathbb{E}\left\{\frac{|I'|}{k}\right\} \approx \frac{R}{\rho}$, whereas now we have:

$$\mathbb{E}\left\{\frac{|I'|}{k}\right\} \approx \frac{R}{2\rho}.$$

In other words, in order to avoid the previous attack we had to take ρ significantly smaller than R and now, we have to take ρ significantly smaller than $R/2$. For all the parameters proposed in the past, it turns out that $d_{\mathrm{GV}}(|I'|, k)$ is almost always equal to 1, which makes the attack generally successful in just one iteration by choosing $p = 1$.

Moreover, if another valid signature σ' is obtained and by taking the union $J_\sigma \cup J_{\sigma'}$ of the supports, then about $3/4$ of the positions of J will be revealed. We can start again the process of finding other message/signature pairs by choosing $K + k + l$ positions among $\{1, 2, \ldots, N\} \setminus (J_\sigma \cup J_{\sigma'})$ to form the sets I. This approach can be iterated as explained in Algorithm 2. This process will quickly reveal the whole set J and from this, the private key is easily extracted as detailed in [COV07].

Finally, let us focus on the variant proposed in [BMJ11]. In this case, we have slightly less information than in the original KKS scheme. This can be explained by the following reasoning. In this case too, we choose S again as $[1 \cdots N] \setminus J_\sigma$, where as before J_σ is defined as $J_\sigma \stackrel{\text{def}}{=} \{i : \sigma_i = 1\}$. However this time, by defining n' again as $n' \stackrel{\text{def}}{=} |I'|$, we have

$$\mathbb{E}\{n' \,|J_\sigma\} = \frac{|J'_\sigma|}{N - |J_\sigma|}(k + K + l)$$

where

$$J'_\sigma = J \setminus J_\sigma.$$

However, this time due to the noise which is added, $|J_\sigma|$ is expected to be larger than before (namely of order $\frac{n}{2} + \frac{(N-n)n}{N}$).

Algorithm 2. Recovering the private key from $t \geqslant 1$ signatures.

PARAMETERS:

r : number of iterations
l : small integer ($l \leqslant 40$)
p : very small integer ($1 \leqslant p \leqslant 4$).

INPUT:

$\hat{\boldsymbol{H}}$: public matrix as defined in Figure 3
$\{\sigma_1, \ldots, \sigma_t\}$: list of $t \geqslant 1$ valid signatures

OUTPUT: $J \subset [1 \cdots N]$ of cardinality n

$J \leftarrow \cup_{i=1}^{t} \mathsf{supp}(\sigma_i)$
repeat
  $S \leftarrow [1 \cdots N] \setminus J$
  $\mathcal{L} \leftarrow \mathsf{KKSforge}(r,l,p,S,\hat{\boldsymbol{H}})$
  for all $\sigma \in \mathcal{L}$ **do**
    $J \leftarrow J \cup \mathsf{supp}(\sigma)$
  end for
until $|J| = n$
return J

5 Analysis of the Attack

The purpose of this section is to provide a very crude upper-bound on the complexity of the attack. We assume here that the code $\mathscr{C}_{\text{rand}}$ of length n which is equal to the restriction on J of $\mathscr{C}_{\text{sec}}$:

$$\mathscr{C}_{\text{rand}} \stackrel{\text{def}}{=} \left\{ (x_j)_{j \in J} \ : \ \boldsymbol{x} = (x_1, \ldots, x_{N+k}) \in \mathscr{C}_{\text{sec}} \right\}$$

behaves like a random code. More precisely we assume that it has been chosen by picking a random parity-check matrix $\boldsymbol{H}_{\text{rand}}$ of size $(n-k) \times n$ (by choosing its entries uniformly at random among $\mathbb{F}_2$). This specifies a code $\mathscr{C}_{\text{rand}}$ of length n as $\mathscr{C}_{\text{rand}} = \{\boldsymbol{x} \in \mathbb{F}_2^n : \boldsymbol{H}_{\text{rand}} \boldsymbol{x}^T = 0\}$. We first give in the following section some quite helpful lemmas about codes of this kind.

5.1 Preliminaries about Random Codes

We are interested in this section in obtaining a lower bound on the probability that a certain subset X of $\mathbb{F}_2^n$ has a non empty intersection with $\mathscr{C}_{\text{rand}}$. For this purpose, we first calculate the two following probabilities. The probabilities are taken here over the random choices of $\boldsymbol{H}_{\text{rand}}$.

Lemma 1. *Let $\boldsymbol{x}$ and $\boldsymbol{y}$ be two different and nonzero elements of $\mathbb{F}_2^n$. Then*

$$\mathbf{prob}(\boldsymbol{x} \in \mathscr{C}_{rand}) = 2^{k-n} \tag{10}$$

$$\mathbf{prob}(\boldsymbol{x} \in \mathscr{C}_{rand}, \boldsymbol{y} \in \mathscr{C}_{rand}) = 2^{2(k-n)} \tag{11}$$

To prove this lemma, we will introduce the following notation and lemma. For $\boldsymbol{x} = (x_i)_{1 \leqslant i \leqslant s}$ and $\boldsymbol{y} = (y_i)_{1 \leqslant i \leqslant s}$ being two elements of $\mathbb{F}_2^s$ for some arbitrary s, we define $\boldsymbol{x} \cdot \boldsymbol{y}$ as

$$\boldsymbol{x} \cdot \boldsymbol{y} = \sum_{1 \leqslant i \leqslant s} x_i y_i,$$

the addition being performed over $\mathbb{F}_2$.

Lemma 2. *Let $\boldsymbol{x}$ and $\boldsymbol{y}$ be two different and nonzero elements of $\mathbb{F}_2^n$ and choose $\boldsymbol{h}$ uniformly at random in $\mathbb{F}_2^n$, then*

$$\mathbf{prob}(\boldsymbol{x} \cdot \boldsymbol{h} = 0) = \frac{1}{2} \tag{12}$$

$$\mathbf{prob}(\boldsymbol{x} \cdot \boldsymbol{h} = 0, \boldsymbol{y} \cdot \boldsymbol{h} = 0) = \frac{1}{4} \tag{13}$$

Proof. To prove Equation (12) we just notice that the subspace $\{\boldsymbol{h} \in \mathbb{F}_2^n : \boldsymbol{x} \cdot \boldsymbol{h} = 0\}$ is of dimension $n - 1$. There are therefore 2^{n-1} solutions to this equation and

$$\mathbf{prob}(\boldsymbol{x} \cdot \boldsymbol{h} = 0) = \frac{2^{n-1}}{2^n} = \frac{1}{2}.$$

On the other hand, the hypothesis made on $\boldsymbol{x}$ and $\boldsymbol{y}$ implies that $\boldsymbol{x}$ and $\boldsymbol{y}$ generate a subspace of dimension 2 in $\mathbb{F}_2^n$ and that the dual space, that is $\{\boldsymbol{h} \in \mathbb{F}_2^n : \boldsymbol{x} \cdot \boldsymbol{h} = 0, \boldsymbol{y} \cdot \boldsymbol{h} = 0\}$ is of dimension $n - 2$. Therefore

$$\mathbf{prob}(\boldsymbol{x} \cdot \boldsymbol{h} = 0, \boldsymbol{y} \cdot \boldsymbol{h} = 0) = \frac{2^{n-2}}{2^n} = \frac{1}{4}$$

□

Proof (of Lemma 1). Let $\boldsymbol{h}_1, \ldots, \boldsymbol{h}_{n-k}$ be the $n - k$ rows of $\boldsymbol{H}_{\text{rand}}$. Then

$$\begin{aligned}
\mathbf{prob}(\boldsymbol{x} \in \mathscr{C}_{\text{rand}}) &= \mathbf{prob}(\boldsymbol{H}_{\text{rand}} \boldsymbol{x}^T = 0) \\
&= \mathbf{prob}(\boldsymbol{h}_1 \cdot \boldsymbol{x} = 0, \ldots, \boldsymbol{h}_{n-k} \cdot \boldsymbol{x} = 0) \\
&= \mathbf{prob}(\boldsymbol{h}_1 \cdot \boldsymbol{x} = 0) \ldots \mathbf{prob}(\boldsymbol{h}_{n-k} \cdot \boldsymbol{x} = 0) &\quad (14) \\
&= 2^{k-n} &\quad (15)
\end{aligned}$$

where Equation (14) follows by the independence of the events and Equation (15) uses Lemma 2. Equation (11) is obtained in a similar fashion. □

Lemma 3. *Let X be some subset of $\mathbb{F}_2^n$ of size m and let f be the function defined by $f(x) \stackrel{\text{def}}{=} \max\left(x(1 - x/2), 1 - \frac{1}{x}\right)$. We denote by x the quantity $\frac{m}{2^{n-k}}$, then*

$$\mathbf{prob}(X \cap \mathscr{C}_{rand} \neq \emptyset) \geq f(x).$$

Proof. For $\boldsymbol{x}$ in X we define $E_{\boldsymbol{x}}$ as the event "$\boldsymbol{x}$ belongs to $\mathscr{C}_{\text{rand}}$" and we let $q \stackrel{\text{def}}{=} 2^{k-n}$. We first notice that

$$\mathbf{prob}(X \cap \mathscr{C}_{\text{rand}} \neq \emptyset) = \mathbf{prob}\left(\bigcup_{\boldsymbol{x} \in X} E_{\boldsymbol{x}}\right).$$

By using the Bonferroni inequality [Com74, p. 193] on the probability of the union of events we obtain

$$\begin{aligned}
\mathbf{prob}\left(\bigcup_{\boldsymbol{x}\in X} E_{\boldsymbol{x}}\right) &\geq \sum_{\boldsymbol{x}\in X} \mathbf{prob}(E_{\boldsymbol{x}}) - \sum_{\{\boldsymbol{x},\boldsymbol{y}\}\subset X} \mathbf{prob}(E_{\boldsymbol{x}} \cap E_{\boldsymbol{y}}) && (16)\\
&\geq mq - \frac{m(m-1)}{2}q^2 && (17)\\
&\geq mq - \frac{m^2q^2}{2}\\
&\geq mq(1 - mq/2),
\end{aligned}$$

where (17) follows from Lemma 1. This bound is rather sharp for small values of mq. On the other hand for larger values of mq, another lower bound on $\mathbf{prob}(X \cap \mathscr{C}_{\text{rand}} \neq \emptyset)$ is more suitable [dC97]. It gives

$$\begin{aligned}
\mathbf{prob}\left(\bigcup_{\boldsymbol{x}\in X} E_{\boldsymbol{x}}\right) &\geq \sum_{\boldsymbol{x}\in X} \frac{\mathbf{prob}(E_{\boldsymbol{x}})^2}{\sum_{\boldsymbol{y}\in X} \mathbf{prob}(E_{\boldsymbol{x}} \cap E_{\boldsymbol{y}})} && (18)\\
&\geq \frac{mq^2}{q + (m-1)q^2} && (19)\\
&\geq \frac{mq^2}{q + mq^2} && (20)\\
&\geq \frac{1}{1 + \frac{1}{mq}}\\
&\geq 1 - \frac{1}{mq},
\end{aligned}$$

By taking the maximum of both lower bounds, we obtain our lemma. □

5.2 Estimating the Complexity of Algorithm 1

Here we estimate how many iterations have to be performed in order to break the scheme when no signature is known and when $S = [1 \cdots N]$. For this purpose, we start by lower-bounding the probability that an iteration is successful. Let us bring the following random variables for $i \in \{1, 2\}$:

$$I'_i \stackrel{\text{def}}{=} I_i \cap J \quad \text{and} \quad W_i \stackrel{\text{def}}{=} |I'_i|\,.$$

By using Lemma 1, we know that an iteration finds a valid signature when there is an $\boldsymbol{x}$ in $\mathscr{C}_{\text{sec}}$ such that

$$|\boldsymbol{x}_{I'_1}| = |\boldsymbol{x}_{I'_2}| = p.$$

Therefore the probability of success P_{succ} is lower bounded by

$$P_{\text{succ}} \geq \sum_{w_1,w_2:w_1+w_2\leqslant n} \mathbf{prob}(W_1 = w_1, W_2 = w_2) \quad (21)$$
$$\mathbf{prob}\left\{\exists \boldsymbol{x} \in \mathscr{C}_{\text{sec}} : |\boldsymbol{x}_{I'_1}| = |\boldsymbol{x}_{I'_2}| = p | W_1 = w_1, W_2 = w_2\right\}$$

On the other hand, by using Lemma 3 with the set

$$X \stackrel{\text{def}}{=} \left\{\boldsymbol{x} = (x_j)_{j\in J} : |\boldsymbol{x}_{I'_1}| = |\boldsymbol{x}_{I'_2}| = p\right\}$$

which is of size $\binom{w_1}{p}\binom{w_2}{p}2^{n-w_1-w_2}$, we obtain

$$\mathbf{prob}\left\{\exists \boldsymbol{x} \in \mathscr{C}_{\text{sec}} : |\boldsymbol{x}_{I'_1}| = |\boldsymbol{x}_{I'_2}| = p | W_1 = w_1, W_2 = w_2\right\} \geq f(x). \quad (22)$$

with

$$x \stackrel{\text{def}}{=} \frac{\binom{w_1}{p}\binom{w_2}{p}2^{n-w_1-w_2}}{2^{n-k}} = \binom{w_1}{p}\binom{w_2}{p}2^{k-w_1-w_2}$$

The first quantity is clearly equal to

$$\mathbf{prob}(W_1 = w_1, W_2 = w_2) = \frac{\binom{n}{w_1}\binom{n-w_1}{w_2}\binom{N-n}{(K+k+l)/2-w_1}\binom{N-n-(K+k+l)/2+w_1}{(K+k+l)/2-w_2}}{\binom{N}{(K+k+l)/2}\binom{N-(K+k+l)/2}{(K+k+l)/2}}. \quad (23)$$

Plugging in the expressions obtained in (22) and (23) in (21) we have an explicit expression of a lower bound on P_{succ}. The number of iterations for our attack to be successful is estimated to be of order $\frac{1}{P_{\text{succ}}}$. We obtain therefore an upper-bound on the expected number of iterations, what we denote by `UpperBound`. Table 2 shows for various KKS parameters, p and l the expected number of iterations.

Table 2. KKS Parameters with the corresponding value of $\frac{1}{P_{\text{succ}}}$

Article	ρ	n	l	p	$n' \stackrel{\text{def}}{=} \mathbb{E}\{\lvert I'\rvert\}$	R	N	UpperBound
[KKS97]	$\frac{60}{1023} \approx 0.059$	1,023	8	1	91	$\frac{192}{3000} \approx 0.064$	3,000	111.26
	$\frac{60}{1023} \approx 0.059$	1,023	14	2	91	$\frac{192}{3000} \approx 0.064$	3,000	14.17
[KKS05]	$\frac{48}{255} \approx 0.188$	255	8	1	66	$\frac{273}{1200} \approx 0.228$	1,200	26.41
	$\frac{48}{255} \approx 0.188$	255	14	2	66	$\frac{273}{1200} \approx 0.228$	1,200	4.37
[KKS97]	$\frac{48}{180} \approx 0.267$	180	8	1	65	$\frac{335}{1100} \approx 0.305$	1,100	6.07
	$\frac{48}{180} \approx 0.267$	180	15	2	65	$\frac{335}{1100} \approx 0.305$	1,100	1.82
[BMJ11]	1/2	320	12	1	165	1/2	11,626	1.24
[BMJ11]	1/2	448	13	1	230	1/2	16,294	1.34
[BMJ11]	1/2	512	13	1	264	1/2	18,586	1.39
[BMJ11]	1/2	768	13	1	395	1/2	27,994	1.61
[BMJ11]	1/2	1,024	14	1	527	1/2	37,274	1.85

5.3 Number of Operations of One Iteration

The complexity of one iteration of Algorithm 1 is $C(p,l) = C_{\text{Gauss}} + C_{\text{hash}} + C_{\text{check}}$ where C_{Gauss} is the complexity of a Gaussian elimination, C_{hash} is the complexity of hashing all the $\boldsymbol{x}_1$'s and C_{check} is the complexity of checking all the $\boldsymbol{x}_2$'s with the following expressions:

$$C_{\text{Gauss}} = O\Big((N+k)(N-k)(N-k-l)\Big) = O(N^3) \tag{24}$$

$$C_{\text{hash}} = O\left(\binom{(K+k+l)/2}{p}\right) \tag{25}$$

$$C_{\text{check}} = O\left(\frac{1}{2^l}(N-K-l)^2 \binom{(K+k+l)/2}{p}^2\right) \tag{26}$$

The last expression giving C_{check} comes from the fact that the algorithm considers $\binom{(K+k+l)/2}{p}$ elements $\boldsymbol{x}_2$, and for each of these candidates, we check about $O\left(\frac{1}{2^l}\binom{(K+k+l)/2}{p}\right)$ elements $\boldsymbol{x}_1$'s, which involves a matrix multiplication in Step 9. Let us note that l will be chosen such that C_{hash} and C_{check} are roughly of the same order, say $2^l \approx \binom{(K+k+l)/2}{p}$.

6 Experimental Results

The attack described in Section 4 was implemented in C and was run on a laptop MacBook Pro with an Intel Core i7 of 2.66 GHz to validate the analysis developed in Section 5. Table 3 presents the average number of iterations that were necessary to obtain a codeword of weight in the range $[t_1 \cdots t_2]$. The average is computed with 4000 tests most of the time, with the exceptions of the penultimate entry (only 1000 tests) and the last entry (only 500 tests). The values of t_1 and t_2 are taken from [KKS97] and [BMJ11]. The algorithm halts whenever it finds a word in the prescribed set. Note that for [BMJ11], we have taken $t_1 = n/2 - \frac{3}{2}\sqrt{n}$ and $t_2 = n/2 + \frac{3}{2}\sqrt{n}$ as advocated by the authors. All the codes that we considered during our simulations were randomly chosen. This setting does not completely comply with the recommendations made by the authors for the schemes given in [KKS97]. In one case, it is suggested to use binary BCH codes of length $n = 255$ and dimension $k = 48$, and in another case a binary code of length $n = 180$ and dimension $k = 48$ that was constructed by means of 12 random binary equidistant codes of length 15, dimension 4 and minimum distance 8. However, we emphasize that these specific constraints are irrelevant because the attack is generic and only requires public data ($\boldsymbol{F}$ and $\boldsymbol{H}$) and aims at forging a valid signature. We can see in Table 3 that the number of iterations are in accordance with the theoretical upper-bound `UpperBound` on the value of $\frac{1}{P_{\text{succ}}}$ obtained in the previous section, which is an upper bound on the average number of iterations.

Table 3. Average number of iterations of Algorithm 1

Article	ρ	n	l	p	R	N	UpperBound	t_1	t_2	Iterations	Average time of an attack (in s)
[KKS97]	$\frac{60}{1023} \approx 0.059$	1,023	8	1	$\frac{192}{3000} \approx 0.064$	3,000	111.26	352	672	102.34	19.50
	$\frac{60}{1023} \approx 0.059$	1,023	14	2	$\frac{192}{3000} \approx 0.064$	3,000	14.17	352	672	9.22	1.754
[KKS05]	$\frac{48}{255} \approx 0.188$	255	8	1	$\frac{273}{1200} \approx 0.228$	1,200	26.41	48	208	24.32	0.384
	$\frac{48}{255} \approx 0.188$	255	14	2	$\frac{273}{1200} \approx 0.228$	1,200	4.37	48	208	3.13	0.051
[KKS97]	$\frac{48}{180} \approx 0.267$	180	8	1	$\frac{335}{1100} \approx 0.305$	1,100	6.07	96	96	5.58	0.061
	$\frac{48}{180} \approx 0.267$	180	15	2	$\frac{335}{1100} \approx 0.305$	1,100	1.82	96	96	1.23	0.017
[BMJ11]	1/2	320	12	1	1/2	11,626	1.24	133	187	1.13	6.425
[BMJ11]	1/2	448	13	1	1/2	16,294	1.34	192	256	1.18	18.90
[BMJ11]	1/2	512	13	1	1/2	18,586	1.39	222	290	1.26	32.23
[BMJ11]	1/2	768	13	1	1/2	27,994	1.61	342	426	1.51	119.5
[BMJ11]	1/2	1,024	14	1	1/2	37,274	1.85	464	560	1.73	350.8

7 Concluding Remarks

Design principles. As explained in Section 3, the parameters of the KKS scheme were chosen in order to make decoding of $\mathscr{C}_{\text{known}}$ intractable when the weight of errors is in the range $[t_1 \cdots t_2]$, where $\mathscr{C}_{\text{known}}$ denotes the code defined by the parity-check matrix $\boldsymbol{H}$. In [BMJ11], it is further required that $\mathscr{C}_{\text{known}}$ is of minimum distance greater than $4n$. Both requirements are clearly insufficient to ensure that the scheme is secure as demonstrated by this paper. We suggest here to replace all these requirements by choosing the parameters such as to make our attack impracticable. This algorithm is exponential in nature when the parameters are well chosen. If we want to avoid that the knowledge of a message-signature pair allows to recover the secret key, this implies for instance that the rate R of $\mathscr{C}_{\text{known}}$ should be significantly larger than 2ρ, that is twice the rate of the secret code $\mathscr{C}_{\text{hidden}}$. This would change the parameters of the scheme significantly and give much larger key sizes than has been proposed in [KKS97, KKS05, BMJ11]. Choosing these parameters requires however to analyze properly the complexity of the attack when one message-signature is known (here we just analyzed the complexity of the attack which does not make use of any message-signature pair). The analysis we performed in our case can be carried over to the case when a message-signature pair is known but this is beyond the scope of this paper and will appear in a full version of this paper.

Relating the security to the problem of decoding a linear code. The attack which has been suggested here is nothing but a well known algorithm for finding low weight codewords or for decoding a generic linear code. It just happens that this algorithm is much more powerful here than for a random linear code due to the peculiar nature of the code it is applied to. However as mentioned above, this attack is exponential in nature and can easily be defeated by choosing the parameters appropriately. It would be interesting to analyze the relationship of the problem of breaking the KKS scheme with decoding problems in more depth, or to prove that the problem which has to be solved is indeed NP hard.

Non-binary codes. Obviously there is a non binary version of the KKS scheme which would deal with codes defined over larger alphabets. The benefits of the generalized scheme are questionable. The attack presented here generalizes easily to higher order fields. What is more, moving to non-binary fields seems to be a poor idea in terms of security. For instance, whereas a message/signature pair reveals only half the positions of J in the binary case, in the q-ary case we expect to obtain roughly a fraction $\frac{q-1}{q}$ of positions of J, which is significantly larger.

Decoding one out of many. Another approach could have been used for attacking the scheme. Let us denote by $\boldsymbol{s}_1, \cdots, \boldsymbol{s}_k$ the columns of $\boldsymbol{F}$. These vectors can be considered as k syndromes of codewords of $\mathscr{C}_{\text{hidden}}$ with respect to the parity-check matrix $\boldsymbol{H}$. If we want to obtain one message/pair we can try to find an error $\boldsymbol{e}_i$ of weight in the range $[t_1 \cdots t_2]$ such that $\boldsymbol{H}\boldsymbol{e}_i^T = \boldsymbol{s}_i$. This suggests to use "the decoding one out of many" approach [Sen11], that is we have k words to decode and we want to decode at least one of them. This problem can be solved more efficiently than just decoding one instance. We can even refine this approach by considering all possible syndromes obtained by all possible (non-zero) combinations $\sum_i \alpha_i \boldsymbol{s}_i$. In this case, we would have to solve "a decoding one out of many" problem with $2^k - 1$ instances. However a naive use of the results of [Sen11] would be far from indicating that all the parameters of [KKS97, KKS05, BMJ11] are easily broken by this approach.

References

[BF02] Barg, A., Forney, G.D.: Random codes: Minimum distances and error exponents. IEEE Transactions on Information Theory 48(9), 2568–2573 (2002)

[BLP11] Bernstein, D.J., Lange, T., Peters, C.: Smaller Decoding Exponents: Ball-Collision Decoding. In: Rogaway, P. (ed.) CRYPTO 2011. LNCS, vol. 6841, pp. 743–760. Springer, Heidelberg (2011)

[BMJ11] Barreto, P.S.L.M., Misoczki, R., Simplicio Jr., M.A.: One-time signature scheme from syndrome decoding over generic error-correcting codes. Journal of Systems and Software 84(2), 198 (2011)

[CFS01] Courtois, N.T., Finiasz, M., Sendrier, N.: How to Achieve a McEliece-Based Digital Signature Scheme. In: Boyd, C. (ed.) ASIACRYPT 2001. LNCS, vol. 2248, pp. 157–174. Springer, Heidelberg (2001)

[Com74] Comtet, L.: Advanced Combinatorics. Reidel, Dordrecht (1974)

[COV07] Cayrel, P.-L., Otmani, A., Vergnaud, D.: On Kabatianskii-Krouk-Smeets Signatures. In: Carlet, C., Sunar, B. (eds.) WAIFI 2007. LNCS, vol. 4547, pp. 237–251. Springer, Heidelberg (2007)

[dC97] de Caen, D.: A lower bound on the probability of a union. Discrete Mathematics 169, 217–220 (1997)

[Dum91] Dumer, I.: On minimum distance decoding of linear codes. In: Proc. 5th Joint Soviet-Swedish Int. Workshop Inform. Theory, Moscow, pp. 50–52 (1991)

[Dum96] Dumer, I.: Suboptimal decoding of linear codes: partition techniques. IEEE Transactions on Information Theory 42(6), 1971–1986 (1996)

[FGO+10] Faugère, J.-C., Gauthier, V., Otmani, A., Perret, L., Tillich, J.-P.: A distinguisher for high rate McEliece cryptosystems. Cryptology ePrint Archive, Report 2010/331 (2010), `http://eprint.iacr.org/`

[FS09] Finiasz, M., Sendrier, N.: Security Bounds for the Design of Code-Based Cryptosystems. In: Matsui, M. (ed.) ASIACRYPT 2009. LNCS, vol. 5912, pp. 88–105. Springer, Heidelberg (2009)

[KKS97] Kabatianskii, G., Krouk, E., Smeets, B.: A Digital Signature Scheme Based on Random Error-Correcting Codes. In: Darnell, M.J. (ed.) Cryptography and Coding 1997. LNCS, vol. 1355, pp. 161–167. Springer, Heidelberg (1997)

[KKS05] Kabatiansky, G., Krouk, E., Semenov, S.: Error Correcting Coding and Security for Data Networks: Analysis of the Superchannel Concept. John Wiley & Sons (2005)

[MS86] MacWilliams, F.J., Sloane, N.J.A.: The Theory of Error-Correcting Codes, 5th edn. North–Holland, Amsterdam (1986)

[Sen11] Sendrier, N.: Decoding one out of many (preprint, 2011)

[Ste88] Stern, J.: A Method for Finding Codewords of Small Weight. In: Wolfmann, J., Cohen, G. (eds.) Coding Theory 1988. LNCS, vol. 388, pp. 106–113. Springer, Heidelberg (1989)

XMSS – A Practical Forward Secure Signature Scheme Based on Minimal Security Assumptions

Johannes Buchmann, Erik Dahmen, and Andreas Hülsing*

Cryptography and Computeralgebra
Department of Computer Science
TU Darmstadt
{buchmann,dahmen,huelsing}@cdc.informatik.tu-darmstadt.de

Abstract. We present the hash-based signature scheme XMSS. It is the first provably (forward) secure and practical signature scheme with minimal security requirements: a pseudorandom and a second preimage resistant (hash) function family. Its signature size is reduced to less than 25% compared to the best provably secure hash based signature scheme.

Keywords: digital signature, practical, minimal security assumptions, hash-based signatures, forward security, provable security.

1 Introduction

Digital signatures are a very important cryptographic tool. The signature schemes currently used in practice are RSA, DSA, and ECDSA. Their security depends on the security of certain trapdoor one-way functions which, in turn, relies on the hardness of factoring integers and computing discrete logarithms, respectively. However, it is unclear whether those computational problems remain hard in the future. In fact, it has been shown by Shor [28] that quantum computers can solve them in polynomial time. Other algorithmic breakthroughs are always possible in the future. In view of the importance of digital signatures it is necessary to come up with alternative practical signature schemes that deliver maximum security. In particular, quantum computers must not be able to break them. They are called post-quantum signature schemes.

In this paper we propose the hash-based signature scheme XMSS (eXtended Merkle Signature Scheme). It is based on the Merkle Signature Scheme [24] and the Generalized Merkle Signature Scheme (GMSS) [10]. We show that XMSS is an efficient post-quantum signature scheme with minimal security assumptions.

This is done as follows. XMSS requires a hash function family $\mathcal{H}$ and another function family F. We prove:

> **(Security)** XMSS is existentially unforgeable under adaptively chosen message attacks in the standard model, provided $\mathcal{H}$ is second preimage resistant and F is pseudorandom.

* Supported by grant no. BU 630/19-1 of the German Research Foundation (www.dfg.de).

B.-Y. Yang (Ed.): PQCrypto 2011, LNCS 7071, pp. 117–129, 2011.

(Efficiency) XMSS is efficient, provided that $\mathcal{H}$ and F are efficient. This claim is supported by experimental results.

The first assertion shows that the security requirements for XMSS are minimal. This follows from [27], [26], [19] and [17] where the existence of a secure signature scheme is proved to imply the existence of a second preimage resistant hash function family and a pseudorandom function family (see Section 3).

The second assertion shows that XMSS is practical as there are many ways to construct very efficient (hash) function families that are believed to be second preimage resistant or pseudorandom, respectively, even in the presence of quantum computers. For example, cryptographic hash functions and block ciphers can be used to construct such families. In particular, there are such constructions based on hard problems in algebra or coding theory. The huge number of instantiations of XMSS guarantees the long-term availability of secure and efficient signature schemes.

The idea of hash-based signatures was introduced by Merkle [24]. The results in [8,9,10,11,12,13,14,15,16,20,21,29] improve the Merkle idea in many respects by providing new algorithmic ideas and security proofs. XMSS incorporates many of those ideas and goes one step further. It is the first practical (forward) secure signature scheme with minimal security requirements in the above sense. All other constructions, but MSS-SPR [14], require a collision resistant hash function family whose existence is not known to follow from the existence of a secure digital signature scheme. Compared to MSS-SPR, which is the currently best hash-based signature scheme that carries a proof of security, we can reduce the signature size by more than 25 % for the same level of security. Furthermore MSS-SPR is not forward-secure. The improved signatur size is very important as the signature size is considered the main drawback of hash-based signatures.

In this paper we show only how to sign fixed length messages. The scheme can easily be extended to sign messages of arbitrary length using TCR hash and sign as proposed in [14]. This requires a target collision resistant hash function family. Target collision resistant hash function families are known to exist if any one-way function exists [27]. Therefore this preserves the minimal security assumptions property.

The paper is organized as follows. In Section 2 we describe the construction of XMSS. Its security and forward security is discussed in Sections 3 and 4. The XMSS-efficiency is shown in Section 5. Section 6 contains a description of our implementation and a presentation of the performance results.

2 The eXtended Merkle Signature Scheme XMSS

In this section we describe XMSS. Like the Merkle signature scheme [24] it uses a one-time signature scheme (OTS) that can only sign one message with one key. To overcome the limitation to one message per key, a hash tree is used to reduce the authenticity of many OTS verification keys to one public XMSS key. To minimize storage requirements, pseudorandom generators (PRG) are used. They generate the OTS signature keys as needed.

The parameters of XMSS are the following:

- $n \in \mathbb{N}$, the security parameter,
- $w \in \mathbb{N}, w > 1$, the Winternitz parameter,
- $m \in \mathbb{N}$, the message length in bits,
- $F(n) = \{f_K : \{0,1\}^n \rightarrow \{0,1\}^n | K \in \{0,1\}^n\}$ a function family,
- $H \in \mathbb{N}$, the tree height, XMSS allows to make 2^H signatures using one keypair,
- h_K, a hash function, chosen randomly with the uniform distribution from the family $\mathcal{H}(n) = \{h_K : \{0,1\}^{2n} \rightarrow \{0,1\}^n | K \in \{0,1\}^n\}$,
- $x \in \{0,1\}^n$, chosen randomly with the uniform distribution. The string x is used to construct the one-time verification keys.

Those parameters are publicly known.

We keep the following description of XMSS and its components short by including references to more detailed descriptions. We write log for $\log_2$.

Winternitz OTS As OTS we use the Winternitz OTS (W-OTS) first mentioned in [24]. We use a slightly modified version proposed in [9]. For $K, x \in \{0,1\}^n$ and $e \in \mathbb{N}$ we define $f_K^e(x)$ as follows. We set $f_K^0(x) = K$ and for $e > 0$ we define $K' = f_K^{e-1}(x)$ and $f_K^e(x) = f_{K'}(x)$. In contrast to previous versions of W-OTS this is a (random) walk through the function family instead of an iterated evaluation of a hash function. This modification allows to eliminate the need for a collision resistant hash function family.

Also, define

$$\ell_1 = \left\lceil \frac{m}{\log(w)} \right\rceil, \quad \ell_2 = \left\lfloor \frac{\log(\ell_1(w-1))}{\log(w)} \right\rfloor + 1, \quad \ell = \ell_1 + \ell_2.$$

The secret signature key of W-OTS consists of ℓ n-bit strings sk_i, $1 \leq i \leq \ell$ chosen uniformly at random. The public verification key is computed as

$$\mathsf{pk} = (\mathsf{pk}_1, \ldots, \mathsf{pk}_\ell) = (f_{\mathsf{sk}_1}^{w-1}(x), \ldots, f_{\mathsf{sk}_\ell}^{w-1}(x)),$$

with f^{w-1} as defined above.

W-OTS signs messages of binary length m. They are processed in base w representation. They are of the form $M = (M_1 \ldots M_{\ell_1})$, $M_i \in \{0, \ldots, w-1\}$. The checksum $C = \sum_{i=1}^{\ell_1}(w - 1 - M_i)$ in base w representation is appended to M. It is of length ℓ_2. The result is $(b_1, \ldots, b_\ell)$. The signature of M is

$$\sigma = (\sigma_1, \ldots, \sigma_\ell) = (f_{\mathsf{sk}_1}^{b_1}(x), \ldots, f_{\mathsf{sk}_\ell}^{b_\ell}(x)).$$

It is verified by constructing $(b_1 \ldots, b_\ell)$ and checking

$$(f_{\sigma_1}^{w-1-b_1}(\mathsf{pk}_0), \ldots, f_{\sigma_\ell}^{w-1-b_\ell}(\mathsf{pk}_0)) \stackrel{?}{=} (\mathsf{pk}_1, \ldots, \mathsf{pk}_\ell).$$

The sizes of signature, public, and secret key are ℓn. For more detailed information see [9].

XMSS Tree The XMSS tree is a modification of the Merkle Hash Tree proposed in [14]. It utilizes the hash function h_K. The XMSS tree is a binary tree. Denote its height by H. It has $H+1$ levels. The leaves are on level 0. The root is on level H. The nodes on level j, $0 \leq j \leq H$, are denoted by $\text{NODE}_{i,j}$, $0 \leq i < 2^{H-j}$. The construction of the leaves is explained below. Level j, $0 < j \leq H$, is constructed using a bitmask $(b_{l,j}||b_{r,j}) \in \{0,1\}^{2n}$ chosen uniformly at random. The nodes are computed as

$$\text{NODE}_{i,j} = h_K((\text{NODE}_{2i,j-1} \oplus b_{l,j})||(\text{NODE}_{2i+1,j-1} \oplus b_{r,j}))$$

for $0 < j \leq H$. The usage of the bitmasks is the main difference to the other Merkle tree constructions. It is borrowed from [5] and allows to replace the collision resistant hash function family. Figure 1 shows the construction of the XMSS tree.

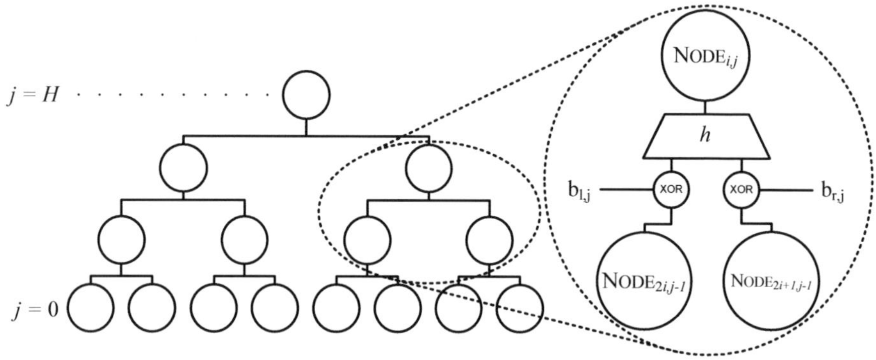

Fig. 1. The XMSS tree construction

We explain the computation of the leaves of the XMSS tree. The XMSS tree is used to authenticate 2^H W-OTS verification keys, each of which is used to construct one leaf of the XMSS tree. The construction of the keys is explained at the end of this section. In the construction of a leaf another XMSS tree is used. It is called L-tree. The first ℓ leaves of an L-tree are the ℓ bit strings $(\mathsf{pk}_0, \ldots, \mathsf{pk}_\ell)$ from the corresponding verification key. As ℓ might not be a power of 2 there are not sufficiently many leaves. Therefore the construction is modified. A node that has no right sibling is lifted to a higher level of the L-tree until it becomes the right sibling of another node. In this construction, the same hash function as above but new bitmasks are used. The bitmasks are the same for each of those trees. As L-trees have height $\lceil \log \ell \rceil$, additional $\lceil \log \ell \rceil$ bitmasks are required.

The XMSS public key PK contains the bitmasks and the root of the XMSS tree.

To sign the ith message, the ith W-OTS key pair is used. The signature $\mathsf{SIG} = (i, \sigma, \text{AUTH})$ contains the index i, the W-OTS signature σ, and the authentication path for the leaf $\text{NODE}_{0,i}$. It is the sequence $\text{AUTH} = (\text{AUTH}_0, \ldots, \text{AUTH}_{H-1})$ of the siblings of all nodes on the path from $\text{NODE}_{0,i}$ to the root. Figure 2 shows

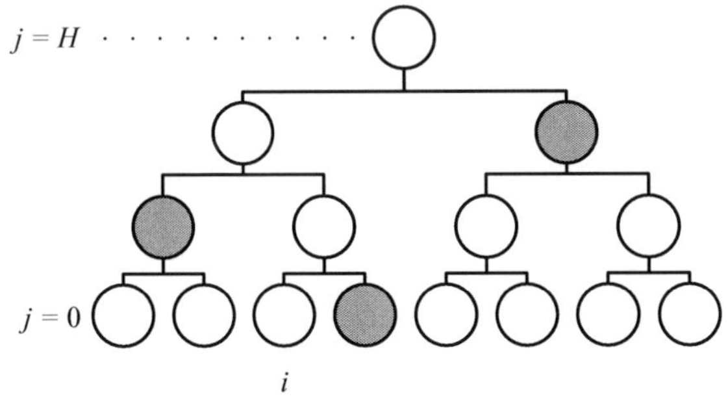

Fig. 2. The authentication path for leaf i

the authentication path for leaf i. To compute the authentication path we use the tree traversal algorithm from [11] as it allows for optimal balanced runtimes using very little memory.

To verify the signature $\mathsf{SIG} = (i, \sigma, \text{AUTH})$, the string $(b_0, \ldots, b_\ell)$ is computed as described in the W-OTS signature generation. Then the ith verification key is computed using the formula

$$(\mathsf{pk}_1, \ldots, \mathsf{pk}_\ell) = (f_{\sigma_1}^{w-1-b_1}(x), \ldots, f_{\sigma_\ell}^{w-1-b_\ell}(x)).$$

The corresponding leaf $\text{NODE}_{0,i}$ of the XMSS tree is constructed using the L-tree. This leaf and the authentication path are used to compute the path $(p_0, \ldots, p_H)$ to the root of the XMSS tree, where $p_0 = \text{NODE}_{0,i}$ and

$$p_j = \begin{cases} h_K((p_{j-1} \oplus b_{l,j})||(\text{AUTH}_{j-1} \oplus b_{r,j})), & \text{if } \lfloor i/2^j \rfloor \equiv 0 \mod 2 \\ h_K((\text{AUTH}_{j-1} \oplus b_{l,j})||(p_{j-1} \oplus b_{r,j})), & \text{if } \lfloor i/2^j \rfloor \equiv 1 \mod 2 \end{cases}$$

for $0 \leq j \leq H$. If p_H is equal to the root of the XMSS tree in the public key, the signature is accepted. Otherwise, it is rejected.

Signature key generation The W-OTS secret signature keys are computed using a seed $\text{SEED} \in \{0,1\}^n$, the pseudorandom function family $F(n)$, and the pseudorandom generator GEN which for $\lambda, \mu \in \{0,1\}^n$ yields

$$\mathsf{GEN}_\lambda(\mu) = f_\mu(1)|| \ldots ||f_\mu(\lambda).$$

For $i \in \{1, \ldots, 2^H\}$ the i-th W-OTS signature key is

$$\mathsf{sk}_i \leftarrow \mathsf{GEN}_\ell(f_{\text{SEED}}(i)).$$

The XMSS secret key contains SEED and the index of the last signature i.

The bit length of the XMSS public key is $(2(H + \lceil \log \ell \rceil) + 1)n$, an XMSS signature has length $(\ell + H)n$, and the length of the XMSS secret signature key is $< 2n$.

3 Security

In this section we discuss the security of XMSS. We use the notations of Section 2 and sketch the proof of the following theorem.

Theorem 1. *If $\mathcal{H}(n)$ is a second preimage resistant hash function family and $F(n)$ a pseudorandom function family, then XMSS is existentially unforgeable under chosen message attacks.*

Before sketching the proof of Theorem 1 we show that it implies the minimality of the security requirements of XMSS. For this purpose, we show how to construct second preimage resistant hash function families and pseudorandom function families from secure signature schemes. Those constructions imply that there is a secure instance of XMSS if there is a secure signature scheme. This implies that the security requirements for XMSS are minimal. Now we present the constructions.

In [27] it is shown that a one-way function can be constructed from any secure signature scheme. Also the construction of a target collision resistant hash function family from a one-way function is presented. Since target collision resistant hash function families are second preimage resistant (see [26]), this implies that second preimage resistant hash function families can be constructed from secure digital signature schemes. In [19] the construction of a pseudorandom generator from a one-way function is presented. In [17] pseudorandom function families are obtained from pseudorandom generators. It follows that secure signature schemes yield pseudorandom function families.

Next, we sketch the proof of Theorem 1. First the notion of existentially unforgeability under chosen message attacks (EU-CMA) [18] has to be adapted to the given setting. XMSS is a signature scheme where the private signature key is not constant as in RSA, but changes over time, whereas the definition of EU-CMA assumes a constant signature key. The definition of EU-CMA security can be described by an experiment, where the adversary has access to a signature oracle, initialized with the constant signature key. The adversary is allowed to send messages to the oracle. The oracle returns the corresponding signatures under the secret signature key. At the end of the experiment, the adversary has to come up with a signature for a message she did not send to the oracle before. For one-time signature schemes like W-OTS the number of queries is limited to one. In the XMSS-experiment the adversary is given access to an XMSS oracle which changes the secret signature key after each signature according to the XMSS construction. The adversary may query this oracle 2^H times.

The proof of Theorem 1 proceeds in two steps. First, it is shown that W-OTS, which is known to be EU-CMA-secure from [9] remains EU-CMA secure in the XMSS construction, where the secret key is generated using a pseudorandom function family.

Second, we modify the security proof in [14]. This is necessary since in contrast to the signature scheme in [14] XMSS generates its signature keys pseudorandomly.

In fact, both steps use the following result. Denote by Dss one of the signature schemes XMSS and W-OTS. In both schemes the key generation algorithm uses n bits of random input and expands them to the λn bits using the pseudorandom function family $F(n)$. Denote by $\text{DSS}^\star$ the modification of DSS, that is obtained by choosing the λn bits randomly. Denote by $\text{InSec}^{\text{EU-CMA}}(\text{DSS}; t, q)$ the maximum success probability over all adversaries running in time $\leq t$ and making at most q_{Sign} oracle queries in the EU-CMA experiment. $\text{InSec}^{\text{EU-CMA}}(\text{DSS}^\star; t, q)$ is defined analogously. Also, define by $\text{InSec}^{\text{PRF}}(F(n); t, q)$ the maximum success probability over all adversaries in distinguishing a random element from $F(n)$ from a random function, when the runtime is bounded by t and she can query an oracle for up to q function values. Then the following is true

$$\begin{aligned}\text{InSec}^{\text{EU-CMA}}(\text{DSS}; t, q) = {} & \text{InSec}^{\text{PRF}}(F(n); (t' + \lambda), \lambda) \\ & + \text{InSec}^{\text{EU-CMA}}(\text{DSS}^\star; t, q)\end{aligned}$$

$t' = t + t_{\mathsf{Kg}} + q t_{\mathsf{Sign}} + t_{\mathsf{Vf}}$.

This is shown by contradiction. Assume there exists an adversary A returning a valid forgery for DSS with success probability

$$\text{Succ}^{\text{EU-CMA}}(\text{DSS}; \mathsf{A}) > \text{InSec}^{\text{EU-CMA}}(\text{DSS}; t, q)$$

in time t. We build a distinguisher Dis for $F(n)$ that runs A and outputs 1, if A returns a valid forgery. Then we show that either Dis can distinguish $F(n)$ with success probability

$$\text{Succ}^{\text{PRF}}(F(n); \mathsf{Dis}) > \text{InSec}^{\text{PRF}}(F; t', q)$$

or A is a forger for $\text{DSS}^\star$ with success probability

$$\text{Succ}^{\text{EU-CMA}}(\text{DSS}^\star; \mathsf{A}) > \text{InSec}^{\text{EU-CMA}}(\text{DSS}^\star; t, q).$$

Both lead to a contradiction of the initial assumptions.

Applying this result to W-OTS shows that W-OTS as used in the XMSS construction is EU-CMA-secure if $F(n)$ is pseudorandom. Together with the assumption that $\mathcal{H}(n)$ is second preimage resistant it follows from [14] that XMSS is EU-CMA-secure if the seeds for the W-OTS signature keys are chosen at random. Finally, applying the above result again, we obtain the following formula which is an exact version of Theorem 1 and therefore concludes the proof.

Denote by $\text{InSec}^{\text{SPR}}(\mathcal{H}(n); t)$ the maximum success probability over all adversaries running in time $\leq t$ for finding a second preimage in $\mathcal{H}(n)$. Furthermore, denote the number of key collisions of $F(n)$ by κ according to [9]. Then the maximum success probability over all adversaries running in time $\leq t$, making at most 2^H oracle queries, against the XMSS EU-CMA security is bounded by

$$\begin{aligned}&\text{InSec}^{\text{EU-CMA}}\left(XMSS\ ; t, q = 2^H\right) \\ &\leq \text{InSec}^{\text{PRF}}\left(F(n); (t' + 2^H), q = 2^H\right) \\ &+ 2 \cdot \max \left\{ \begin{array}{l} (2^{H+\log \ell} - 1) \cdot \text{InSec}^{\text{SPR}}(\mathcal{H}(n); t'), \\ 2^H \Big(\text{InSec}^{\text{PRF}}(F(n); (t' + \ell), q = \ell) \\ \quad + (\ell^2 w^2 \kappa^{w-1} \frac{1}{\left(\frac{1}{\kappa} - \frac{1}{2^n}\right)}) \cdot \text{InSec}^{\text{PRF}}(F(n); (t'), q = 2) \Big) \end{array} \right\}\end{aligned}$$

where $t' = t + 2^H \cdot t_{\mathsf{Sign}} + t_{\mathsf{Vf}} + t_{\mathsf{Kg}}$.

Note that, assuming only generic attacks on $\mathcal{H}(n)$ and $F(n)$ the symmetric bit security of XMSS is

$$\begin{aligned} b &= \log\left(\frac{t}{\mathrm{InSec}^{\text{EU-CMA}}\left(XMSS\;; t, q = 2^H\right)}\right) \\ &\leq \min\left\{n-1, n-H-2-w-2\log(\ell w)\right\} - 1 \end{aligned}$$

4 Forward Security

Given the above result we can go even further. In [1] Anderson introduced the idea of forward security for signature schemes (FSSIG) which was later formalized in [4]. It says that even after a key compromise all signatures created before remain valid. Obviously, this notion is only meaningful for signature schemes that change their secret signature key over time. From an attack based point of view this translates to: If an attacker learns the actual secret key sk_i, she is still not able to forge a signature under a secret key sk_j, $j < i$. This is a desirable property, especially in the context of long term secure signatures, as it allows to remove the need for timestamps and an online trusted third party.

In this section we show that XMSS is forward secure if we slightly modify the key generation process based on an idea from [22]. We describe the modifications. To make XMSS forward secure we use a forward secure PRG FsGen when generating the seeds for the W-OTS secret keys. A forward secure PRG is a stateful PRG that starts from a random initial state. Given a state, it outputs a new state and some output bits. Even if an adversary manages to learn the secret state of a forward secure PRG, she is not able to distinguish the former outputs from random bit strings. In the modified XMSS, the W-OTS seeds are generated by FsGen. Starting from a random input $\textsc{Seed} = \textsc{State}_0$ of length n, FsGen uses $F(n)$ and the previous state $\textsc{State}_{i-1}$ to generate n bits of pseudorandom output $\textsc{Out}_i$ and a new state $\textsc{State}_i$ of length n:

$$(\textsc{State}_i || \textsc{Out}_i) = (f_{\textsc{State}_{i-1}}(0) || f_{\textsc{State}_{i-1}}(1))$$

The generation of the W-OTS secret keys from the seeds still utilizes GEN_ℓ. The secret key of the resulting forward secure XMSS contains the actual state $\textsc{State}_i$ instead of $\textsc{Seed}$. In contrast to the construction from Section 2, the seeds for the W-OTS signature keys are not easily accessible from $\textsc{State}_i$ using one evaluation of $F(n)$. To compute the authentication path, the tree traversal algorithm needs to compute several W-OTS keys before they are needed. This is very expensive using FsGen. This problem is already addressed in [11]. We use their solution that requires to store $2H$ states of FsGen. This results in a secret signature key size of $2Hn$.

For this modified XMSS we proof the following security theorem.

Theorem 2. *If $\mathcal{H}(n)$ is a second preimage resistant hash function family and $F(n)$ a pseudorandom function family, then XMSS with a modified key generation described below is a forward secure digital signature scheme.*

Again we will only sketch the proof and refer the reader to the full version for more details and the formal definitions. We keep the notions from the last two Sections.

We sketch the proof. Our proof is a modification of the proof from [22]. In the proof of Theorem 1 it is shown that XMSS is EU-CMA-secure if the seeds for the W-OTS signature keys are chosen at random. In [6] it is shown that if $F(n)$ is a pseudorandom function family, then $\mathrm{InSec}^{\mathrm{FSPRG}}(\mathsf{FsGen}; t)$, the maximum success probability of any adversary running in time $\leq t$, attacking the forward security of FsGen is:

$$\mathrm{InSec}^{\mathrm{FSPRG}}(\mathsf{FsGen}; t) = 2\widetilde{n} \cdot \mathrm{InSec}^{\mathrm{PRF}}(F(n); (t + 2\widetilde{n}), 2)$$

where $\widetilde{n}$ denotes the maximum number of outputs produced by FsGen. We use these results in the proof from [22] to conclude the following formula which is an exact version of Theorem 2.

The maximum success probability over all adversaries running in time $\leq t$, making at most 2^H oracle queries, in attacking the forward security of the modified XMSS, $\mathrm{InSec}^{\mathrm{FSSIG}}(XMSS; t, q = 2^H)$, is bounded by

$$\begin{aligned}
&\mathrm{InSec}^{\mathrm{FSSIG}}\left(XMSS; t, q = 2^H\right) \\
&\leq 2^{2H+1} \cdot \mathrm{InSec}^{\mathrm{PRF}}(F(n); (t' + 2), q = 2) \\
&+ 2 \cdot \max \left\{ \begin{array}{l} (2^{H+\log \ell} - 1) \cdot \mathrm{InSec}^{\mathrm{SPR}}(\mathcal{H}(n); t'), \\ 2^H \Big(\mathrm{InSec}^{\mathrm{PRF}}(F(n); (t' + \ell), q = \ell) \\ \quad + (\ell^2 w^2 \kappa^{w-1} \frac{1}{\left(\frac{1}{\kappa} - \frac{1}{2^n}\right)}) \cdot \mathrm{InSec}^{\mathrm{PRF}}(F(n); (t'), q = 2) \Big) \end{array} \right\}
\end{aligned}$$

$t' = t + 2^H \cdot t_{\mathsf{Sign}} + t_{\mathsf{Vf}} + t_{\mathsf{Kg}}$.

5 Efficiency

In this Section we discuss the efficiency of XMSS. We will show that XMSS and its the forward secure variant are efficient if $\mathcal{H}(n)$ is an efficient second preimage resistant hash function family and $F(n)$ an efficient pseudorandom function family. Efficiency here refers to the runtimes and space requirements for sufficiently secure parameters. It is expressed as a function of the security parameter n. In the Section 6 we will propose parameters that are secure according to [23] and present experimental results that support the efficiency of XMSS.

The runtime of all three algorithms of XMSS is dominated by the number $\#call_F$ of calls to $F(n)$ and the number $\#call_{\mathcal{H}}$ of calls to $\mathcal{H}(n)$. We ignore the negligible computational overhead for adding the bitmasks, control flow and computing the base w representation of the message. Using a simple counting argument we obtain the following result:

For one call to the XMSS signature algorithm, the number of calls to $\mathcal{H}(n)$ and $F(n)$ is bounded by

$$\#call_{\mathcal{H}} \leq \frac{H+2}{2} * (H + \ell), \quad \#call_F \leq \frac{H+2}{2} * (\ell(w+1)) + 4H.$$

For one call to the XMSS signature verification algorithm, the number of calls to $\mathcal{H}(n)$ and $F(n)$ is bounded by

$$\#call_{\mathcal{H}} \leq H + \ell, \quad \#call_F \leq \ell w.$$

For one call to the XMSS key generation algorithm, the number of calls to $\mathcal{H}(n)$ and $F(n)$ is bounded by

$$\#call_{\mathcal{H}} \leq 2^H(\ell + 1), \quad \#call_F \leq 2^H(2 + \ell(w + 1)).$$

The space requirements for the internal state of Sign and Kg (including sk) are at most $6H * n$ bits. Vf needs no internal state. Hence, the space used by XMSS is at most $6H * n$ bits.

6 Implementation

We have implemented XMSS to evaluate its practical performance. The implementation was done in C, using the AES and SHA-2 implementation of OpenSSL[1]. The implementation is straightforward, except for the construction of $\mathcal{H}(n)$ and $F(n)$ for which we implemented constructions based on hash functions and block ciphers.

First we discuss the hash function based constructions. In our implementation any hash function from the OpenSSL library can be used that uses the Merkle-Darmgard (M-D) construction [25]. The family $F(n)$ is constructed as follows.

Given a hash function Hash with block length b and output size n that uses the M-D construction we build the function family $F(n)$ as

$$f_K(M) = \texttt{Hash}(\texttt{Pad}(K)||\texttt{Pad}(M)),$$

for key $K \in \{0,1\}^n$, message $M \in \{0,1\}^n$ and $\texttt{Pad}(x) = (x||10^{b-|x|-1})$ for $|x| < b$.

We show that this is a pseudorandom function family if Hash is a good cryptographic hash function. In [2] it is assumed, that the compression function of a good M-D hash function is a pseudorandom function family if keyed using the input. In [3], it is assumed, that the compression function of a good M-D hash function is a pseudorandom function family if keyed on the chaining input. Further it is shown, that a fixed input length M-D hash function, keyed using the initialization vector (IV) is a pseudorandom function family for fixed length inputs. In our construction the internal compression function of hash is evaluated twice: First on the IV and the padded key, second on the resulting chaining value and the padded message. Due to the pseudorandomness of the compression function when keyed on the message input, the first evaluation works as a pseudorandom key generation. As we have a fixed message length the second iteration is a pseudorandom function family keyed using the IV input.

[1] http://www.openssl.org/

Table 1. XMSS performance for $H = 20$, $m = 256$. b denotes the bit security. * Using AES-NI. ** Although the authors of [14] mention the possibility to generate the secret key using a pseudorandom generator, this is not covered by their security proof. For the provided values a secret key of size $2^H \cdot n$ is assumed. A secret key size of 152 bits is possible, slightly reducing the bit security. Hence we exclude this value from the comparison for fairness.

		Timings (ms)			Sizes (bit)			
Function	w	Sign	Verify	Keygen	Signature	Public key	Secret key	b
AES-128*	4	1.72	0.11	109,610.45	19,608	7,296	152	82
AES-128	4	2.87	0.22	158,208.49	19,608	7,296	152	82
SHA-256	4	6.30	0.51	408,687.43	39,192	13,568	280	210
SHA-256	16	7.00	0.52	466,236.55	22,296	13,568	280	196
SHA-256	64	15.17	1.02	1,099,377.18	16,664	13,568	280	146
SHA-256	108	33.47	2.34	2,288,355.24	15,384	13,568	280	100
RSA 2048		3.08	0.09	-	≤ 2048	≤ 4096	≤ 4096	87
DSA 2048		0.89	1.06	-	≤ 2048	≤ 4096	≤ 4096	87
MSS-SPR (n=128)					68,096	7680	-**	98

For $\mathcal{H}(n)$ we use `Hash` without modifications, as we only need a randomly chosen element of $\mathcal{H}(n)$ and not the whole family. We follow the standard assumption for the security of keyless hash functions. It assumes that a keyless hash function is an element of a family of hash functions, chosen uniformly at random.

Next we present the constructions using a block cipher $\mathrm{E}(K, M)$ with block and key length n bit. This is of special interest in case of AES, because many smartcard crypto co-processors and also most actual Intel processors provide hardware acceleration for AES. For $F(n)$ we use E without modification, as a standard assumption states that a good block cipher can be modelled as pseudorandom permutation. $\mathcal{H}(n)$ is constructed as $h_K(M) = C_2$ for $M = M_1||M_2$, with

$$C_i = \mathrm{E}_{C_{i-1}}(M_i) \oplus M_i, \quad C_0 = K, \quad 0 \leq i \leq 2$$

in M-D mode. In [7] the authors give a black box proof for the security of this construction. We do not use M-D strengthening, as our domain has fixed size.

Table 1 shows our results on an Intel(R) Core(TM) i5 CPU M540 @ 2.53GHz with Infineon AES-NI[2] for XMSS. For the forward secure construction the signature key size grows to 10.240 bits (5.120 bits) for SHA-256 (AES-128), respectively. We used a tree height $H = 20$. This leads to instances usable for about one million signatures. Further we assumed a message length of $m = 256$ bit. The last column of the table shows the bit security of the configuration. Following the heuristic of Lenstra and Verheul [23] the AES configuration with bit security 82 is secure until 2015. The SHA-256 configurations with bit security 100 (146, 196, 210) are secure until 2039 (2099, 2164, 2182). According to [23], RSA as well as

[2] `http://software.intel.com/en-us/articles/intel-advanced-encryption-standard-instructions-aes-ni`

DSA using a 2048-bit key are assumed to be secure until 2022. The timings for RSA and DSA where taken using the OpenSSL `speed` command. As this does not provide timings for key generation, we had to leave this field blank. The results show that XMSS is comparable to existing signature schemes. Only the key generation takes a lot of time. But as key generation is an offline task, it can be scheduled.

The last row of table 1 shows the signature size and public key size for MSS-SPR [14]. To make the results from [14] comparable, we computed the signature and public key size for message length $m = 256$ bit, using their formulas. [14] does not provide runtimes, therefore we had to leave these fields blank. Comparing XMSS using SHA-256 and $w = 108$ with MSS-SPR shows that even for a slightly higher bit security we achieve a signature length of less than 25 % of the signature length of MSS-SPR. We also tried to compare XMSS with GMSS [10], but as the authors do not provide a security proof, a fair comparison is not possible without presenting a security proof for GMSS.

References

1. Anderson, R.: Two remarks on public key cryptology. In: Manuscript. Relevant material presented by the author in an invited lecture at the 4th ACM Conference on Computer and Communications Security, CCS, pp. 1–4. Citeseer (1997)
2. Bellare, M., Canetti, R., Krawczyk, H.: Keying Hash Functions for Message Authentication. In: Koblitz, N. (ed.) CRYPTO 1996. LNCS, vol. 1109, pp. 1–15. Springer, Heidelberg (1996)
3. Bellare, M., Canetti, R., Krawczyk, H.: Pseudorandom functions revisited: The cascade construction and its concrete security. In: Proceedings of 37th Annual Symposium on Foundations of Computer Science, pp. 514–523. IEEE (1996)
4. Bellare, M., Miner, S.K.: A Forward-Secure Digital Signature Scheme. In: Wiener, M. (ed.) CRYPTO 1999. LNCS, vol. 1666, pp. 431–448. Springer, Heidelberg (1999)
5. Bellare, M., Rogaway, P.: Collision-Resistant Hashing: Towards Making UOWHFs Practical. In: Kaliski Jr., B.S. (ed.) CRYPTO 1997. LNCS, vol. 1294, pp. 470–484. Springer, Heidelberg (1997)
6. Bellare, M., Yee, B.S.: Forward-Security in Private-Key Cryptography. In: Joye, M. (ed.) CT-RSA 2003. LNCS, vol. 2612, pp. 1–18. Springer, Heidelberg (2003)
7. Black, J.A., Rogaway, P., Shrimpton, T.: Black-Box Analysis of the Block-Cipher-Based Hash-Function Constructions from PGV. In: Yung, M. (ed.) CRYPTO 2002. LNCS, vol. 2442, pp. 103–118. Springer, Heidelberg (2002)
8. Bleichenbacher, D., Maurer, U.M.: Optimal Tree-based One-time Digital Signature Schemes. In: Puech, C., Reischuk, R. (eds.) STACS 1996. LNCS, vol. 1046, pp. 363–374. Springer, Heidelberg (1996)
9. Buchmann, J., Dahmen, E., Ereth, S., Hülsing, A., Rückert, M.: On the Security of the Winternitz One-Time Signature Scheme. In: Nitaj, A., Pointcheval, D. (eds.) AFRICACRYPT 2011. LNCS, vol. 6737, pp. 363–378. Springer, Heidelberg (2011)
10. Buchmann, J., Dahmen, E., Klintsevich, E., Okeya, K., Vuillaume, C.: Merkle Signatures with Virtually Unlimited Signature Capacity. In: Katz, J., Yung, M. (eds.) ACNS 2007. LNCS, vol. 4521, pp. 31–45. Springer, Heidelberg (2007)
11. Buchmann, J., Dahmen, E., Schneider, M.: Merkle Tree Traversal Revisited. In: Buchmann, J., Ding, J. (eds.) PQCrypto 2008. LNCS, vol. 5299, pp. 63–78. Springer, Heidelberg (2008)

12. Buchmann, J., Dahmen, E., Szydlo, M.: Hash-based Digital Signature Schemes. In: Bernstein, D.J., Buchmann, J., Dahmen, E. (eds.) Post-Quantum Cryptography, pp. 35–93. Springer, Heidelberg (2009)
13. Buchmann, J., García, L.C.C., Dahmen, E., Döring, M., Klintsevich, E.: CMSS – An Improved Merkle Signature Scheme. In: Barua, R., Lange, T. (eds.) INDOCRYPT 2006. LNCS, vol. 4329, pp. 349–363. Springer, Heidelberg (2006)
14. Dahmen, E., Okeya, K., Takagi, T., Vuillaume, C.: Digital Signatures Out of Second-Preimage Resistant Hash Functions. In: Buchmann, J., Ding, J. (eds.) PQCrypto 2008. LNCS, vol. 5299, pp. 109–123. Springer, Heidelberg (2008)
15. Dods, C., Smart, N.P., Stam, M.: Hash Based Digital Signature Schemes. In: Smart, N.P. (ed.) Cryptography and Coding 2005. LNCS, vol. 3796, pp. 96–115. Springer, Heidelberg (2005)
16. García, L.C.C: On the security and the efficiency of the Merkle signature scheme. Technical Report Report 2005/192, Cryptology ePrint Archive - Report 2005/192 (2005), `http://eprint.iacr.org/2005/192/`
17. Goldreich, O., Goldwasser, S., Micali, S.: How to construct random functions. Journal of the ACM 33(4), 792–807 (1986)
18. Goldwasser, S., Micali, S., Rivest, R.L.: A digital signature scheme secure against adaptive chosen-message attacks. SIAM J. Comput. 17(2), 281–308 (1988)
19. Håstad, J., Impagliazzo, R., Levin, L.A., Luby, M.: A pseudorandom generator from any one-way function. SIAM J. Comput. 28, 1364–1396 (1999)
20. Hevia, A., Micciancio, D.: The Provable Security of Graph-Based One-Time Signatures and Extensions to Algebraic Signature Schemes. In: Zheng, Y. (ed.) ASIACRYPT 2002. LNCS, vol. 2501, pp. 379–396. Springer, Heidelberg (2002)
21. Jakobsson, M., Leighton, T., Micali, S., Szydlo, M.: Fractal Merkle Tree Representation and Traversal. In: Joye, M. (ed.) CT-RSA 2003. LNCS, vol. 2612, pp. 314–326. Springer, Heidelberg (2003)
22. Krawczyk, H.: Simple forward-secure signatures from any signature scheme. In: CCS 2000: Proceedings of the 7th ACM Conference on Computer and Communications Security, pp. 108–115. ACM, New York (2000)
23. Lenstra, A.K., Verheul, E.R.: Selecting cryptographic key sizes. Journal of Cryptology 14, 255–293 (2001)
24. Merkle, R.C.: A Certified Digital Signature. In: Brassard, G. (ed.) CRYPTO 1989. LNCS, vol. 435, pp. 218–238. Springer, Heidelberg (1990)
25. Merkle, R.C.: One Way Hash Functions and DES. In: Brassard, G. (ed.) CRYPTO 1989. LNCS, vol. 435, pp. 428–446. Springer, Heidelberg (1990)
26. Rogaway, P., Shrimpton, T.: Cryptographic Hash-Function Basics: Definitions, Implications, and Separations for Preimage Resistance, Second-Preimage Resistance, and Collision Resistance. In: Roy, B.K., Meier, W. (eds.) FSE 2004. LNCS, vol. 3017, pp. 371–388. Springer, Heidelberg (2004)
27. Rompel, J.: One-way functions are necessary and sufficient for secure signatures. In: STOC 1990: Proceedings of the Twenty-Second Annual ACM Symposium on Theory of Computing, pp. 387–394. ACM Press, New York (1990)
28. Shor, P.W.: Algorithms for quantum computation: Discrete logarithms and factoring. In: Proceedings of the 35th Annual IEEE Symposium on Foundations of Computer Science (FOCS 1994), pp. 124–134. IEEE Computer Society Press (1994)
29. Szydlo, M.: Merkle Tree Traversal in Log Space and Time. In: Cachin, C., Camenisch, J.L. (eds.) EUROCRYPT 2004. LNCS, vol. 3027, pp. 541–554. Springer, Heidelberg (2004)

On the Differential Security of Multivariate Public Key Cryptosystems

Daniel Smith-Tone[1,2]

[1] Department of Mathematics, University of Louisville, Louisville, Kentucky, USA
[2] National Institute of Standards and Technology, Gaithersburg, Maryland, USA
daniel.smith@nist.gov

Abstract. Since the discovery of an algorithm for factoring and computing discrete logarithms in polynomial time on a quantum computer, the cryptographic community has been searching for an alternative for security in the approaching post-quantum world. One excellent candidate is multivariate public key cryptography. Though the speed and parameterizable nature of such schemes is desirable, a standard metric for determining the security of a multivariate cryptosystem has been lacking. We present a reasonable measure for security against the common differential attacks and derive this measurement for several modern multivariate public key cryptosystems.

Keywords: Matsumoto-Imai, multivariate public key cryptography, differential, symmetry.

1 Introduction

In recent years a great deal of focus has been directed towards post-quantum cryptology. This increased attention is indicative of a paradigm shift which has been occurring since, in [1], Peter Shor developed algorithms for factoring and computing discrete logarithms in polynomial time on a quantum computing device. In the face of mounting evidence that quantum computing is not a physical impossibility but merely an engineering challenge, it is more important than ever that we develop secure systems relying on problems of greater difficulty than the classical number theoretic schemes.

Multivariate Public Key Cryptography(MPKC) has emerged as one of a few serious candidates for security in the post-quantum world. This emergence is due to several facts. First, the problem of solving a system of quadratic equations is known to be NP-hard, and seems to be hard even in the average case. No great reduction of the complexity of this problem has been found in the quantum model of computing, and, indeed, if this problem is discovered to be solvable in the quantum model, we can solve all NP problems, which seems particularly wishful. Second, multivariate systems are very efficient, often having speeds dozens of times faster than RSA, [2–4]. Finally, it is easy to parameterize

B.-Y. Yang (Ed.): PQCrypto 2011, LNCS 7071, pp. 130–142, 2011.

many multivariate systems in such a way that vastly different schemes are derived with potentially vastly different resistances to specialized attacks.

One of the great challenges facing MPKC is the task of deriving security proofs. In fact, there currently is no widely accepted quantification for indistinguishability between systems of multivariate equations. One reason for the absence of such a quantification is the fact that even with a great deal of structure in the construction of a multivariate cryptosystem, the coefficients can appear to have a uniform distribution. In fact, history has shown that once a way to distinguish a system of structured multivariate equations from a collection of random equations is discovered, a method of solving this system is often quickly developed.

Recently, several cryptanalyses of various multivariate cryptosystems have pointed out weaknesses in the predominant philosophy for the construction of multivariate public key cryptosystems. Several systems, SFLASH, Square, for example, which are based on simple modifications of the prototypical Matsumoto-Imai public key cryptosystem, have been broken by very similar differential attacks exploiting some symmetry which is inherent to the field structure these systems utilize. See [5–8]. In fact, even various attacks on other multivariate schemes, for example the oil-vinegar attack, see [9], can be viewed as a dual attack, finding a differential invariant.

In [10], a classification of field maps exhibiting the multiplicative symmetry was presented. In this article we are interested in the dual problem, that is, identifying all possible initial general linear differential symmetries a field map can possess. Such a characterization will lead to a fuller understanding of the theory, potentially establish a foundation for modeling more general security proofs, and establish a reasonable and quantitative criterion for the development of future multivariate schemes.which we may model

The paper is organized as follows. The next section illustrates the ubiquitous nature of the differential attack by recasting the attack on the balanced oil and vinegar scheme in the differential setting. In the following section, we focus on differential symmetry, presenting the general linear symmetry and discussing the general structure of the space of linear maps exhibiting this symmetry. The subsequent section restricts the analysis of this space to the case in which the hidden field map of the cryptosystem is a C^* monomial. Next the specific case of the squaring map used in Square is analyzed. The space of linear maps is then determined for projected systems such as the projected SFLASH analogue, pSFLASH. Finally, we review these results and analyze the dimension of this space of linear maps as a metric for determining differential security.

2 Differential Symmetries and Invariants

Differential attacks play a crucial role in multivariate public key cryptography. Such attacks have not only broken many of the so called "big field" schemes, they have directed the further development of the field by inspiring modifiers — Plus (+), Minus (-), Projection (p), Perturbation (P), Vinegar (v) — and the creation of newer more robust techniques.

The differential of a field map, f, is defined by $Df(a,x) = f(a+x) - f(a) - f(x) + f(0)$. The use of this discrete differential appears to occur in very many cryptanalyses of post-quantum multivariate schemes. In fact, we can even consider Patarin's initial attack, in [11], on Imai and Matsumoto's C^* scheme, see [12], as the exploitation of a trivial differential symmetry. Suppose $f(x) = x^{q^\theta+1}$ and let $y = f(x)$. Since the differential of f, Df, is a symmetric bilinear function, $0 = Df(y,y) = Df(y, x^{q^\theta+1}) = yx^{q^{2\theta}+q^\theta} + y^{q^\theta}x^{q^\theta+1} = x^{q^\theta}(yx^{q^{2\theta}} + y^{q^\theta}x)$. Dividing by x^{q^θ} we have Patarin's linear relation, $yx^{q^{2\theta}} = y^{q^\theta}x$; see [11] for details.

Differential methods provide powerful tools for decomposing a multivariate scheme. To illustrate the nearly universal nature of differential attacks, we review the attack of Kipnis and Shamir, see [9], on a non-big-field system, the oil and vinegar scheme. Though they use differing terminology, the attack exploits a symmetry hidden in the differential structure of the scheme.

Recall that the oil and vinegar scheme is based on a hidden quadratic system of equations, $f : k^n \to k^o$, in two types of variables, $x_1, ..., x_o$, the oil variables, and $x_{o+1}, ..., x_{o+v=n}$, the vinegar variables. We focus on the balanced oil and vinegar scheme, in which $o = v$. Let $c_1, ..., c_v$ be random constants. The map f has the property that $f(x_1, ..., x_v, c_1, ..., c_v)$ is affine in $x_1, ..., x_v$. The encryption map, $\overline{f}$ is the composition of f with an n-dimensional invertible affine map, L.

Let O represent the subspace generated by the first v basis vectors, and let V denote the cosummand of O. Notice that the discrete differential given by $Df(a,x) = f(x+a) - f(x) - f(a) + f(0)$ has the property that for all a and x in O, $Df(a,x) = 0$. Thus for each coordinate, i, the differential coordinate form Df_i can be represented:

$$Df_i = \begin{bmatrix} 0 & Df_{i1} \\ Df_{i1}^T & Df_{i2} \end{bmatrix}.$$

Let M_1 and M_2 be two invertible matrices in the span of the Df_i. Then $M_1^{-1}M_2$ is an O-invariant transformation of the form:

$$\begin{bmatrix} A & B \\ 0 & C \end{bmatrix}.$$

Now the Df_i are not known, but $D(f \circ L)_i = L^T Df_i L$, so the $L^T Df_i L$ are known. Notice that if M is in the span of the Df_i, then $L^T M L$ is in the span of the $L^T Df_i L$. Also, since $(L^T M_1 L)^{-1}(L^T M_2 L) = L^{-1}M_1^{-1}M_2 L$, there is a large space of matrices leaving $L^{-1}O$ invariant, which Kipnis and Shamir are able to exploit to effect an attack against the balanced oil and vinegar scheme; see [9] for details. Making the oil and vinegar scheme unbalanced, see [13], corrects this problem by making any subspace which is invariant under a general product $M_1^{-1}M_2$ very small, see [14].

While the differential analysis of the oil and vinegar systems is a very specific case of utilizing an invariant related to the differential structure of the hidden map, several general attacks on big field schemes rely on a type of linear symmetry. The following sections focus on a systematic study of this type of symmetry, and conditions in which such a symmetry can be utilized for a differential attack.

3 Properties of General Linear Symmetries

Let k be an extension field of $\mathbb{F}_q$, the field with q elements. Dubois et al. completed a successful attack against the SFLASH signature scheme, see [8], by utilizing a multiplicative symmetry of the form:

$$Df(\sigma a, x) + Df(a, \sigma x) = (\sigma^{q^\theta} + \sigma)Df(a, x), \tag{1}$$

where $f : k \to k$ is a C^* monomial map, and $\sigma \in k$.

Consider the more general initial linear symmetric relation as suggested by Dubois et al., in [8], of the form:

$$Df(La, x) + Df(a, Lx) = \Lambda_L Df(a, x), \tag{2}$$

where $f : k \to k$ is a function, and $L, \Lambda_L : k \to k$ are $\mathbb{F}_q$-linear. This definition is perfectly appropriate, since we are guaranteed a solution space of dimension at least n for C^* monomial maps, f. In addition, it is clear that we have additive closure, in general. Let S_G denote the set of all linear maps, L, satisfying (2). Notice:

$$\begin{aligned} Df((L+M)a, x) + Df(a, (L+M)x) &= Df(La, x) + Df(a, Lx) \\ &\quad + Df(Ma, x) + Df(a, Mx) \\ &= \Lambda_L Df(a, x) + \Lambda_M Df(a, x) \\ &= (\Lambda_L + \Lambda_M)\, Df(a, x). \end{aligned} \tag{3}$$

For a more general function, f, however, we have no guarantee of such a large space of solutions as possessed by C^* monomials; however, in characteristic two, the discovery of one such symmetric relation allows the generation of a space of maps satisfying the symmetry which has both an additive and square structure. It is worth exploring to see how much structure such a symmetry holds.

Note that if L is in S_G:

$$\begin{aligned} Df(L^2 a, x) + Df(a, L^2 x) &= Df(L^2 a, x) + Df(La, Lx) \\ &\quad + Df(La, Lx) + Df(a, L^2 x) \\ &= \Lambda_L Df(La, x) + \Lambda_L Df(a, Lx) \\ &= \Lambda_L \left(Df(La, x) + Df(a, Lx)\right) \\ &= \Lambda_L^2 Df(a, x). \end{aligned} \tag{4}$$

Notice that for odd characteristic, there is no way to add the needed terms of the form $Df(La, Lx)$. We do not have, in general, multiplicative closure, but for any polynomial function, p, with terms of degree zero or a power of two, if

$L \in S_G$ then $p(L) \in S_G$. Thus, the existence of a single linear map L satisfying the initial general linear symmetry guarantees the existence of a relatively large space of maps satisfying the symmetry.

Therefore S_G is the $\mathbb{F}_q$-vector space sum of rings of the form $\mathbb{F}_q$ or $\mathbb{F}_q\left[L^{2^i}\right]$. Given just a few elements of S_G, we can potentially generate a large subspace of S_G, which is a very appealing situation for an adversary.

This situation is exactly the scenario which has resulted in the breaking of SFLASH and other C^* variants. In [8], it was shown that $k < S_G$ when f is a C^* monomial, and thus S_G is so large that an element can be detected using the relation (2) even when up to one half of the public equations are removed.

Thus the task of constructing a differentially secure multivariate cryptosystem must necessarily include an analysis of the space of linear maps, S_G, illustrating the symmetry. If S_G is very small, then recovering an element from this subspace may be an infeasible task, and the differential attack is doomed.

4 Properties Relative to C^* Monomials

If we restrict our attention to the case in which f is a C^* monomial map of the form $f(x) = x^{q^\theta+1}$, we can derive some additional properties of S_G indicating why so many C^* variants have fallen to differential attacks. Immediately, we know that there is an injective map $g : k \to S_G$, since f has the multiplicative symmetry. Furthermore, by considering the linearized polynomial form of an arbitrary linear map, $L \in GL(\mathbb{F}_q, n)$, we can continue, revealing the exact multiplicative structure of S_G.

Theorem 1. *If f is a C^* monomial, then S_G, equipped with standard multiplication is a k-algebra, and consequently has a large dimension as an $\mathbb{F}_q$-vector space. Furthermore, if $3\theta \neq n$, $S_G \cong k$.*

Proof. Consider the linearized polynomial form of $M \in S_G$, $Mx = \sum_{i=0}^{n-1} m_i x^{q^i}$. We will find conditions on the coefficients, m_i of this linearized polynomial form. For the generic C^* monomial map, $f(x) = x^{q^\theta+1}$, we have that the discrete differential, $Df(a,x) = a^{q^\theta}x + ax^{q^\theta}$. Thus:

$$\begin{aligned}
Df(Ma,x) + Df(a,Mx) &= \sum_{i=0}^{n-1}\left(m_i^{q^\theta} a^{q^{i+\theta}} x + m_i a^{q^i} x^{q^\theta}\right) \\
&\quad + \sum_{i=0}^{n-1}\left(m_i a^{q^\theta} x^{q^i} + m_i^{q^\theta} a x^{q^{i+\theta}}\right) \\
&= \sum_{i=0}^{n-1} m_i^{q^\theta}\left(a^{q^{i+\theta}} x + a x^{q^{i+\theta}}\right) \\
&\quad + \sum_{i=0}^{n-1} m_i \left(a^{q^i} x^{q^\theta} + a^{q^\theta} x^{q^i}\right).
\end{aligned} \tag{5}$$

Since $M \in S_G$, there is an $\mathbb{F}_q$-linear map, $\Lambda_M(x) = \sum_{i=0}^{n-1} \lambda_i x^{q^i}$ such that the equation $Df(Ma, x) + Df(a, Mx) = \Lambda_M Df(a, x)$ holds. Therefore we have:

$$\sum_{i=0}^{n-1} m_i^{q^\theta}\left(a^{q^{i+\theta}}x + ax^{q^{i+\theta}}\right) + m_i\left(a^{q^i}x^{q^\theta} + a^{q^\theta}x^{q^i}\right) = \sum_{i=0}^{n-1}\lambda_i\left(a^{q^\theta}x + ax^{q^\theta}\right)^{q^i}$$
$$= \sum_{i=0}^{n-1}\lambda_i\left(a^{q^{i+\theta}}x^{q^i} + a^{q^i}x^{q^{i+\theta}}\right). \tag{6}$$

We can collect the coefficients of each monomial, $a^i x^j$, and set each to zero, obtaining relations on the coefficients of the linearized form of M and Λ_M.

If $q^\theta + 1$ shares a nontrivial factor with $q^n - 1$, then f is not strictly speaking a C^* monomial, since it is not a permutation polynomial. Thus we treat the case $\theta \notin \{0, \frac{n}{2}, \frac{n}{4}\}$, encompassing all C^* monomials, as well as many functions which are not C^* monomials. If we collect the coefficients of the monomial ax^{q^θ}, we get the relation $\lambda_0 = m_0 + m_0^{q^\theta}$. The coefficients of monomials of the form ax^{q^i}, for $i \notin \{0, \pm\theta\}$, generate the relations $m_{i-\theta} = 0$. Thus $m_i = 0$ for all $i \notin \{0, -\theta, -2\theta\}$. Collecting the coefficients of the monomials of the form $a^{q^\theta}x^{q^i}$ for $i \notin \{0, \theta, 2\theta\}$, we have $m_i = 0$.

Therefore, if a nonzero coefficient exists other than m_0, then either $-\theta = \theta$, which implies $\theta = \frac{n}{2}$, $-\theta = 2\theta$, implying $3\theta = n$, or $-2\theta = 2\theta$, which implies $\theta = \frac{n}{4}$. Of these cases, only $3\theta = n$ represents a possible C^* monomial. Thus, if $3\theta \neq n$, then for all $i \neq 0$, $m_i = 0$, and in this case, $Mx = m_0 x$ is multiplication by an element in k; consequently, $S_G \cong k$.

If $3\theta = n$, then m_0, $m_{\frac{n}{3}}$, and $m_{\frac{2n}{3}}$ can possibly be nonzero. To prove that S_G is still a ring in this case, notice that given two linear maps, M and L, each with all coefficients zero except possibly m_0, $m_{\frac{n}{3}}$, $m_{\frac{2n}{3}}$, l_0, $l_{\frac{n}{3}}$, and $l_{\frac{2n}{3}}$, we have:

$$\begin{aligned} LMx = {} & (l_0 m_0 + l_{\frac{n}{3}} m_{\frac{2n}{3}}^{q^{\frac{n}{3}}} + l_{\frac{2n}{3}} m_{\frac{n}{3}}^{q^{\frac{2n}{3}}})x \\ & + (l_0 m_{\frac{n}{3}} + l_{\frac{n}{3}} m_0^{q^{\frac{n}{3}}} + l_{\frac{2n}{3}} m_{\frac{2n}{3}}^{q^{\frac{2n}{3}}})x^{q^{\frac{n}{3}}} \\ & + (l_0 m_{\frac{2n}{3}} + l_{\frac{n}{3}} m_{\frac{n}{3}}^{q^{\frac{n}{3}}} + l_{\frac{2n}{3}} m_0^{q^{\frac{2n}{3}}})x^{q^{\frac{2n}{3}}}, \end{aligned} \tag{7}$$

which is, again a linear map with all coefficients zero except for the 0-th, $\frac{n}{3}$-th, and $\frac{2n}{3}$-th. Thus S_G has multiplicative closure, and is a 3-dimensional k-algebra.

In the above theorem we didn't mention anything about characteristic. Strictly speaking, a C^* monomial is linearly equivalent to a quadratic permutation polynomial of the form $f(x) = x^{q^\theta+1}$. This is only possible, however, when q is even, since trivially, $2|(q^\theta + 1, q^n - 1)$. Some cryptosystems, however, do use this form of core map in odd characteristic, choosing a map which is 2-to-1, or few-to-1. Such systems never use $\theta \in \{\frac{n}{2}, \frac{n}{4}\}$, since such maps would have exponential collisions. It is for this reason that in the above theorem we relaxed the constraints and allowed any map with $\theta \notin \{0, \frac{n}{2}, \frac{n}{4}\}$. We have completely characterized the symmetries in these cases.

5 Symmetries for Non-permutation Polynomials

In [5, 6], two notable systems, Square and Square-Vinegar, introduced the idea of utilizing a quadratic map over a field of odd characteristic. The C^* form of the core map of Square is $f(x) = x^{q^\theta+1}$ where $\theta = 0$. The theorem of the preceding section doesn't apply to the case $\theta = 0$, therefore we will treat this case separately, and completely characterize S_S, the space of linear maps, L, satisfying (2).

Theorem 2. *Let q be odd. Then $S_S \cong k$.*

Proof. First, $Df(a, x) = 2ax$. Therefore, by the symmetric application of the linear function $M(x) = \sum_{i=0}^{n-1} m_i x^{q^i}$, we have:

$$Df(Ma, x) + Df(a, Mx) = 2\left(\sum_{i=0}^{n-1} m_i a^{q^i}\right)x + 2a\left(\sum_{i=0}^{n-1} m_i x^{q^i}\right). \quad (8)$$

Setting this quantity equal to $\Lambda_M Df(a, x)$ we have:

$$2\left(\sum_{i=0}^{n-1} m_i a^{q^i}\right)x + 2a\left(\sum_{i=0}^{n-1} m_i x^{q^i}\right) = \sum_{i=0}^{n-1} \lambda_i 2^{q^i} a^{q^i} x^{q^i}. \quad (9)$$

We can collect the coefficients of each monomial $a^{q^i} x^{q^j}$ and set each equal to zero to determine relations between M and Λ_M. Collecting coefficients for monomials of the form ax^{q^i}, for $i \neq 0$, we get the relations, $2m_i = 0$. Thus $m_i = 0$ for all $i \neq 0$, and M is multiplication by m_0 in k; consequently, $S_S \approx k$.

It is important to note that the Square systems have been broken by a differential attack in [7] which recovers the multiplicative structure of k by utilizing a symmetry Square exhibits under left composition. This method of finding a terminal symmetry under left composition was discovered for two reasons: first, the Square systems did not preclude such an attack by employing the minus modifier or an alternative precaution; and second, the designers were able to mask the initial multiplicative symmetry of the core map of Square by projecting the input of the C^* monomial into a subspace, making an attack using a symmetry of the form (2) infeasible. If we include the minus modifier, i.e. consider Square-, then the attack of [7] fails, and the question of which symmetries exist over a subspace becomes more critical.

6 Symmetries over Subspaces

In [15], Ding et al. began the work of classifying the initial general linear symmetries for C^* monomial maps over subspaces. Their result was imprecisely stated, but they successfully proved that "almost always" if a field map has an initial general linear symmetry over a subspace then that symmetry is a multiplicative symmetry.

As stated, the claim indicated that for a bijective C^* monomial, f, given any hyperplane, $H = \pi(k)$, if we have:

$$Df(Ma, \pi x) + Df(\pi a, Mx) = \Lambda_M Df(\pi a, \pi x), \tag{10}$$

for all $a, x \in k$, then $M = M_\sigma \circ \pi$ and $\Lambda_M = M_{\sigma+\sigma^{q^\theta}}$, for some $\sigma \in k$.

To prove that the statement as given in [15] is in err, let us define the space saving notation $S_f(A, B)(a, x) = Df(Aa, Bx) + Df(Ba, Ax)$, and take the following example. Let $k = GF(64)$ over $\mathbb{F}_2$, $f(x) = x^5$, $\pi x = x + x^2$, and $Mx = x^4 + x^8$. By a simple calculation,

$$\begin{aligned} S_f(M, \pi)(a, x) &= (a^{16} + a^{32})(x + x^2) + (a + a^2)(x^{16} + x^{32}) \\ &= a^{16}x + ax^{16} + a^{32}x + ax^{32} + a^{16}x^2 + a^2x^{16} + a^{32}x^2 + a^2x^{32} \\ &= \left(ax^4 + a^4x + a^2x^4 + a^4x^2 + ax^8 + a^8x + a^2x^8 + a^8x^2\right)^{16} \\ &= Df(\pi a, \pi x)^{16}. \end{aligned} \tag{11}$$

(Here we note that two terms of the form $(a^4+a^8)(x^4+x^8)$ cancelled each other in the first line above.) Thus, we have found a counterexample with $\Lambda_M x = x^{16}$ and $Mx = (\pi x)^4$, which is certainly not the composition of a multiplication map and π. Here the fact that $2\,(codim(H) + \theta) = n$ created some extra symmetries in the relations between the coefficients of M and Λ_M. Informally, θ was an exceptional choice which permits the existence of a linear map allowing collisions between monomials generated from $Df(Ma, \pi x)$ and $Df(\pi a, Mx)$. Since the arithmetic of k has characteristic 2, collision corresponds with annihilation.

We can resolve the minor issues with the result of Ding et al. and generalize the statement somewhat by providing a more detailed analysis of the symmetry:

$$Df(Ma, \pi x) + Df(\pi a, Mx) = \Lambda_M Df(\pi a, \pi x), \tag{12}$$

for more general linear maps, π. In particular, a more precise formulation of the result of Ding et al. is the special case of $d = 1$ in the following theorem.

Theorem 3. *Let $f(x) = x^{q^\theta+1}$ be a C^* map, and let M and $\pi x = \sum_{i=0}^{d} x^{q^i}$ be linear. Suppose $Df(Ma, \pi x) + Df(\pi a, Mx) = \Lambda_M Df(\pi a, \pi x)$. If $\theta + d < \frac{n}{2}$, $|n - 3\theta| > d$, and $0 < d < \theta - 1$, then $M = M_\sigma \pi$ for some $\sigma \in k$.*

Proof. Our strategy for the proof will be to determine relations between the coefficients of the linearized polynomial forms of M and Λ_M. We will zigzag back and forth between solving for coefficients of M and of Λ_M, further resolving the relationship between the two maps with each step. We will extensively use the "space of indices," the torus consisting of the pairs (r, s) (mod n) which correspond to monomials of the form $a^{q^r}x^{q^s}$. The geometry of this space of indices will be useful in determining relations on the coefficients of the corresponding monomials in the expansions of (12).

Expanding the right hand side of (12) repeatedly, using the bilinearity of Df, we obtain:

$$\begin{aligned}\Lambda_M Df(\pi a,\pi x) &= \sum_{i=0}^{n-1}\lambda_i Df(\pi a,\pi x)^{q^i}\\ &= \sum_{i=0}^{n-1}\lambda_i Df(\sum_{j=0}^{d} a^{q^j},\sum_{l=0}^{d} x^{q^l})^{q^i}\\ &= \sum_{i=0}^{n-1}\sum_{j=0}^{d}\sum_{l=0}^{d}\lambda_i Df(a^{q^j},x^{q^l})^{q^i}\\ &= \sum_{i=0}^{n-1}\sum_{j=0}^{d}\sum_{l=0}^{d}\lambda_i\left(a^{q^{\theta+j}}x^{q^l}+a^{q^j}x^{q^{\theta+l}}\right)^{q^i}.\end{aligned} \tag{13}$$

Notice that for each monomial term in this expression, the difference between the exponent of q in the power of a and the exponent of q in the power of x is $l-\theta-j \pmod n$ or $l+\theta-j \pmod n$. Also, there is the restriction that $0\le l,j\le d$. From these facts we can determine which monomials never occur in the right side of (12).

The monomial $a^{q^r}x^{q^s}$ may only occur in the right side of (12) if the difference between the coordinates, $(s-r)\ (mod\ n)\in[-\theta-d,-\theta+d]\cup[\theta-d,\theta+d]$, where we require $2\theta+2d<n$, avoiding overlap. Also, implicitly, we have the restriction that for such an interval, (u,v), the positive residues u and v satisfy $0\le v-u\le n-1$. For all other pairs, (r,s), $a^{q^r}x^{q^s}$ certainly has a coefficient of zero in the right side of (12). Therefore, we will study the set of pairs of indices,

$$E=\{(r,s)|s-r\in(-\theta+d,\theta-d)\cup(\theta+d,-\theta-d)\}.$$

This set is the diagonal band in the space of indices for which the corresponding coefficients have no contribution from the right side of (12); refer to the shaded region in the figure below.

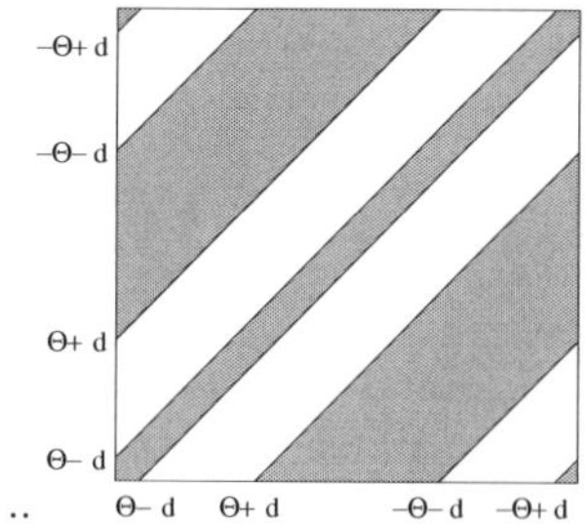

Fig. 1. The space of indices with the shaded region corresponding to monomials which cannot occur on the right side of (12)

Expanding the left hand side of (12), similarly:

$$\begin{aligned}
S_f(M,\pi)(a,x) &= Df(Ma,\pi x) + Df(\pi a, Mx)\\
&= Df(\sum_{i=0}^{n-1} m_i a^{q^i}, \pi x) + Df(\pi a, \sum_{i=0}^{n-1} m_i x^{q^i})\\
&= \sum_{i=0}^{n-1} \left(Df(m_i a^{q^i}, \pi x) + Df(\pi a, m_i x^{q^i})\right)\\
&= \sum_{i=0}^{n-1} \left(Df(m_i a^{q^i}, \sum_{j=0}^{d} x^{q^j}) + Df(\sum_{j=0}^{d} a^{q^j}, m_i x^{q^i})\right)\\
&= \sum_{i=0}^{n-1}\sum_{j=0}^{d} \left(Df(m_i a^{q^i}, x^{q^j}) + Df(a^{q^j}, m_i x^{q^i})\right)\\
&= \sum_{i=0}^{n-1}\sum_{j=0}^{d} \left(m_{i-\theta}^{q^\theta} a^{q^i} x^{q^j} + m_i a^{q^i} x^{q^{\theta+j}} + m_i a^{q^{\theta+j}} x^{q^i} + m_{i-\theta}^{q^\theta} a^{q^j} x^{q^i}\right).
\end{aligned} \tag{14}$$

Now, to analyze which monomials of the form $a^{q^r} x^{q^s}$, have nontrivial coefficients for the pair of "indices" (r,s), we construct four index sets, A, B, C, and D, relative to the four monomials in the above expression, respectively. We have:

$$\begin{aligned}
A &= [0, n-1] \times [0, d]\\
B &= [0, n-1] \times [\theta, \theta + d]\\
C &= [\theta, \theta + d] \times [0, n-1]\\
D &= [0, d] \times [0, n-1]\,.
\end{aligned} \tag{15}$$

We can see that only the pairs (A,C), (A,D), (B,C), and (B,D) have non trivial intersections. Isolating the index pairs occurring in only one of these index spaces we can find relations on the coefficients of M and Λ_M which involve only one m_i. If, furthermore, the index pair occurs in E, then the corresponding coefficient from Λ_M is zero. Let $*$ denote the operation of taking one of these sets minus the union of the other three. We notice that:

$$\begin{aligned}
A^* &= ([d+1, \theta-1] \cup [\theta+d+1, n-1]) \times [0, d]\\
B^* &= ([d+1, \theta-1] \cup [\theta+d+1, n-1]) \times [\theta, \theta+d]
\end{aligned} \tag{16}$$

if $d < \theta$.

For both A^* and B^*, the first coordinate is the index associated with the coefficient of M in (14); we are, therefore, interested in which values of the first coordinate are possible in $A^* \cap E$ and $B^* \cap E$. Equivalently, we want to discover $\pi_1(A^* \cap E)$ and $\pi_1(B^* \cap E)$, where π_1 is the projection mapping onto the first coordinate. By a simple calculation, we have:

$$\begin{aligned}
\pi_1(A^* \cap E) &= [\theta+d+1, -\theta-1] \cup [-\theta+d+1, n-1] \cup [d+1, \theta-1]\\
\pi_1(B^* \cap E) &= [d+1, \theta-1] \cup [\theta+d+1, 2\theta-1] \cup [2\theta+d+1, n-1]\,,
\end{aligned} \tag{17}$$

see the figure below.

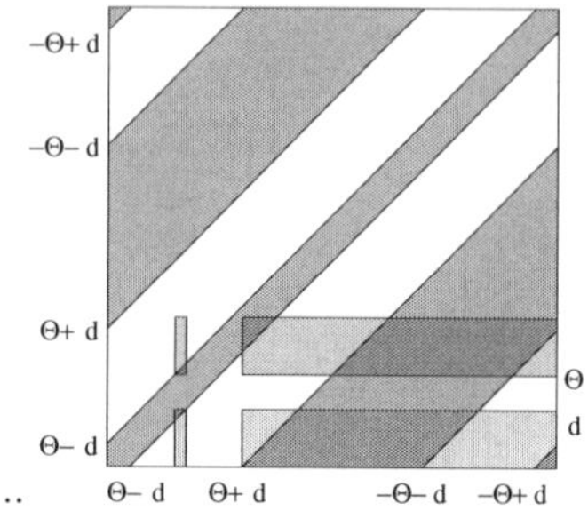

Fig. 2. The intersection of A^* and B^* with E

Since the coefficient of M associated with (r,s) in A^* is $m_{r-\theta}$, and the coefficient associated with (r,s) in B^* is m_r, we know that $m_r = 0$ for every r in the union, $((\pi_1(A^* \cap E) - \theta) \cup \pi_1(B^* \cap E))$, where:

$$\pi_1(A^* \cap E) - \theta = [d+1, -2\theta-1] \cup [-2\theta+d+1, -\theta-1] \cup [-\theta+d+1, n-1]. \tag{18}$$

Notice that $\pi_1(B^* \cap E)$ and $\pi_1(A^* \cap E) - \theta$ are symmetric with respect to $[d+1, n-1]$, and therefore their union is $[d+1, n-1]$ if and only if the first "gap", $[\theta, \theta+d]$, of $\pi_1(B^* \cap E)$ is contained in the first or second subinterval of $\pi_1(A^* \cap E) - \theta$. This occurs when either $\theta + d \leq n - 2\theta - 1$, which is equivalent to $3\theta + d < n$, or $n - 2\theta + d + 1 \leq \theta$, which is equivalent to $n < 3\theta - d$; thus, since by hypothesis $|n - 3\theta| > d$, we have $m_r = 0$ for all $r \in [d+1, n-1]$.

Furthermore, since the boundary of E, ∂E, corresponds to regions at which the coefficient of the right side of (12) is a single λ_i, we can use the complementary technique, checking the coefficients corresponding to $\partial E - (A \cup B \cup C \cup D)$, to reveal that $\lambda_i = 0$ for $i \in [d+1, \theta-1] \cup [\theta+1, n-\theta-1] \cup [n-\theta+1, n-d-1]$. Moreover, we can compare coefficients at the intersection of ∂E and one of A^*, B^*, C^*, or D^*. For $\partial E \cap A^*$, we get the relations $\lambda_i = m_{d+i}$ for $i \in [1, d-1]$, and for $\partial E \cap B^*$, we get $\lambda_i = m_i$ for $i \in [n-d+1, n-1]$. Since we have already shown that such coefficients of M are zero, λ can only be nonzero for the values λ_0, λ_d, λ_θ, $\lambda_{n-\theta}$, and λ_{n-d}.

Using this information we can greatly simplify (13), and as a consequence, get further information about the coefficients of M. In particular, from collecting coefficients for monomials with indices (θ, i), for $i \in [0, d]$, we get the relations $m_0^{q^\theta} + m_i = \lambda_0 + \lambda_{-d}$. Thus, $m_i = m_0$ for $i \leq d$, and, finally, we see that $M = m_0 \pi$.

The preceding theorem gives us precise criteria for when the space of linear maps, S_G, consists of only projected multiplication maps. Furthermore, it was stated in [10] and [15] that these multiplication maps satisfy the relation (2) only if the multiplication commutes with the projection, which happens precisely when the image of the projection is a subspace over an intermediate extension field of $\mathbb{F}_q$. Clearly, in the case $d = 1$, $d + \theta < \frac{n}{2}$, and $|n - 3\theta| > 1$, πk is a hyperplane, and thus $S_G \cong \mathbb{F}_q$, which is optimal.

7 Conclusion

Multivariate public key cryptography has several desirable traits as a potential candidate for post-quantum security. Unfortunately, a standard metric by which we can judge the security of a multivariate scheme has yet to be determined. One consequence of this current status of the field is the similar cryptanalyses of several promising ideas.

We offer the size of the space of linear maps, S_G, illustrating the initial differential symmetries of the core map, f, as a benchmark for the judgement of differential security in modern multivariate public key cryptosystems. As evidence of the feasibility and utility of this method as a measurement of differential security, we measure these spaces for several key players in the evolution of the recent big-field schemes. In the cases of schemes which have been broken, we find that these spaces are large, at least as large as the size of the big field. In the cases of currently considered secure variants, such as the projected SFLASH scheme, pSFLASH, we find that we can make this space as small as possible.

References

1. Shor, P.W.: Polynomial-time algorithms for prime factorization and discrete logarithms on a quantum computer. SIAM J. Sci. Stat. Comp. 26, 1484 (1997)
2. Chen, A.I.-T., Chen, M.-S., Chen, T.-R., Cheng, C.-M., Ding, J., Kuo, E.L.-H., Lee, F.Y.-S., Yang, B.-Y.: SSE Implementation of Multivariate PKCs on Modern X86 CPUs. In: Clavier, C., Gaj, K. (eds.) CHES 2009. LNCS, vol. 5747, pp. 33–48. Springer, Heidelberg (2009)
3. Chen, A.I.-T., Chen, C.-H.O., Chen, M.-S., Cheng, C.-M., Yang, B.-Y.: Practical-Sized Instances of Multivariate PKCs: Rainbow, TTS, and ℓIC-Derivatives. In: Buchmann, J., Ding, J. (eds.) PQCrypto 2008. LNCS, vol. 5299, pp. 95–108. Springer, Heidelberg (2008)
4. Yang, B.-Y., Cheng, C.-M., Chen, B.-R., Chen, J.-M.: Implementing Minimized Multivariate PKC on Low-Resource Embedded Systems. In: Clark, J.A., Paige, R.F., Polack, F.A.C., Brooke, P.J. (eds.) SPC 2006. LNCS, vol. 3934, pp. 73–88. Springer, Heidelberg (2006)
5. Clough, C., Baena, J., Ding, J., Yang, B.-Y., Chen, M.-s.: Square, a New Multivariate Encryption Scheme. In: Fischlin, M. (ed.) CT-RSA 2009. LNCS, vol. 5473, pp. 252–264. Springer, Heidelberg (2009)
6. Baena, J., Clough, C., Ding, J.: Square-Vinegar Signature Scheme. In: Buchmann, J., Ding, J. (eds.) PQCrypto 2008. LNCS, vol. 5299, pp. 17–30. Springer, Heidelberg (2008)
7. Billet, O., Macario-Rat, G.: Cryptanalysis of the Square Cryptosystems. In: Matsui, M. (ed.) ASIACRYPT 2009. LNCS, vol. 5912, pp. 451–468. Springer, Heidelberg (2009)

8. Dubois, V., Fouque, P.-A., Shamir, A., Stern, J.: Practical Cryptanalysis of SFLASH. In: Menezes, A. (ed.) CRYPTO 2007. LNCS, vol. 4622, pp. 1–12. Springer, Heidelberg (2007)
9. Kipnis, A., Shamir, A.: Cryptanalysis of the Oil & Vinegar Signature Scheme. In: Krawczyk, H. (ed.) CRYPTO 1998. LNCS, vol. 1462, pp. 257–266. Springer, Heidelberg (1998)
10. Smith-Tone, D.: Properties of the Discrete Differential with Cryptographic Applications. In: Sendrier, N. (ed.) PQCrypto 2010. LNCS, vol. 6061, pp. 1–12. Springer, Heidelberg (2010)
11. Patarin, J.: Cryptanalysis of the Matsumoto and Imai Public Key Scheme of Eurocrypt '88. In: Coppersmith, D. (ed.) CRYPTO 1995. LNCS, vol. 963, pp. 248–261. Springer, Heidelberg (1995)
12. Matsumoto, T., Imai, H.: Public Quadratic Polynomial-Tuples for Efficient Signature-Verification and Message-Encryption. In: Günther, C.G. (ed.) EUROCRYPT 1988. LNCS, vol. 330, pp. 419–453. Springer, Heidelberg (1988)
13. Kipnis, A., Patarin, J., Goubin, L.: Unbalanced Oil and Vinegar Signature Schemes. In: Stern, J. (ed.) EUROCRYPT 1999. LNCS, vol. 1592, pp. 206–222. Springer, Heidelberg (1999)
14. Patarin, J.: The oil and vinegar algorithm for signatures. Presented at the Dagsthul Workshop on Cryptography (1997)
15. Ding, J., Dubois, V., Yang, B.Y., Chen, C.H.O., Cheng, C.M.: Could SFLASH be Repaired? In: Aceto, L., Damgård, I., Goldberg, L.A., Halldórsson, M.M., Ingólfsdóttir, A., Walukiewicz, I. (eds.) ICALP 2008, Part II. LNCS, vol. 5126, pp. 691–701. Springer, Heidelberg (2008)

Implementation of McEliece Based on Quasi-dyadic Goppa Codes for Embedded Devices

Stefan Heyse

Horst Görtz Institute for IT Security
Ruhr University Bochum
44780 Bochum, Germany
heyse@crypto.rub.de

Abstract. Most public-key cryptosystems frequently implemented have been proven secure on the basis of the presumed hardness of two mathematical problems: factoring the product of two large primes (FP) and computing discrete logarithms (DLP). At present, both problems are believed to be computationally infeasible with an ordinary computer. However, a quantum-computer having the ability to perform computations on a few thousand qbits could solve both problems using Shor's algorithm [23]. Although a quantum computer of this dimension has not been reported, development and cryptanalysis of alternative public-key cryptosystems seem suitable. To achieve acceptance and attention in practice, they have to be implemented efficiently. Furthermore, the implementations have to perform fast while keeping memory requirements low for security levels comparable to conventional schemes. The McEliece encryption and decryption do not require computationally expensive multiple precision arithmetic. Hence, it is predestined for an implementation on embedded devices. The major disadvantage of the McEliece public-key cryptosystem(PKC) is its very large public key of several hundred thousands bits. For this reason, the McEliece PKC has achieved little attention in the practice. Another disadvantage of the McEliece scheme, like many other schemes, is that it is not semantically secure. The quasi-dyadic McEliece variant proposed by Barreto and Misoczki addresses both problems. In this work we provide an implementation of this alternative public-key cryptosystem, which is semantically secure and uses a 40 times smaller public key and a five times smaller secret key compared to a previously published implementation [6].

Keywords: McEliece, Goppa Code, Quasi-Dyadic, Embedded Device, Post-Quantum.

1 Introduction

Only few implementations of the original McEliece public-key cryptosystem have been reported. For instance, there exist two software implementations for 32-bit

B.-Y. Yang (Ed.): PQCrypto 2011, LNCS 7071, pp. 143–162, 2011.

architectures: an i386 assembler implementation [20] and a C-implementation [21]. Two implementations of the McEliece PKC on an 8-bits AVR microcontroller and an FPGA have been provided by [6]. The microcontroller implementation encrypts with 3,889 bits/second and decrypts with 2,835 bits/second at a clock frequency of 32 MHz clock frequency. The main disadvantage of this implementation is the use of external memory for encryption. As explained above, the public-key of the McEliece PKC in [6] is 437.75 Kbytes in size such that external memory has to be used to store the key. The quasi-dyadic variant should solve the problem of large public keys, increasing the practicability of the McEliece public-key cryptosystem. To the best of our knowledge, no implementations of the quasi-dyadic McEliece variant have been proposed targeting an embedded device.

The remainder of this work is organized as follows. Section 2 introduces the classical McEliece public key scheme. In further progress we describe how binary dyadic and quasi-dyadic Goppa codes are constructed. Section 3 gives the scheme definition of the quasi-dyadic McEliece variant and describes the Kobara-Imai's specific conversion γ. In Section 4, our implementation of the McEliece PKC with quasi-dyadic Goppa codes on an 8-bits AVR microcontroller is explained. We provide the results of our implementation with respect to memory requirements and performance in Section 5 and conclude in Section 6.

2 Background on the McEliece Cryptosystem

The McEliece cryptosystem [15] was developed by Robert McEliece in 1978 and was the first proposed public-key cryptosystem (PKC) based on error-correcting codes.

The idea behind this scheme is to pick randomly a code from a family of codes with an existing efficient decoding algorithm and to use the description of this code as private key. To obtain the public key the private key is disguised as a general linear code by means of several secret transformations. The decoding of general linear codes is known to be $\mathcal{NP}$-hard. Hence, the purpose of these transformations is to hide any visible structure of the private key which might be used to identify the underlying code.

The common system parameters for the McEliece PKC are parameters of the underlying $[n, k, d]$ binary Goppa code defined by an (irreducible) polynomial of degree t over $GF(2^m)$ called Goppa polynomial. Corresponding to each such polynomial there exist a binary Goppa code of length $n = 2^m$, dimension $k \geq n - mt$ and minimum distance $d = 2t + 1$ where t is the number of errors correctable by an efficient decoding algorithm. The public key is $K_{pub} = (\hat{G}, t)$, where $\hat{G} = S \cdot G \cdot P$. The private key is $K_{pr} = (S, G, P)$, where G is a $k \times n$ generator matrix for the code $\mathcal{C}$, S is a $k \times k$ scrambling matrix and P is a $n \times n$ permutation matrix. The McEliece encryption is done by multiplying a k-bit message vector by the recipient's public generator matrix $\hat{G}$ and adding a random error vector e with Hamming weight at most t. The decoding problem is the problem of decoding a linear code $\hat{\mathcal{C}}$ equivalent to a binary Goppa code $\mathcal{C}$.

The knowledge of the permutation P is necessary to solve this problem. After reversing the permutation transformation, the decoder for $\mathcal{C}$ can be used to decode the permuted ciphertext $\hat{c}$ to a message $\hat{m} = S \cdot m$. The original message m is then obtained from $\hat{m}$ by $m = \hat{m}S^{-1}$.

2.1 Recommended Parameters and Key Sizes

The parameters influencing the security of the McEliece PKC are the code length n, the code dimension k, and the number of added errors t. In his original paper [15] McEliece suggests using $[n = 2^m, k = n - mt, d = 2t + 1] = [1024, 524, 101]$ Goppa codes over $GF(2^m)$ where $m = 10$ and $t = 50$. In [4] the authors present an improved attack on the McEliece scheme. This new attack reduces the number of operations needed to break the McEliece scheme with original parameters to about 2^{60} instead of 2^{80} which was assumed before. To achieve 80-bit, 128-bit, and 256-bit security level the authors suggest using [2048,1751,55], [2960,2288,113], and [6624,5129,231] binary Goppa codes, respectively.

Table 1 summarizes all suggested parameters as well as the resulting key sizes for specific security levels. It is very common to give the public key in systematic form as a $(n - k) \times k$ matrix. But all published implementations targeting embedded devices choose to store the full $(n \times k)$ public key. This has the advantage of a smaller secret key, which cannot be stored in external memory. If the public key is non systematic, the matrix S in the secret key is completely random and can be generated at runtime form a small seed. For this reason column four gives the size of non-systematic public keys.

Table 1. Recommended parameters and key sizes for the original McEliece PKC

Security Level	**[n,k,d]-Code**	**Added errors**	**Size of K_{pub} in Kbits**	**Size of $K_{pr} = (G(x), P, S)$ in Kbits**
hardly 80-bit	[1632,1269,67]	34	2022	(0.34,15.94,1573)
80-bit	[2048,1751,55]	27	3502	(0.30,22,2994)
128-bit	[2960,2288,113]	56	6614	(0.61,31.80,5112)
256-bit	[6624,5129,231]	117	33178	(1.38,77.63,25690)

The major disadvantage of the McEliece public-key cryptosystem is its very large public key of several hundred thousand bits. The complete public generator matrix $\hat{G}$ of an (n, k) linear code occupies $n \cdot k$ bits storage space. For this reason, the McEliece PKC has achieved little attention in the practice. Particularly with regard to bounded memory capabilities of embedded devices, it is essential to improve the McEliece cryptosystem by finding a way to reduce the public key size.

2.2 Goppa Codes

Goppa codes were introduces by V. D. Goppa in 1970 [9]. Binary Goppa codes form a family of binary linear codes generated by a Goppa polynomial

$G(x) = \sum_{i=0}^{t} g_i x^i$ of degree t with coefficients taken in a finite field $\mathbb{F}_q$ where $q = 2^m$ and a subset $L = (L_0, \ldots, L_{n-1}) \in \mathbb{F}_q^n$, whose elements L_i are not roots of $G(x)$. Lower bounds on their dimension and minimum distance are known, as well as an efficient polynomial-time decoding algorithm.

Theorem 1. *Let L be a sequence $L = (L_0, \ldots, L_{n-1}) \in \mathbb{F}_q^n$ of distinct elements and $G(x)$ a Goppa polynomial of degree t where $G(L_i) \neq 0$, $\forall\, 0 \leq i \leq n-1$. For any vector $c = (c_0, \ldots, c_{n-1}) \in \mathbb{F}_p^n$ we define the* syndrome *of c by*

$$S_c(x) = -\sum_{i=0}^{n-1} \frac{c_i}{G(L_i)} \frac{G(x) - G(L_i)}{x - L_i} \mod G(x) \equiv \sum_{i=0}^{n-1} \frac{c_i}{x - L_i} \mod G(x).$$

The binary Goppa code $\Gamma(L, G(x))$ *is defined as the following subspace of $\mathbb{F}_p^n$.*

$$\Gamma(L, G(x)) = \{c \in \mathbb{F}_p^n \mid S_c(x) \equiv 0 \mod G(x)\}$$

An alternative way to define Goppa codes is to treat them as subfield subcodes of Generalized Reed-Solomon codes. In that special case Goppa codes are also called alternant codes.

Definition 1. *Given a sequence $L = (L_0, \ldots, L_{n-1}) \in \mathbb{F}_q^n$ of distinct elements and a sequence $D = (D_0, \ldots, D_{n-1}) \in \mathbb{F}_q^n$ of nonzero elements, the Generalized Reed-Solomon code $GRS_t(L, D)$ is the [n,k,t+1] linear error-correcting code defined by the parity-check matrix $H_{L,D} = vdm(t, L) \cdot Diag(D)$ where $vdm(t, L)$ denotes the $t \times n$ Vandermonde matrix with elements $vdm_{ij} = L_j^i$.*

$$H_{L,D} := \begin{pmatrix} D_0 & D_1 & \cdots & D_{n-1} \\ D_0 L_0 & D_1 L_1 & \cdots & D_{n-1} L_{n-1} \\ \vdots & \vdots & \ddots & \vdots \\ D_0 L_0^{t-1} & D_1 L_1^{t-1} & \cdots & D_{n-1} L_{n-1}^{t-1} \end{pmatrix}$$

In the original McEliece cryptosystem binary irreducible Goppa codes are used. A Goppa code is *irreducible* if the used Goppa polynomial $G(x)$ is irreducible over $\mathbb{F}_q$. In this case the Goppa code can correct up to t errors.

If $G(x) = \prod_{i=0}^{t-1}(x - z_i)$ is a monic polynomial with t *distinct* roots all in $\mathbb{F}_q$ then it is called *separable* over $\mathbb{F}_q$. In case of $q = 2^m$ the Goppa code can also correct t errors. A Goppa code generated by a separable polynomial over $\mathbb{F}_q$ admits a parity-check matrix in Cauchy form [14].

Definition 2. *Given two disjoint sequences $z = (z_0, \ldots, z_{t-1}) \in \mathbb{F}_q^t$ and $L = (L_0, \ldots, L_{n-1}) \in \mathbb{F}_q^n$ of distinct elements, the Cauchy matrix $C(z, L)$ is the $t \times n$ matrix with elements $C_{ij} = 1/(z_i - L_j)$.*

Theorem 2. *The Goppa code generated by a monic polynomial $G(x) = (x - z_0) \cdots (x - z_{t-1})$ without multiple zeros admits a parity-check matrix of the form $H = C(z, L)$, i.e. $H_{ij} = 1/(z_i - L_j)$, $0 \leq i < t, 0 \leq j < n$.*

2.3 Dyadic Goppa Codes

In [16] Barreto and Misoczki show how to build binary Goppa codes which admit a parity-check matrix in dyadic form. The family of dyadic Goppa codes offers the advantage of having a compact and simple description. In their proposal the authors make extensive use of the fact that using Goppa polynomials separable over $\mathbb{F}_q$ the resulting Goppa code admits a parity-check matrix in Cauchy form by Theorem 2. Hence, it is possible to construct parity-check matrices which are in Cauchy and dyadic form, simultaneously.

Definition 3. *Let $\mathbb{F}_q$ denote a finite field and $h = (h_0, h_1, \ldots, h_{n-1}) \in \mathbb{F}_q$ a sequence of $\mathbb{F}_q$ elements. The* dyadic matrix *$\Delta(h) \in \mathbb{F}_q^n$ is the symmetric matrix with elements $\Delta_{ij} = h_{i \oplus j}$, where $\oplus$ is the bitwise exclusive-or. The sequence h is called* signature *of $\Delta(h)$ and coincides with the first row of $\Delta(h)$. Given $t > 0$, $\Delta(h,t)$ denotes $\Delta(h)$ truncated to its first t rows.*

When n is a power of 2 every 1×1 matrix is a dyadic matrix, and for $k > 0$ any $2^k \times 2^k$ matrix $\Delta(h)$ is of the form $\Delta(h) := \begin{pmatrix} A & B \\ B & A \end{pmatrix}$ where A and B are dyadic $2^{k-1} \times 2^{k-1}$ matrices.

Theorem 3. *Let $H \in \mathbb{F}_q^{n \times n}$ with $n > 1$ be a dyadic matrix $H = \Delta(h)$ for some signature $h \in \mathbb{F}_q^n$ and a Cauchy matrix $C(z, L)$ for two disjoint sequences $z \in \mathbb{F}_q^n$ and $L \in \mathbb{F}_q^n$ of distinct elements, simultaneously. It follows that*

- *$\mathbb{F}_q$ is a field of characteristic 2*
- *h satisfies $\frac{1}{h_{i \oplus j}} = \frac{1}{h_i} + \frac{1}{h_j} + \frac{1}{h_0}$*
- *the elements of z are defined as $z_i = \frac{1}{h_i} + \omega$, and*
- *the elements of L are defined as $L_i = \frac{1}{h_j} + \frac{1}{h_0} + \omega$ for some $\omega \in \mathbb{F}_q$*

It is obvious that a signature h describing such a dyadic Cauchy matrix cannot be chosen completely at random. Hence, the authors suggest only choosing nonzero distinct h_0 and h_i at random, where i scans all powers of two smaller than n, and to compute all other values for h by $h_{i \oplus j} = \frac{1}{\frac{1}{h_i} + \frac{1}{h_j} + \frac{1}{h_0}}$ for $0 < j < i$.

Algorithm 1 in [16] shows how binary dyadic Goppa codes are constructed. It takes as input three integers: q, N, and t. The first integer $q = p^d = 2^m$ where $m = s \cdot d$ defines the finite field $\mathbb{F}_q$ as degree d extension of $\mathbb{F}_p = \mathbb{F}_{2^s}$. The code length N is a power of two such that $N \leq q/2$. The integer t denotes the number of errors correctable by the Goppa code. The algorithm outputs the support L, a separable polynomial $G(x)$, as well as the dyadic parity-check matrix $H \in \mathbb{F}_q^{t \times N}$ for the binary Goppa code $\Gamma(L, G(x))$ of length N and designed minimum distance $2t + 1$.

Furthermore, Algorithm 1 in [16] generates the essence η of the signature h of H where $\eta_r = \frac{1}{h_{2^r}} + \frac{1}{h_0}$ for $r = 0, \ldots, \lfloor \lg N \rfloor - 1$ with $\eta_{\lfloor \lg N \rfloor} = \frac{1}{h_0}$, so that, for $i = \sum_{k=0}^{\lfloor \lg N \rfloor - 1} i_k 2^k$, $\frac{1}{h_i} = \eta_{\lfloor \lg N \rfloor} + \sum_{k=0}^{\lfloor \lg N \rfloor - 1} i_k \eta_k$. The first $\lfloor \lg t \rfloor$ elements of η together with $\lfloor \lg N \rfloor$ completely specify the roots of the Goppa polynomial $G(x)$, namely, $z_i = \eta_{\lfloor \lg N \rfloor} + \sum_{k=0}^{\lfloor \lg t \rfloor - 1} i_k \eta_k$.

The number of possible dyadic Goppa codes which can be produced by these algorithm is the same as the number of distinct essences of dyadic signatures corresponding to Cauchy matrices. This is about $\prod_{i=0}^{\lfloor \lg N \rfloor}(q-2^i)$. The algorithm also produces equivalent essences where the elements corresponding to the roots of the Goppa polynomial are only permuted. That leads to simple reordering of those roots. As the Goppa polynomial itself is defined by its roots regardless of their order, the actual number of possible Goppa polynomials is $\left(\prod_{i=0}^{\lfloor \lg N \rfloor}(q-2^i)\right)/(\lfloor \lg N \rfloor !)$.

2.4 Quasi-Dyadic Goppa Codes

A cryptosystem cannot be securely defined using completely dyadic Goppa codes which admit a parity-check matrix in Cauchy form. By solving the overdefined linear system $\frac{1}{H_{ij}} = z_i + L_j$ with nt equations and $n+t$ unknowns the Goppa polynomial $G(x)$ would be revealed immediately. Hence, Barreto and Misoczki propose using binary Goppa codes in quasi-dyadic form for cryptographic applications.

Definition 4. *A quasi-dyadic matrix is a possibly non-dyadic block matrix whose component blocks are dyadic submatrices.*

A quasi-dyadic Goppa code over $\mathbb{F}_p = \mathbb{F}_{2^s}$ for some s is obtained by constructing a dyadic parity-check matrix $H_{dyad} \in \mathbb{F}_q^{t \times n}$ over $\mathbb{F}_q = \mathbb{F}_{p^d} = \mathbb{F}_{2^m}$ of length $n = lt$ where n is a multiple of the desired number of errors t, and then computing the co-trace matrix $H'_{Tr} = Tr'(H_{dyad}) \in \mathbb{F}_p^{dt \times n}$. The resulting parity-check matrix for the quasi-dyadic Goppa code is a non-dyadic matrix composed of blocks of dyadic submatrices [16].

3 Scheme Definition of QD-McEliece

The main difference between the original McEliece scheme and the quasi-dyadic variant is the key generation algorithm 1 shown below. It takes as input the system parameters t, n, and k and outputs a binary Goppa code in quasi-dyadic form over a subfield $\mathbb{F}_p$ of $\mathbb{F}_q$, where $p = 2^s$ for some s, $q = p^d = 2^m$ for some d with $m = ds$. The code length n must be a multiple of t such that $n = lt$ for some $l > d$.

The key generation algorithm proceeds as follows. It first runs **Algorithm 1 in [16]** to produce a dyadic code C_{dyad} of length $N >> n$, where N is a multiple of t not exceeding the largest possible length $q/2$. The resulting code admits a $t \times N$ parity-check matrix $H_{dyad} = \left[B_0|\cdots|B_{N/t-1}\right]$ which can be viewed as a composition of N/t dyadic blocks B_i of size $t \times t$ each. In the next step the key generation algorithm uniformly selects l dyadic blocks of H_{dyad} of size $t \times t$ each. This procedure leads to the same result as puncturing the code C_{dyad} on a random set of block coordinates T_t of size $(N-n)/t$ first, and then permuting the remaining l blocks by changing their order. The block

Algorithm 1. QD-McEliece: Key generation algorithm

Input: Fixed common system parameters: t, $n = l \cdot t$, $k = n - dt$
Output: private key K_{pr}, public key K_{pub}

1: $(L_{dyad},\ G(x)\ H_{dyad},\ \eta) \leftarrow$ **Algorithm 1 in [16]** $(2^m, N, t)$, where $N >> n$, $N = l' \cdot t < q/2$
2: Select uniformly at random l distinct blocks $\left[B_{i_0}|\cdots|B_{i_{l-1}}\right]$ in any order from H_{dyad}
3: Select l dyadic permutations $\Pi^{j_0}, \cdots, \Pi^{j_{l-1}}$ of size $t \times t$ each
4: Select l nonzero scale factors $\sigma_0, \ldots, \sigma_{l-1} \in \mathbb{F}_p$. If $p = 2$, then all scale factors are equal to 1.
5: Compute $H = \left[B_{i_0}\Pi^{j_0}|\cdots|B_{i_{l-1}}\Pi^{j_{l-1}}\right] \in (\mathbb{F}_q^{t\times t})^l$
6: Compute $\Sigma = Diag(\sigma_0 I_t, \ldots, \sigma_{l-1} I_t) \in (\mathbb{F}_p^{t\times t})^{l\times l}$
7: Compute the co-trace matrix $H'_{Tr} = Tr'(H\Sigma) = Tr'(H)\Sigma \in (\mathbb{F}_p^{t\times t})^{l\times l}$
8: Bring H'_{Tr} in systematic form $\hat{H} = [Q|I_{n-k}]$, e.g. by means of Gaussian elimination
9: Compute the public generator matrix $\hat{G} = [I_k|Q^T]$
10: **return** $K_{pub} = (\hat{G}, t)$, $K_{pr} = (H_{dyad}, L_{dyad}, \eta, G(x), (i_0, \ldots, i_{l-1}), (j_0, \ldots, j_{l-1}), (\sigma_0, \ldots, \sigma_{l-1}))$

permutation sequence $(i_0, \ldots, i_l)$ is the first part of the trapdoor information. It can also be described as an $N \times n$ permutation matrix P_B. Then the selection and permutation of $t \times t$ blocks can be done by right-side multiplication $H_{dyad} \times P_B$. Further transformations performed to disguise the structure of the private code are dyadic inner block permutations.

Definition 5. *A* dyadic permutation Π^j *is a dyadic matrix whose signature is the j-th row of the identity matrix. A dyadic permutation is an involution, i.e.* $(\Pi^j)^2 = I$. *The j-th row (or equivalently the j-th column) of the dyadic matrix defined by a signature h can be written as* $\Delta(h)_j = h\Pi^j$.

The key generation algorithm first chooses a sequence of integers $(j_0, \ldots, j_{l-1})$ defining the positions of ones in the signatures of the l dyadic permutations. Then each block B_i is multiplied by a corresponding dyadic permutation Π^j to obtain a matrix H which defines a permutation equivalent code $\mathcal{C}_H$ to the punctured code $\mathcal{C}_{dyad}^{T_t}$. Since the dyadic inner-block permutations can be combined to an $n \times n$ permutation matrix $P_{dp} = Diag(\Pi^{j_0}, \cdots, \Pi^{j_{l-1}})$ we can write $H = H_{dyad} \cdot P_B \cdot P_{dp}$. The last transformation is scaling. Therefore, first a sequence $(\sigma_0, \ldots, \sigma_{l-1}) \in F_p$ is chosen, and then each dyadic block of H is multiplied by a diagonal matrix $\sigma_i I_t$ such that $H' = H \cdot \Sigma = H_{dyad} \cdot P_B \cdot P_{dp} \cdot \Sigma$. Finally, the co-trace construction derives from H' the parity-check matrix H'_{Tr} for a binary quasi-dyadic permuted subfield subcode over $\mathbb{F}_p$. Bringing H'_{Tr} in systematic form, e.g. by means of Gaussian elimination, we obtain a systematic parity-check matrix $\hat{H}$ for the public code. $\hat{H}$ is still a quasi-dyadic matrix composed of dyadic submatrices which can be represented by a signature of length t each and which are no longer associated to a Cauchy matrix. The generator matrix $\hat{G}$ obtained from $\hat{H}$ defines the public code $\mathcal{C}_{pub}$ of length n and dimension k over $\mathbb{F}_p$, while $\hat{H}$ defines a dual code $\mathcal{C}_{pub}^{\perp}$ of length n and dimension $k = n - dt$. The trapdoor information consisting of the essence η of the signature h_{dyad}, the sequence $(i_0, \ldots, i_{l-1})$ of blocks, the sequence $(j_0, \ldots, j_{l-1})$ of dyadic permutation identifiers, and the sequence of scale factors $(\sigma_0, \ldots, \sigma_{l-1})$ relates the public code defined by $\hat{H}$ with the private code defined by H_{dyad}. The public code defined by $\hat{G}$ admits a further parity-check matrix $V_{L^*,G} = \mathtt{vdm}(L^*, G(x)) \cdot \mathtt{Diag}(G(L_i^*)^{-1})$ where L^* is the permuted support obtained from L_{dyad} by $L^* = L_{dyad} \cdot P_B \cdot P_{db}$. Bringing $V_{L^*,G}$ in systematic form leads to the same quasi-dyadic parity-check matrix $\hat{H}$ for the code $\mathcal{C}_{pub}$. The matrix $V_{L^*,G}$ is permutation equivalent to the parity-check matrix $V_{L,G} = \mathtt{vdm}(L, G(x)) \cdot \mathtt{Diag}(G(L_i)^{-1})$ for the shortened private code $\mathcal{C}_{pr} = \mathcal{C}_{dyad}^{T_t}$ obtained by puncturing the large private code $\mathcal{C}_{dyad}$ on the set of block coordinates T_t. The support L for the code $\mathcal{C}_{pr}$ is obtained by deleting all components of L_{dyad} at the positions indexed by T_t. Classical irreducible Goppa codes use support sets containing all elements of $\mathbb{F}_q$. Thus, the support corresponding to such a Goppa code can be published while only the Goppa polynomial and the (support) permutation are parts of the secret key. In contrast, the support sets L and L^* for $\mathcal{C}_{pr}$ and $\mathcal{C}_{pub}$, respectively, are not full but just subsets of $\mathbb{F}_q$ where L^* is a permuted version of L. Hence, the support sets contain additional information and have to be kept secret.

The encryption algorithm of the QD-McEliece variant is the same as that of the original McEliece cryptosystem. First a message vector is multiplied by the systematic generator matrix $\hat{G}$ for the quasi-dyadic public code $\mathcal{C}_{pub}$ to obtain the corresponding codeword. Then a random error vector of length n and hamming weight at most t is added to the codeword to obtain a ciphertext.

The decryption algorithm of the QD-McEliece version is essentially the same as that of the classical McEliece cryptosystem. The following decryption strategies are conceivable.

Permute the ciphertext and undo the inner block dyadic permutation as well as the block permutation to obtain an extended permuted ciphertext of length N such that $ct_{perm} = ct \cdot P_B \cdot P_{dp}$. Then use the decoding algorithm of the large private code $\mathcal{C}_{dyad}$ to obtain the corresponding codeword. Multiplying ct_{perm} by the parity-check matrix for $\mathcal{C}_{dyad}$ yields the same syndrome as reversing the dyadic permutation and the block permutation without extending the length of the ciphertext and using a parity-check matrix for the shortened private code $\mathcal{C}_{pr}$. A better method is to decrypt the ciphertext directly using the equivalent parity-check matrix $V_{L^*,G}$ for syndrome computation. Patterson's decoding algorithm can be used to detect the error and to obtain the corresponding codeword. Since $\hat{G}$ is in systematic form, the first k bits of the resulting codeword correspond to the encrypted message.

3.1 Parameter Choice and Key Sizes

For an implementation on an embedded microcontroller the best choice is to use Goppa codes over the base field $\mathbb{F}_2$. In this case the matrix vector multiplication can be performed most efficiently. Hence, the subfield $\mathbb{F}_p = \mathbb{F}_{2^s}$ should be chosen to be the base field itself where $s = 1$ and $p = 2$. Furthermore, as the register size of embedded microcontrollers is restricted to 8 bits it is advisable to construct subfield subcodes of codes over $\mathbb{F}_{2^8}$ or $\mathbb{F}_{2^{16}}$. But the extension field $\mathbb{F}_{2^8}$ is too small to derive secure subfield subcodes from codes defined over it.

Over the base subfield $\mathbb{F}_2$ of $\mathbb{F}_{2^{16}}$ [16] suggests using the parameters summarized in Table 2.

Table 2. Suggested parameters for McEliece variants based on quasi-dyadic Goppa codes over $\mathbb{F}_2$

level	**t**	**n = l·t**	**k = n - m·t**	**key size ($m \cdot k$ bits)**
80	2^6	$36 \cdot 2^6 = 2304$	$20 \cdot 2^6 = 1280$	$20 \cdot 2^{10}$ bits $= 20$ Kbits
112	2^7	$28 \cdot 2^7 = 3584$	$12 \cdot 2^7 = 1536$	$12 \cdot 2^{11}$ bits $= 24$ Kbits
128	2^7	$32 \cdot 2^7 = 4096$	$16 \cdot 2^7 = 2048$	$16 \cdot 2^{11}$ bits $= 32$ Kbits
192	2^8	$28 \cdot 2^8 = 7168$	$12 \cdot 2^8 = 3072$	$12 \cdot 2^{12}$ bits $= 48$ Kbits
256	2^8	$32 \cdot 2^8 = 8192$	$16 \cdot 2^8 = 4096$	$16 \cdot 2^{12}$ bits $= 64$ Kbits

As the public generator matrix $\hat{G}$ is in systematic form, only its non-trivial part Q of length $n - k = m \cdot t$ has to be stored. This part consists of $m(l - m)$ dyadic submatrices of size $t \times t$ each. Storing only the t-length signatures of Q, the resulting public key size is $m(l - m)t = m \cdot k$ bits in size. Hence, the public key size is a factor of t smaller compared to the generic McEliece version where the key even in systematic form is $(n - k) \cdot k$ bits in size.

3.2 Security of QD-McEliece

A recent work [7] presents an efficient attack recovering the private key in specific instances of the quasi-dyadic McEliece variant. Due to the structure of a quasi-dyadic Goppa code additional linear equations can be constructed. These equations reduce the algebraic complexity of solving a multidimensional system of equations using Groebner bases [1]. In the case of the quasi-dyadic McEliece variant there are $l - m$ linear equations and $l - 1$ unknowns Y_i. The dimension of the vector space solution for the $Y_i's$ is $m - 1$. Once the unknowns Y_i are found all other unknowns X_i can be obtained by solving a system of linear equations. In our case there are 35 unknowns Y_i, 20 linear equations, and the dimension of the vector space solution for the $Y_i's$ is 15. The authors remark that the solution space is manageable in practice as long as $m < 16$. The attack was not successful with $m = 16$. Hence, up to now the McEliece variant using subfield subcodes over the base field of large codes over $\mathbb{F}_{2^{16}}$ is still secure.

3.3 Conversions for CCA2-Secure McEliece Variants

In [13] Kobara and Imai considered conversions for achieving CCA2-security in a restricted class of public-key cryptosystems. The authors reviewed these conversions for applicability to the McEliece public key cryptosystem and showed two of them to be convenient. These are Pointcheval's generic conversion [19] and Fujisaki-Okamoto's generic conversion [8]. Both convert partially trapdoor one-way functions (PTOWF)[1] to public key cryptosystems fulfilling the CCA2 indistinguishability.

The main disadvantage of both conversions is their high redundancy of data. Hence, Kobara and Imai developed three further specific conversions (α, β, γ) decreasing data overhead of the generic conversions even below the values of the original McEliece PKCs for large parameters. Their work shows clearly that the Kobara-Imai's specific conversion γ (KIC-γ) provides the lowest data redundancy for large parameters n and k. In particular, for parameters $n = 2304$ and $k = 1280$ used in this work for the construction of the quasi-dyadic McEliece-type PKC the data redundancy of the converted variant is even below that of the original scheme without conversion.

4 Implementational Aspects

In this section we discuss aspects of our implementation of the McEliece variant based on quasi-dyadic Goppa codes of length $n = 2304$, dimension $k = 1280$, and correctable number of errors $t = 64$ over the subfield $\mathbb{F}_2$ of $\mathbb{F}_{2^{16}}$ providing a security level of 80 bit. Target platform is the ATxmega256A1, a RISC microcontroller frequently used in embedded systems. This microcontroller operates at a clock frequency of up to 32 MHz, provides 16 Kbytes SRAM and 256 Kbytes Flash memory.

4.1 Field Arithmetic

To implement the field arithmetic on an embedded microcontroller most efficiently both representations of the field elements of $\mathbb{F}_q$, polynomial and exponential, should be precomputed and stored as *log-* and *antilog* table, respectively. Each table occupies $m \cdot 2^m$ bits of storage. Unfortunately, we cannot store the whole log- and antilog tables for $\mathbb{F}_{2^{16}}$ because each table is 128 Kbytes in size. Neither the SRAM memory of the ATXmega256A1 (16 Kbytes) nor the Flash memory (256 Kbytes) would be enough to implement the McEliece PKC when completely storing both tables. Hence, we make use of *tower field arithmetic*. Efficient algorithms for arithmetic over tower fields are proposed in [2], [17], and [18]. It is possible to view the field $\mathbb{F}_{2^{2k}}$ as a field extension of degree 2 over $\mathbb{F}_{2^k}$. Thus, we can consider the finite field $\mathbb{F}_{2^{16}} = \mathbb{F}_{(2^8)^2}$ as a tower of $\mathbb{F}_{2^8}$ constructed by an irreducible polynomial $p(x) = x^2 + x + p_0$ where $p_0 \in \mathbb{F}_{2^8}$. If β is a root of $p(x)$ in $\mathbb{F}_{2^{16}}$ then $\mathbb{F}_{2^{16}}$ can be represented as a two dimensional vector space over $\mathbb{F}_{2^8}$ and an element $A \in \mathbb{F}_{2^{16}}$ can be written as $A = a_1\beta + a_0$ where $a_1, a_0 \in \mathbb{F}_{2^8}$. To perform field arithmetic over $\mathbb{F}_{2^{16}}$ we store the log- and antilog tables for $\mathbb{F}_{2^8}$ and use them for fast mapping between exponential and polynomial representations

[1] A PTOWF is a function $F(x, y) \rightarrow z$ for which no polynomial time algorithm exists recovering x or y from their image z alone, but the knowledge of a secret enables a partial inversion, i.e., finding x from z.

of elements of $\mathbb{F}_{2^8}$. Each table occupies only 256 bytes, therefore both tables can smoothly be copied into the fast SRAM memory of the microcontroller at startup time. The next question is how to realize the mapping $\varphi\colon A \rightarrow (a_1, a_0)$ of an element $A \in \mathbb{F}_{2^{16}}$ to two elements $(a_1, a_0) \in \mathbb{F}_{2^8}$, and the inverse mapping $\varphi^{-1}\colon a_1, a_0 \rightarrow A$ such that $A = a_1\beta + a_0$. Both mappings can be implemented by means of a special transformation matrix and its inverse, respectively [18]. As the input and output for the McEliece scheme are binary vectors, field elements are only used in the scheme internally. Hence, we made an informed choice against the implementation of both mappings. Instead, we represent each field element A of $\mathbb{F}_{2^{16}}$ as a structure of two uint8_t values describing the elements of $\mathbb{F}_{2^8}$ and perform all operations on these elements directly.

4.2 Implementation of the QD-McEliece Variant

Encryption. The first step of the McEliece encryption is codeword computation. This is performed through multiplication of a plaintext p by the public generator matrix $\hat{G}$ which serves as public key. In our case the public generator matrix $\hat{G} = [I_k|M]$ is systematic. Hence, the first k bits of the codeword are the plaintext itself, and only the submatrix M of $\hat{G}$ is used for the computation of the parity-check bits. $M \in (\mathbb{F}_2^{t\times t})^{d\times(l-d)}$ can be considered as a composition of $d\cdot(l-d)$ dyadic submatrices $\Delta(h_{xy})$ of size $t \times t$ each, represented by a signature h_{xy} of length t each. It also can be seen as a composition of $l-d$ dyadic matrices $\Delta(h_x, t)$ of size $dt \times t$ each, represented by a signature of length $dt = n-k$ each.

$$M := \begin{pmatrix} \mathbf{m_{0,0}} & \cdots & \mathbf{m_{0,n\text{-}k\text{-}1}} \\ \vdots & \ddots & \vdots \\ m_{t-1,0} & \cdots & m_{t-1,n-k-1} \\ \hline \mathbf{m_{t,0}} & \cdots & \mathbf{m_{t,n\text{-}k\text{-}1}} \\ \vdots & \ddots & \vdots \\ m_{2t-1,0} & \cdots & m_{2t-1,n-k-1} \\ \hline \vdots & \ddots & \vdots \\ \hline \mathbf{m_{(l\text{-}d\text{-}1)t,0}} & \cdots & \mathbf{m_{(l\text{-}d\text{-}1)t,n\text{-}k\text{-}1}} \\ \vdots & \ddots & \vdots \\ m_{(l-d)t-1,0} & \cdots & m_{(l-d)t-1,n-k-1} \end{pmatrix} \begin{matrix} \left.\vphantom{\begin{matrix}a\\b\\c\end{matrix}}\right\} \Delta(h_0, t) \\ \left.\vphantom{\begin{matrix}a\\b\\c\end{matrix}}\right\} \Delta(h_1, t) \\ \\ \left.\vphantom{\begin{matrix}a\\b\\c\end{matrix}}\right\} \Delta(h_{l-d}, t) \end{matrix}$$

In both cases the compressed representation of M serving as public key K_{pub} for the McEliece encryption is

$$K_{pub} = [(m_{0,0}, \cdots, m_{0,n-k-1}), \cdots, (m_{(l-d-1)t,0}, \cdots, m_{(l-d)t-1,n-k-1})].$$

The public key is 2.5 KBytes in size and can be copied into the SRAM of the microcontroller at startup time for faster encryption. The plaintext $p = (p_0, \cdots, p_{t-1}, p_t, \cdots, p_{2t-1}, \cdots, p_{(l-d-1)t}, \cdots, p_{(l-d)t-1})$ is a binary vector of length $k = 1280 = 20 \cdot 64 = (l-d)t$. Hence, the codeword computation is done by adding the rows of M corresponding to the non-zero bits of p. As we do not store M but just its compressed representation, only the bits p_{it} for all $0 \leq i \leq (l-d-1)$ can be encrypted directly by adding the corresponding signatures. To encrypt all other bits of p the corresponding rows of M have to be reconstructed from K_{pub} first. The components $h_{i,j}$ of a dyadic matrix $\Delta(h,t)$ are normally computed as $h_{i,j} = h_{i \oplus j}$ which is a simple reordering of the elements of the signature h. Unfortunately, we cannot use this equation directly because the public key is stored as an array of $(n-k)(l-d)/8$ elements of type `uint8_t`. Furthermore, for every $t = 64$ bits long substring of the plaintext a different length-$(n-k)$ signature has to be used for encryption.

Decryption. For decryption we use the equivalent shortened Goppa code $\Gamma(L^*, G(x))$ defined by the Goppa polynomial $G(x)$ and a (permuted) support sequence $L^* \subset \mathbb{F}_{2^{16}}$. The support sequence consists of $n = 2304$ elements of $\mathbb{F}_{2^{16}}$ and is 4.5 KBytes in size. We store the support sequence in an array of type `gf16_t` and size 2304. The Goppa polynomial is a monic separable polynomial of degree $t = 64$. As t is a power of 2, the Goppa polynomial is sparse and of the form $G(x) = G_0 + \sum_{i=0}^{6} G_{2^i} x^{2^i}$. Hence, it occupies just $8 \cdot 16$ bits storage space. We can store both the support sequence and the Goppa polynomial in the SRAM of the microcontroller. Furthermore, we precompute the sequence $Diag(G(L_0^*)^{-1}, \ldots, G(L_{n-1}^*)^{-1})$ for the parity-check matrix $V_{t,n}(L^*, D)$. Due to the construction of the Goppa polynomial $G(x) = \prod_{i=0}^{t-1}(x - z_i)$ where $z_i = 1/h_i + \omega$ with a random offset ω, the following holds for all $G(L_{jt+i}^*)^{-1}$.

$$G(L_{jt+i}^*)^{-1} = \prod_{r=0}^{t-1}(L_{jt+i}^* + z_r)^{-1} = \prod_{r=0}^{t-1}(1/h_{jt+i}^* + 1/h_r + 1/h_0)^{-1} = \prod_{r=0}^{t-1} h_{jt+r}^* = \prod_{r=jt}^{jt+t-1} h_r^*$$

h^* denotes a signature obtained by puncturing and permuting the signature h for the large code C_{dyad} such that $h^* = h \cdot P$ where P is the secret permutation matrix. Hence, the evaluation of the Goppa polynomial on any element of the support block $(L_{jt}^*, \ldots, L_{jt+t-1}^*)$ where $j \in \{0, \ldots, l-1\}$, $i \in \{0, \ldots, t-1\}$ leads to the same result. For this reason, only $n/t = l = 36$ values of type `gf16_t` need to be stored. Another polynomial we need for Patterson's decoding algorithm is $W(x)$ satisfying $W(x)^2 \equiv x \mod G(x)$. As the Goppa polynomial $G(x)$ is sparse, the polynomial $W(x)$ is also sparse and of the form $W(x) = W_0 + \sum_{i=0}^{5} W_{2^i} x^{2^i}$. $W(x)$ occupies $7 \cdot 16$ bits storage space.

Syndrome Computation. The first step of the decoding algorithm is the syndrome computation. Normally, the syndrome computation is performed through solving the equation $S_c(x) = S_e(x) \equiv \sum_{i \in E} \frac{1}{x - L_i^*} \mod G(x)$ where E denotes a set of error positions. The polynomial $\frac{1}{x - L_i^*}$ satisfies the equation

$$\frac{1}{x - L_i^*} \equiv \frac{1}{G(L_i^*)} \sum_{j=s+1}^{t} G_j {L_i^*}^{j-s-1} \mod G(x), \; \forall 0 \leq s \leq t-1 \qquad (1)$$

The coefficients of this polynomial are components of the $i-th$ column of the Vandermonde parity-check matrix for the Goppa code $\Gamma(G(x), L^*)$. Hence, to compute the syndrome of a ciphertext c we perform the on-the-fly computation of the rows of the parity-check matrix. As the Goppa polynomial is a sparse monic polynomial of the form $G(x) = G_0 + \sum_{i=0}^{6} G_{2^i} x^{2^i}$ with $G_{64} = 1$, we can simplify the Equation 1, and thus, reduce the number of operations needed for the syndrome computation. The main advantage of this computation method is that it is performed on-the-fly such that no additional storage space is required. To speed-up the syndrome computation the parity-check matrix can be precomputed at the expense of additional $n(n-k) = 288$ KBytes memory. As the size of the Flash memory of ATxmega256A1 is restricted to 256 Kbytes, we cannot store the whole parity-check matrix. It is just possible to store 52 coefficients of each syndrome polynomial at most, and to compute the remaining coefficients on-the-fly. A better possibility is to work with the systematic quasi-dyadic public parity-check matrix $\hat{H} = [Q^T | I_{n-k}]$ from which the public generator matrix $\hat{G} = [I_k | Q]$ is obtained. To compute a syndrome the vector matrix multiplication $\hat{H} \cdot c^T = c \cdot \hat{H}^T$ is performed. For the transpose parity-check matrix $\hat{H}^T = [Q^T | I_{n-k}]^T$ holds, where Q is the quasi-dyadic part composed of dyadic submatrices. Hence, to compute a syndrome we proceed as follows. The first k bits of the ciphertext are multiplied by the part Q which can be represented by the signatures of the dyadic submatrices. The storage space occupied by this part is 2.5 KBytes. The multiplication is performed in the same way as encryption of a plaintext (see Section 4.2) and results in a binary vector s' of length $n-k$. The last $n-k$ bits of the ciphertext are multiplied by the identity matrix I_{n-k}. Hence, we can omit the multiplication and just add the last $n-k$ bits of c to s'. To obtain a syndrome for the efficiently decodable code the vector s' first has to be multiplied by a scrambling matrix S. We stress that this matrix brings the Vandermonde parity-check matrix for the private code $\Gamma(G(x), L^*)$ in systematic form which is the same as the public parity-check matrix. Hence, S has to be kept secret. We generate S over $\mathbb{F}_2$ and afterwards represent it over $\mathbb{F}_{2^{16}}$. Thus, the multiplication of a binary vector s' by S results in a polynomial $S_c(x) \in \mathbb{F}_{2^{16}}[x]$ which is a valid syndrome. The matrix S is 128 KBytes in size and can be stored in the Flash memory of the microcontroller. The next step, which is computing the error locator polynomial $\sigma(x)$, is implemented straightforward using Patterson's algorithm.

Searching for roots of $\sigma(x)$. The last and the most computationally expensive step of the decoding algorithm is the search for roots of the error locator polynomial $\sigma(x)$. For this purpose, we first planed to implement the Berlekamp trace algorithm [3] which is known to be one of the best algorithm for finding roots of polynomials over finite fields with small characteristic. Considering the complexity of this algorithm we found out that it is absolutely unsuitable for punctured codes over a large field, because of the required computation of traces and gcds. The next root finding method we analyzed is the Chien search [5] which has a theoretical complexity of $\mathcal{O}(n \cdot t)$ if $n = 2^m$. The Chien search scans automatically all $2^m - 1$ field elements, in a more sophisticated manner than the simple polynomial evaluation method. Unfortunately, in our case $n << 2^m$ such that the complexity of the Chien search becomes $\mathcal{O}(2^{16} \cdot t)$ which is enormous compared to the complexity of the simple polynomial evaluation method. Another disadvantage of both the Berlekamp trace algorithm and the Chien search is that after root extraction the found roots have to be located within the support sequence to identify error positions. That is not the case when evaluating the error locator polynomial on the support set directly. In this case we know the positions of the elements L_i^* and can correct errors directly by flipping the corresponding bits in the ciphertext. The only algorithm which actually decreases the computation costs of the simple evaluation method in the case of punctured codes is the Horner scheme [12]. The complexity of the Horner scheme does not depend on the extension degree of the field but on the number of possible root candidates, which is n. In addition, as the Horner scheme evaluates the error locator polynomial on the support set L^*, the root positions within L^* are known such that errors can be corrected more efficiently. Hence, we have implemented this root finding algorithm. After a root L_i^* of $\sigma(x)$ has been found we perform the polynomial division of $\sigma(x)$ by $(x - L_i^*)$. We observed that the polynomial division by $(x - L_i^*)$ can be performed sequentially reusing values computed in previous iteration steps. In the first step we compute the coefficient y_{t-2} of the searched polynomial $y(x)$. In every iteration step j we use the previous coefficient y_{t-j+1} to compute $y_{t-j} = y_{t-j+1}L_i^* + \sigma_{t-j}$. The whole procedure requires $t-3$ multiplications and $t-2$ additions to divide a degree-t polynomial by $x - L_i^*$. The main advantage of performing polynomial division each time a root has been found is that the degree of the error locator polynomial decreases. Hence, the next evaluation steps require less operations.

4.3 Implementation of the KIC-γ

For the implementation of Kobara-Imai's specific conversion γ [13] two parameters have to be chosen: the length of the random value r and the length of the public constant $Const$. The length of r should be equal to the output length of the used hash function. Here we choose the Blue Midnight Wish (BMW) hash function, because of the availability of a fast assembly implementation. As we have $|r| = 256$ and $|Const| = 160$, the message to be encrypted should be of the length $|m| \geq \lfloor \log_2 \binom{n}{t} \rfloor + k + |r| - |Const| = 1281$ bits. Hence, we encrypt messages of length 1288 bits = 161 bytes. In this case the data redundancy is even below of that of the McEliece scheme without conversion: $1288/2304 \leq 1280/2304$.

The first steps of the KIC-γ encryption function are the generation of a random seed r for the function $Gen(r)$, as well as the one-time-pad encryption of the message m padded with the public constant $Const$ and the output of $Gen(r)$. The result is a $1288 + 160 = 1448$ bits = 181 bytes value y_1. In the next step the hash value of y_1 is added to the random seed r by the xor operation to obtain the value y_2. $k = 1280$ bits from $(y_2||Y_1)$ are used as input for McEliece and from the remaining 424 bits the error vector is constructed by the constant weight encoding function $Conv$[22,11].

To decrypt a ciphertext the KIC-γ first stores the first two bytes of the ciphertext in y_5. Then it calls the McEliece decryption function which returns the encrypted plaintext y_3 and the error vector $\delta_j = i_j - i_{j-1} - 1$ where i_r denote the error positions. To obtain part y_4 from the error vector constant weight decoding function is used. Now $(y_2||y_1) = (y_5||y_4||y_3)$ is known and the message m can be obtained.

5 Results

This section presents the results of our implementation of the McEliece variant based on [2304, 1280, 129] quasi-dyadic Goppa codes providing an 80-bit security level for the 8-bits AVR microcontroller. As we use a systematic generator matrix for the Goppa code, we also implemented Kobara-Imai's specific conversion γ developed for CCA2-secure McEliece variants. Due to the parameters chosen for KIC-γ the actual length of the message to be encrypted increases to 1288 bytes while the ciphertext length increases to 2312 bytes. Table 3 summarizes the sizes of all parameters being precomputed and used for the encryption and decryption algorithms.

Table 3. Sizes of tables and values in memory

	Parameter	**Size**
QD-McEliece encryption	K_{pub}	2560 bytes
QD-McEliece decryption	log table for $\mathbb{F}_{2^8}$	256 bytes
	antilog table for $\mathbb{F}_{2^8}$	256 bytes
	Goppa polynomial $G(x)$	16 bytes
	Polynomial $W(x)$	14 bytes
	Support sequence L^*	4608 bytes
	Array with elements $1/G(L_i^*)$	72 bytes
	Matrix S	131072 bytse
KIC-γ	Public constant $Const$	20 bytes

Except for the matrix S which is used only within the syndrome computation method with precomputation, all precomputed values can be copied into the faster SRAM of the microcontroller at startup time resulting in faster encryption and decryption. The performance results of our implementation were obtained from AVR Studio in version 4.18. Table 4 summarizes the clock cycles needed for specific operations and sub-operations for the conversion and encryption of a message. Note that we used fixed random values for the implementation of KIC-γ. The encryption of a 1288 bits message requires 6,358,952 cycles. Hence, when running at 32 MHz, the encryption takes about 0.1987 seconds while the throughput is 6482 bits/second.

Table 4. Performance of the QD-McEliece encryption including KIC-γ on the AVR μC ATxmega256@32 MHz

Operation	Sub-operation	Clock cycles
Hash		15,083
CWencoding		50,667
Other		8,927
QD-McEliece encryption	Vector-matrix multiplication	6,279,662
	Add error vector	4,613

Table 5. Performance of the QD-McEliece decryption on the AVR μC ATxmega256@ 32 MHz

Operation	Sub-operation	Clock cycles
QD-McEliece decryption	Syndrome computation on-the-fly	25,745,284
	Syndrome computation with S	9,118,828
	Syndrome inversion	3,460,823
	Computing $\sigma(x)$	1,625,090
	Error correction (HS)	31,943,688
	Error correction (HS with PD)	19,234,171
CWdecoding		61,479
Hash		15,111
Other		19,785

Table 5 presents the results of the operations and sub-operations of the QD-McEliece decryption function including KIC-γ.

Table 5 shows clearly that the error correction using the Horner scheme with polynomial division (PD) is about 40% faster then the Horner scheme without polynomial division. Considering the fact that the error correction is one of the most computationally expensive functions within the decryption algorithm the polynomial division provides a significant speed gain for this operation. In the case that the syndrome is computed using the precomputed matrix S and the error correction is performed using the Horner scheme with polynomial division decoding of a 2312 bits ciphertext requires 33,535,287 cycles. Running at 32 MHz the decryption takes 1.0480 seconds while the ciphertext rate is 2206 bits/second[2]. Decryption with the on-the-fly syndrome computation method takes 50,161,743 cycles. Hence, running at 32 MHz the decryption of a ciphertext takes 1.5676 seconds in this case while the ciphertext rate is 1475 bits/second. Although the on-the-fly decryption is about 1.5 times slower, no additional Flash memory is required so that a migration to cheaper devices is possible.

[2] Chiphertext rate denotes number of ciphertext bits processed per second.

Table 6 summarizes the resource requirements of our implementation. The third column of the table refers to the decryption method with precomputed matrix S, the fourth to the on-the-fly syndrome decoding method. For a comparison we also provide the resource requirements for the McEliece version based on [2048,1751,55]-Goppa codes [6].

Table 6. Resource requirements of QD-McEliece on the AVR μC ATxmega256@ 32 MHz

	Operation	Flash memory	External memory
QD-McEliece with KIC-γ	Encryption	11 Kbyte	–
	Decryption (with S)	156 Kbyte	–
	Decryption (on-the-fly)	21 Kbyte	–
McEliece[6]	Encryption	684 byte	438 Kbyte
	Decryption	130.4 Kbyte	–

As we can see, the memory requirements of the quasi-dyadic encryption routine including KIC-γ are minimal because of the compact representation of the public key. Hence, much cheaper microcontrollers such as ATxmega32 with only 4 Kbytes SRAM and 32 Kbytes Flash ROM could be used for encryption. In contrast, the implementation of the original McEliece version even requires 438 Kbyte external memory. The implementation of the decryption method with on-the-fly syndrome computation could also be migrated to a slightly cheaper microcontroller such as ATxmega128 with 8 Kbyte SRAM and 128 Kbyte Flash memory.

Table 7 gives a comparison of our implementation of the quasi-dyadic McEliece variant including KIC-γ with the implementation of the original McEliece PKC and the implementations of other public-key cryptosystems providing an 80-bit security level. RSA-1024 and ECC-160 [10] were implemented on a Atmel ATmega128 microcontroller at 8 MHz while the original McEliece version was implemented on a Atmel ATxmega192 microcontroller at 32 MHz. For a fair comparison with our implementation running at 32 MHz, we scale timings at lower frequencies accordingly.

Although we additionally include KIC-γ in the quasi-dyadic McEliece encryption, we were able to out perform both, the original McEliece version and ECC-160, in terms of number of operations per second. In particular, the throughput of our implementation significantly exceeds that of ECC-160.

Unfortunately, we could not out perform the original McEliece scheme neither in throughput nor in number of operations per second for the decryption. The reason is that the original McEliece version is based on Goppa codes with much smaller number of errors $t = 27$. Due to this fact, this McEliece version works with polynomials of smaller degree such that most operations within the decoding algorithm can be performed more efficiently. Another disadvantage of our implementation is that all parameters are defined over the large field $\mathbb{F}_{2^{16}}$.

Table 7. Comparison of the quasi-dyadic McEliece variant including KIC-γ (n'=2312, k'=1288, t=64) with original McEliece PKC (n=2048, k=1751, t=27), ECC-P160, and RSA-1024

Method	**Time** sec	**Throughput** bits/sec
QD-McEliece encryption	0.1987	6482
QD-McEliece decryption (with S)	1.0480	1229
QD-McEliece decryption (on-the-fly)	1.5676	822
McEliece encryption [6]	0.4501	3889
McEliece decryption [6]	0.6172	2835
ECC-160 [10]	0.2025	790
RSA-1024 $2^{16}+1$ [10]	0.1075	9525
RSA-1024 w. CRT [10]	2.7475	373

As we could not store the log- and antilog tables for this field in the Flash memory, we had to implement the tower field arithmetic which significantly reduces performance. For instance, one multiplication over a tower $\mathbb{F}_{(2^8)^2}$ involves 5 multiplications over the subfield $\mathbb{F}_{2^8}$. Hence, much more arithmetic operations have to be performed to decrypt a ciphertext.

Nevertheless, the decryption function is still faster than the RSA-1024 private key operation and exceeds the throughput of ECC-160. Furthermore, although slower, the on-the-fly decoding algorithm requires 81% less memory compared to the original McEliece version such that migration to cheaper devices is possible.

6 Conclusion and Further Research

In this work we have implemented a McEliece variant based on quasi-dyadic Goppa codes on a 8-bits AVR microcontroller. The family of quasi-dyadic Goppa codes offers the advantage of having a compact and simple description. Using quasi-dyadic Goppa codes the public key for the McEliece encryption is significantly reduced. Furthermore, we used a generator matrix for the public code in systematic form resulting in an additional key reduction. As a result, the public key size is a factor t less compared to generic Goppa codes used in the original McEliece PKC. Moreover, the public key can be kept in this compact size not only for storing but for processing as well. However, the systematic coding necessitates further conversion to protect the message. Without any conversions the encrypted message would be revealed immediately from the ciphertext. Hence, we have implemented Kobara-Imai's specific conversion γ: a conversion scheme developed specially for CCA2 secure McEliece variants.

Our implementation out performs the implementations of the original McEliece PKC and ECC-160 in encryption. In particular, the quasi-dyadic McEliece encryption is 2.3 times faster than the original McEliece PKC and exceeds the throughput of both, the original McEliece PKC and ECC-160, by 1.7 and 8.2

times, respectively. In addition, our encryption algorithm requires 96,7% less memory compared to the original McEliece version and can be migrated to much cheaper devices.

The performance of the McEliece decryption algorithm is closely related to the number of errors added within the encryption. In our case the number of errors is 64 which is 2.4 times greater compared to the original McEliece PKC. Hence, the polynomials used are huge and the parity-check matrix is too large to be completely precomputed and stored in the Flash memory. In addition, the error correction requires more time because a polynomial of degree 64 has to be evaluated. We showed in Section 4.2 that none of the frequently used error correction algorithms, such as the Berlekamp trace algorithm and the Chien search, are suitable for punctured and shortened codes obtained from codes over very large fields. Furthermore, the tower field arithmetic significantly reduces the performance of the decoding algorithm. Nevertheless, the decryption algorithms with precomputation and on-the-fly computation are 2.6 and 1.8 times faster than the RSA-1024 private key operation and exceed the throughput of ECC-160. Furthermore, although slower, the on-the-fly decoding algorithm requires 81% less memory compared to the original McEliece version such that migration to cheaper devices is possible.

Acknowledgement. I would like to thank Olga Paustjan and Paulo Barreto for fruitful discussions. Special thanks to an anonymous reviewer for many useful comments.

References

1. Adams, W., Loustaunau, P.: An Introduction to Gröbner Bases, vol. 3 (1994)
2. Afanasyev, V.B.: On the complexity of finite field arithmetic. In: Fifth Joint Soviet-Swedish Intern. Workshop Information Theory, pp. 9–12 (January 1991)
3. Berlekamp, E.R.: Factoring polynomials over large finite fields. Mathematics of Computation 24(111), 713–715 (1970)
4. Bernstein, D.J., Lange, T., Peters, C.: Attacking and Defending the McEliece Cryptosystem. In: Buchmann, J., Ding, J. (eds.) PQCrypto 2008. LNCS, vol. 5299, pp. 31–46. Springer, Heidelberg (2008)
5. Chien, R.: Cyclic decoding procedure for the bose-chaudhuri-hocquenghem codes. IEEE Transactions on Information Theory IT-10(10), 357–363 (1964)
6. Eisenbarth, T., Güneysu, T., Heyse, S., Paar, C.: MicroEliece: McEliece for Embedded Devices. In: Clavier, C., Gaj, K. (eds.) CHES 2009. LNCS, vol. 5747, pp. 49–64. Springer, Heidelberg (2009)
7. Faugère, J.-C., Otmani, A., Perret, L., Tillich, J.-P.: Algebraic Cryptanalysis of McEliece Variants with Compact Keys. In: Gilbert, H. (ed.) EUROCRYPT 2010. LNCS, vol. 6110, pp. 279–298. Springer, Heidelberg (2010)
8. Fujisaki, E., Okamoto, T.: Secure Integration of Asymmetric and Symmetric Encryption Schemes. In: Wiener, M. (ed.) CRYPTO 1999. LNCS, vol. 1666, pp. 537–554. Springer, Heidelberg (1999)
9. Goppa, V.D.: A New Class of Linear Correcting Codes. Probl. Pered. Info. 6(3), 24–30 (1970)

10. Gura, N., Patel, A., Wander, A., Eberle, H., Shantz, S.C.: Comparing Elliptic Curve Cryptography and RSA on 8-Bit CPUs. In: Joye, M., Quisquater, J.-J. (eds.) CHES 2004. LNCS, vol. 3156, pp. 119–132. Springer, Heidelberg (2004)
11. Heyse, S.: Low-Reiter: Niederreiter Encryption Scheme for Embedded Microcontrollers. In: Sendrier, N. (ed.) PQCrypto 2010. LNCS, vol. 6061, pp. 165–181. Springer, Heidelberg (2010)
12. Horner, W.G.: A new method of solving numerical equations of all orders, by continuous approximation. Philosophical Transactions of the Royal Society of London 109, 308–335 (1981)
13. Kobara, K., Imai, H.: Semantically Secure McEliece Public-key Cryptosystems-conversions for McEliece PKC. In: Kim, K.-c. (ed.) PKC 2001. LNCS, vol. 1992, pp. 19–35. Springer, Heidelberg (2001)
14. MacWilliams, F.J., Sloane, N.: The Theory of Error-Correcting Codes. North-Holland Mathematical Library, vol. 16 (1997)
15. McEliece, R.J.: A public-key cryptosystem based on algebraic coding theory. DSN Progress Report 42-44, Jet Propulsion Laboratory (January-February 1978)
16. Misoczki, R., Barreto, P.S.: Compact McEliece Keys from Goppa Codes. In: Jacobson Jr., M.J., Rijmen, V., Safavi-Naini, R. (eds.) SAC 2009. LNCS, vol. 5867, pp. 376–392. Springer, Heidelberg (2009)
17. Morii, M., Kasahara, M.: Efficient construction of gate circuit for computing multiplicative inverses over $gf(2^m)$. Transactions of the IEICE E72, 37–42 (1989)
18. Paar, C.: Efficient VLSI Architectures for Bit-Parallel Computation in Galois Fields. Dissertation, Institute for Experimental Mathematics, Universität Essen (1994)
19. Pointcheval, D.: Chosen-Ciphertext Security for Any One-Way Cryptosystem. In: Imai, H., Zheng, Y. (eds.) PKC 2000. LNCS, vol. 1751, pp. 129–146. Springer, Heidelberg (2000)
20. Preneel, B., Bosselaers, A., Govaerts, R., Vandewalle, J.: A software implementation of the McEliece public-key cryptosystem. In: Proceedings of the 13th Symposium on Information Theory in the Benelux, Werkgemeenschap voor Informatieen Communicatietheorie, pp. 119–126. Springer, Heidelberg (1992)
21. Prometheus. Implementation of McEliece cryptosystem for 32-bit microprocessors (c-source), `http://www.eccpage.com/`
22. Sendrier, N.: Encoding information into constant weight words. In: IEEE Conference, ISIT 2005, pp. 435–438 (September 2005)
23. Shor, P.W.: Polynomial-time algorithms for prime factorization and discrete logarithms on a quantum computer. SIAM J. Comput. 26(5), 1484–1509 (1997)

Efficient Threshold Encryption from Lossy Trapdoor Functions*

Xiang Xie[1,2], Rui Xue[1], and Rui Zhang[1]

[1] The State Key Laboratory of Information Security
Institute of Software, Chinese Academy of Sciences
[2] Graduate University of Chinese Academy of Sciences
{xiexiang,rxue,r-zhang}@is.iscas.ac.cn

Abstract. This paper discusses the problem of building secure threshold public key encryption (TPKE) schemes from lossy trapdoor functions, which can in turn be built from a number of assumptions, e.g. lattices. Our methodology is generic and our concrete instantiation is more efficient than previous construction.

Keywords: threshold public key encryption, lossy trapdoor functions, lattices.

1 Introduction

A threshold public key encryption (TPKE) scheme [8,9,13,17] distributes the decryption power among n decryption servers so that k servers or more working together can do the decryption. Not only TPKE itself provides useful functionalities, it is also an important building block for other cryptographic primitives, such as mix-net (anonymous channel) [6], public key encryption with non-interactive opening [7,15,16]. Currently popular TPKE schemes are often based on discrete-logarithm type or RSA type assumptions [9,29,4,3]. It is worth emphasizing that most known TPKE schemes become insecure in the existence of quantum attackers though they achieve good efficiency. Therefore, it is both necessary and challenging to build alternative efficient constructions of TPKE from other assumptions valid even against quantum attackers.

In this paper, we give a new construction of TPKE from lossy trapdoor functions [23], which can be implemented using a number of assumptions including lattices. Instantiating our methodology, we obtain a TPKE scheme, which is (slightly) more efficient than the only known TPKE based on lattices [2].

Let us briefly explain our ideas here. First recall that in [10], Dodis and Katz proposed a generic construction of TPKE based on so-called multiple encryption techniques [31,10]. However, a prerequisite of their construction is that each encryption component must be adaptively chosen-tag secure. We remark that no efficient scheme was known previously on how to build such primitives using

* This work is partially supported by the Fund of the National Natural Science Foundation of China under Grants No. 60873260 and the National High Technology Research and Development Program of China under Grant No. 2009AA01Z414.

B.-Y. Yang (Ed.): PQCrypto 2011, LNCS 7071, pp. 163–178, 2011.

lattices. However, we revisit their proof, and find that this strong requirement is not necessary. In fact, a strictly weaker component, namely, public key encryption secure against selective-chosen tag attack (sTag), is already sufficient.

Next, towards our goal of TPKE resilient to quantum attackers, we demonstrate a public key encryption from lossy trapdoor functions meets this desirable requirement [23], namely, security against selective-tag and adaptive chosen ciphertext attack. Note that in their paper [23], Peikert and Waters proved that the scheme is passively chosen ciphertext secure (CCA1). However, our result shows actually it also provides adaptive chosen ciphertext security under a selective-tag setting. Combine the above two strategies together, we achieve an efficient construction for TPKE from lossy trapdoor functions.

Instantiating our generic construction with lattices, we obtain a concrete TPKE scheme under hardness of the shortest independent vector problem (SIVP). We then compare our TPKE with some well-known schemes in the literature and conclude our TPKE scheme is efficient.

1.1 Related Work

Desmedt first introduced the concept of threshold encryption [8]. Dodis and Katz gave a generic construction of threshold decryption schemes [10]. The first practical TPKE with chosen ciphertext security (CCA) [21,24,12] was proposed by Shoup and Gennaro [29]. Canetti and Goldwasser [4] gave the first TPKE in the standard model, but their security model was weaker than [29]. Boneh, Boyen and Halevi proposed the first TPKE scheme with threshold decryption without random oracles [3], but their scheme, and the subsequent schemes all require pairings [1]. The first threshold decryption scheme based on lattice was proposed by Bendlin and Damgård [2].

1.2 Our Contributions

First, we show that a generic CCA secure threshold public key encryption scheme in the standard model can be derived from tag-based public key encryption schemes which only require weak security. Dodis and Katz proposed a generic CCA secure construction whose components are tag-based public key encryption schemes. While the our constructions are selective-tag CCA security for the encryption components. We improve their results.

Next, we revisit the security of a CCA1-secure public key scheme from lossy trapdoor functions [23], and prove that this scheme also provides selective-tag security against CCA attacks. Finally, combine the two results together, we design an efficient TPKE scheme from lossy trapdoor functions, which can be implemented using lattices. We also give comparisons among known TPKE schemes.

2 Preliminaries

In this section, we review some useful notations and definitions.

Notations. If x is a string, then $|x|$ denotes its length, while if S is a set then $|S|$ denotes its size. If S is a set then $s \leftarrow S$ denotes the operation of picking an element s of S uniformly at random. Unless otherwise indicated, algorithms are randomized. We write $z \leftarrow \mathcal{A}^{\mathcal{O}_1,\mathcal{O}_2,\ldots}(x, y, \ldots)$ to indicate that $\mathcal{A}$ is an algorithm with inputs $x, y, \ldots$ and access to oracles $\mathcal{O}_1, \mathcal{O}_2, \ldots$ and let z be the output.

Let X and Y be two random variables over some set S. The *statistic distance* between X and Y is defined as $\Delta(X, Y) = \frac{1}{2}\sum_{s\in S}\big|\Pr[X = s] - \Pr[Y = s]\big|$. We say X and Y are statistically indistinguishable if $\Delta(X, Y)$ is negligible. It's routine to see that statistical indistinguishability implies computational indistinguishability.

2.1 Tag-Based Encryption

Informally in a tag-based encryption scheme, the encryption and decryption operations take an additional "tag". we review selective-tag security against chosen-ciphertext attacks [19] of tag-based encryption. More formally, a tag-based encryption scheme $\mathcal{TBE}$=(TGen, TEnc, TDec) consists of three algorithms. TGen takes as input a security parameter λ, and outputs a key pair (pk, dk); TEnc takes as input a public key pk, a plaintext m, a tag τ, and outputs a ciphertext c; TDec takes as input a secret key dk, a ciphertext c, a tag τ, and outputs a plaintext m.

For correctness, we require that $\forall\lambda$, τ and m, $(pk, dk) \leftarrow \mathtt{TGen}(\lambda)$, we have $\mathtt{TDec}(dk, \tau, \mathtt{TEnc}(pk, \tau, m)) = m$. We now define the selective-tag CCA security. Consider the following experiment between a chlangger and an adversary $\mathcal{A}$:

$$
\begin{array}{l}
\textbf{Experiment } \mathbf{Exp}_{\mathcal{TBE},\mathcal{A}}^{tbe\text{-}stag\text{-}cca}(\lambda) \\
\quad (\tau^*, St_0) \leftarrow \mathcal{A}(\lambda, init) \\
\quad (pk, dk) \leftarrow \mathtt{TGen}(\lambda) \\
\quad (m_0, m_1, St) \leftarrow \mathcal{A}^{\mathtt{DEC}(dk,\cdot)}(find, pk, St_0) \\
\quad b \leftarrow \{0,1\},\ c^* \leftarrow \mathtt{TEnc}(pk, \tau^*, m_b) \\
\quad b' \leftarrow \mathcal{A}^{\mathtt{DEC}(dk,\cdot)}(guess, c^*, St) \\
\quad \text{If } b' = b \text{ then return 1 else return 0}
\end{array}
$$

where the oracle $\mathtt{DEC}(dk, (c, \tau))$ returns $m \leftarrow \mathtt{TDec}(dk, \tau, c)$ with the restriction that $\mathcal{A}$ is not allowed to query oracle $\mathtt{DEC}(dk, \cdot)$ for challenge tag τ^*. Both messages must be of the same size. We define the advantage of $\mathcal{A}$ in the above experiment as $\mathbf{Adv}_{\mathcal{TBE},\mathcal{A}}^{tbe\text{-}stag\text{-}cca}(\lambda) = \left|\Pr\left[\mathbf{Exp}_{\mathcal{TBE},\mathcal{A}}^{tbe\text{-}stag\text{-}cca}(\lambda) = 1\right] - \frac{1}{2}\right|$. A tag-based encryption scheme $\mathcal{TBE}$ is said to be selective-tag secure against chosen-ciphertext attacks if the advantage function is negligible for all probabilistic polynomial time (PPT) $\mathcal{A}$. In the above experiment $\mathcal{A}$ is allowed to make decryption queries for any $\tau \neq \tau^*$. This includes queries for the challenge ciphertext c^* with tags $\tau \neq \tau^*$. The challenge tag τ^* has to be output by $\mathcal{A}$ before seeing the public key. Tags in public key encryption have been considered in [28]. We briefly review the definition of full-tag CCA security in [28].

Full-tag CCA security for TBE schemes considered in [28] are almost the same as selective-tag CCA security, except that $\mathcal{A}$ is allowed to choose τ^* at the end of its *find* stage, possibly depending on the public key and its queries. In the *guess* stage, $\mathcal{A}$ is allowed to ask any decryption queries $(\tau, c) \neq (\tau^*, c^*)$. In particular,

$\mathcal{A}$ can make decryption queries for (τ^*, c) with $c \neq c^*$, which is not allowed to ask in selective-tag CCA security. This is a stronger security requirement than selective-tag CCA security. In [10] the components for a threshold scheme should achieve full-tag CCA security, while we point that for our construction, the weaker definition is sufficient.

2.2 Secret Sharing

A secret sharing scheme consists of two algorithms $\mathcal{SS} = (\texttt{Share}, \texttt{Rec})$. $\texttt{Share}(\cdot)$ takes as input a message m, and outputs n secret shares $s_1, ..., s_n$. $\texttt{Rec}(\cdot)$ is a deterministic algorithm which takes as input n shares $s_1, ..., s_n$, and outputs message m or $\perp$. The security of a secret sharing scheme is quantified by several thresholds, we use the definition used in [10].

1. t_p – the *privacy* threshold. Determines the maximum number of shares which reveal "no information" about the massage (the distribution of these shares does not depend on the message.).
2. t_f – the *fault-tolerance* threshold. Determines the minimum number of correct shares which suffice to recover the message, when the other shares are missing.
3. t_r – the *robustness* threshold. Determines the minimum number of correct shares which suffice to recover the message, when the other shares are adversarially set.
4. t_s – the *soundness* threshold. Determines the minimum number of correct shares which ensure that it is impossible to recover an incorrect message $m' \notin \{m, \perp\}$, when the other shares are adversarially set.

The above thresholds must satisfy $t_p + 1 \leq t_f \leq t_r \leq n$ and $t_s \leq t_r$.The security properties corresponding to the thresholds above can all be formalized in a straightforward way. We say the above scheme a (t_p, t_f, t_r, t_s, n)-secure secret sharing scheme. We use $\mathcal{SS}$ instead of a share verification algorithm proposed in [29]. Shamir's scheme [27] is a classical example achieves information-theoretic privacy with $t_f = t_p + 1$. In [10] they presented two methods to transform any (t_p, t_f, n)-secure secret sharing scheme into a *robust* (t_p, t_f, t_r, t_s, n)-secure secret sharing scheme, achieving optimal values $t_s = 0$ and $t_r = t_f$ by using signature schemes and commitment schemes.

2.3 Threshold Encryption

The concept of threshold encryption was first proposed in [8]. In a threshold encryption scheme, there is a single public key, but the corresponding private decryption key is shared among a set of decryption servers. A user who wants to encrypt a message can run the encryption algorithm using the public key. A user who wants to decrypt a ciphertext gives the ciphertext to the servers, requesting a decryption share. When the user collects all the shares, he can apply a combining algorithm to obtain the message.

We first consider the *message privacy* in the model. A (t_p, n)-threshold encryption scheme $\mathcal{THE}$=(ThGen, ThEnc, ThDec, ThCom) consists of four algorithms:

ThGen: takes as input a security parameter λ, the number of decryption servers $n(\geq 1)$, and the threshold parameter t_p $(1 \leq t_p \leq n)$; it outputs

$$(PK, \overrightarrow{SK}) \leftarrow \texttt{ThGen}(\lambda, n, t_p),$$

where PK is the public encryption key, and $\overrightarrow{SK} = (SK_1, ..., SK_n)$ is the list of private keys. For $1 \leq i \leq n$, the private key SK_i is given to server i by a trusted dealer.

ThEnc: takes as input a public key PK and a message m, and outputs a ciphertext $\psi \leftarrow \texttt{ThEnc}(PK, m)$.

ThDec: takes as input a private key SK_i, and a ciphertext ψ, and outputs a decryption share $\delta_i \leftarrow \texttt{ThDec}(SK_i, \psi)$.

ThCom: (deterministic) takes as input all the the decryption shares $S = (\delta_1, ...\delta_n)$, and outputs the message $m = \texttt{ThCom}(S)$.

For correctness, we need that for any plaintext m, any output ψ of $\texttt{ThEnc}(PK, m)$, and all decryptions S of ψ, we have $\texttt{ThCom}(S) = m$.

Data Privacy. We define the security of threshold encryption formally under chosen-ciphertext attack with *static* server corruption. Consider the following experiment between a challenger and an adversary $\mathcal{A}$:

Experiment $\mathbf{Exp}^{the\text{-}cca}_{\mathcal{THE},\mathcal{A}}(\lambda)$
$((i_1, ..., i_{t_p}), St_0) \leftarrow \mathcal{A}(\lambda, Init)$
$(PK, \overrightarrow{SK}) \leftarrow \texttt{ThGen}(\lambda)$
$(m_0, m_1, St) \leftarrow \mathcal{A}^{\texttt{TDEC}(i, SK_i, \cdot)}\Big(find, PK, (SK_{i_1}, ..., SK_{i_{t_p}}), St_0\Big)$
$b \leftarrow \{0,1\}, c^* \leftarrow \texttt{ThEnc}(PK, m_b)$
$b' \leftarrow \mathcal{A}^{\texttt{TDEC}(i, SK_i, \cdot)}(guess, c^*, St)$
If $b = b'$ then return 1 else return 0

where the oracle $\texttt{TDEC}(i, SK_i, c)$ returns $\delta_i = \texttt{ThDec}(SK_i, c)$ with the restriction that $\mathcal{A}$ is not allowed to query the oracle for c^* in the *guess* phase. Both messages must be the same size. We define the advantage of $\mathcal{A}$ in the above experiment as $\mathbf{Adv}^{the\text{-}cca}_{\mathcal{THE},\mathcal{A}}(\lambda) = \left|\Pr\left[\mathbf{Exp}^{the\text{-}cca}_{\mathcal{THE},\mathcal{A}} = 1\right] - \frac{1}{2}\right|$. A threshold encryption $\mathcal{THE}$ is said to be secure against chosen-ciphertext attacks with privacy threshold t_p if the advantage function is negligible for all PPT $\mathcal{A}$.

Decryption Robustness. The correctness property of the threshold encryption only ensures correct decryption when all algorithms are honestly and correctly executed. Just as in the case of secret sharing, however, one may often desire fault-tolerance, robustness, and/or soundness. As in the case of secret sharing, these are parameterized by thresholds t_f, t_r, t_s, whose meaning is completely analogous to their meaning in the case of secret sharing (described earlier). Our constructions can achieve optimal $t_s = 0$, $t_r = t_f$, and any $t_p < t_f$.

2.4 Strongly Unforgeable One-Time Signature

We now review the definition and the security of signature schemes. A signature scheme $\Sigma = (\texttt{Gen}, \texttt{Sign}, \texttt{Ver})$ consists of three PPT algorithms . $\texttt{Gen}$ takes as input a parameter λ, it outputs a verification key vk and a signing key sk, $(vk, sk) \leftarrow \texttt{Gen}(\lambda)$. $\texttt{Sign}$ takes as input a signing key sk, and a message m, it outputs a signature $\sigma \leftarrow \texttt{Sign}(sk, m)$. $\texttt{Ver}$ takes as input a verification key vk,a message m, and a signature σ, it outputs $0/1 \leftarrow \texttt{Ver}(vk, m, \sigma)$.

We define the strong existential unforgeability under one-time chosen message attack. Consider the following experiment between a challenger and an adversary $\mathcal{A}$:

Experiment $\mathbf{Exp}_{\Sigma,\mathcal{A}}^{ots\text{-}suf\text{-}cma}(\lambda)$
$(vk, sk) \leftarrow \texttt{Gen}(\lambda)$
$(m^*, \sigma^*) \leftarrow \mathcal{A}^{\texttt{SIGN}(sk,\cdot)}(find, vk)$
If $\texttt{Ver}(vk, m^*, \sigma^*) = 1$, then return 1 else return 0

where the oracle $\texttt{SIGN}(sk, m)$ returns $\sigma \leftarrow \texttt{Sign}(sk, m)$ and $\mathcal{A}$ may only make at most one query to oracle $\texttt{SIGN}(sk, \cdot)$. $\mathcal{A}$ is said to win the Experiment if (1) $Ver(vk, m^*, \sigma^*) = 1$; (2) $(m, \sigma) \neq (m^*, \sigma^*)$. We define the advantage of $\mathcal{A}$ in the above experiment as $\mathbf{Adv}_{\Sigma,\mathcal{A}}^{ots\text{-}suf\text{-}cma}(\lambda) = \Pr\left[\mathbf{Exp}_{\Sigma,\mathcal{A}}^{ots\text{-}suf\text{-}cma}(\lambda) = 1\right]$. A one-time signature scheme Σ is said to be strongly unforgeable under chosen message attacks if the advantage function is negligible in λ for all PPT $\mathcal{A}$.

3 Main Construction

We describe our construction in this section. In [10], Dodis and Katz provided a formal definition of multiple encryption, and pointed out that threshold encryption is an application of multiple encryption. They also show how to construct a scheme with their highest security, i.e. strong MCCA. In this section we use weaker components to construct threshold schemes.

Let $\mathcal{TBE} = (\texttt{TGen}, \texttt{TEnc}, \texttt{TDec})$ be a tag-based encryption, $SS = (\texttt{Share}, \texttt{Rec})$ be a secret sharing scheme, $\Sigma = (\texttt{Gen}, \texttt{Sign}, \texttt{Ver})$ be a one-time signature scheme. We assume that the shares of $\mathcal{SS}$ are in the message space of $\mathcal{TBE}$, and the verification key is in the tag space of $\mathcal{TBE}$. We construct a generic threshold encryption scheme as follows:

ThGen: For $i = 1, ..., n$, Let $(pk_i, dk_i) \leftarrow \texttt{TGen}(\lambda)$, and set $PK = (pk_1, ..., pk_n)$, and $\overrightarrow{SK} = (dk_1, ..., dk_n)$. Send decryption key dk_i to server i by a trusted dealer.

ThEnc: Given a plaintext m. Let $(s_1, ..., s_n) = \texttt{Share}(m)$, and $(vk, sk) \leftarrow \texttt{Gen}(\lambda)$, use vk as the tag for tag-based encryption. Set $c_i = \texttt{TEnc}(pk_i, vk, s_i)$, then, compute the signature $\sigma = \texttt{Sign}(sk, c_1, ..., c_n)$. It outputs $c = (c_1, ..., c_n, vk, \sigma)$.

ThDec: Given a ciphertext c. For each server i, parses c as $(c_1, ..., c_n, vk, \sigma)$, and checks the signature. If $\texttt{Ver}(vk, c_1, ..., c_n, \sigma) = 0$, output $\perp$; else computes $s_i = \texttt{TDec}(dk_i, vk, c_i)$ using the private key dk_i. It returns the decryption share $\delta_i = s_i$.

`ThCom`: Given a set of all decryption shares $S = (\delta_1, ..., \delta_n)$. It outputs $m = \texttt{Rec}(s_1, ..., s_n)$.

Theorem 1. *$\mathcal{THE}$ constructed above is a CCA secure threshold encryption with thresholds t_p,t_f,t_r,t_s, if $\mathcal{TBE}$ is selective-tag CCA secure, $\mathcal{SS}$ is (t_p, t_f, t_r, t_s, n) secure, and Σ is one-time strongly unforgeable under chosen message attacks.*

Robustness thresholds t_f,t_r,t_s follow immediately from those of the secret sharing scheme. We now argue message privacy. We will use the following useful lemma.

Lemma 1. *Let S_1,S_2 and R be events defined on some probability space. Suppose that the event $S_1 \wedge \neg R$ occurs if and only if $S_2 \wedge \neg R$ occurs. Then*

$$\left|\Pr[S_1] - \Pr[S_2]\right| \leq \Pr[R].$$

We now define a sequence of games between a simulator and an adversary $\mathcal{A}$. Before defining the games, we define two "global" aspects in all the games. First, the simulator chooses a one-time signature key pair $(vk^*, sk^*) \leftarrow \texttt{Gen}(\lambda)$. Second, whenever $\mathcal{A}$ submits m_0, m_1, the simulator uses (vk^*, sk^*) instead of generating a new one-time signature key pair to generate the challenge ciphertext c^*.

Game$_{0'}$: This game is identical to the threshold encryption CCA experiment. At the beginning, $\mathcal{A}$ chooses t_p servers to corrupt. W.l.o.g we assume that $\mathcal{A}$ corrupts servers $\{n - t_p + 1, ..., n\}$. The simulator runs the key generation algorithm `ThGen` to obtain the public key $PK = (pk_1, ..., pk_n)$, and decryption keys $\overrightarrow{SK} = (dk_1, ..., dk_n)$. Then the simulator gives the public key PK, and $dk_{n-t_p+1}, ..., dk_n$ to $\mathcal{A}$. $\mathcal{A}$ can make queries to uncorrupted servers with ciphertext c, and the simulator decrypts c with the corresponding decryption key and returns the decryption share. After that $\mathcal{A}$ chooses two plaintexts m_0,m_1, and gives them to the simulator. The simulator chooses $b \in \{0, 1\}$ at random and returns $c^* = \texttt{ThEnc}(PK, m_b)$ to $\mathcal{A}$. We denote $c^* = (c_1^*, ..., c_n^*, vk^*, \sigma^*)$. After $\mathcal{A}$ obtains c^*, it still can make queries to uncorrupted servers with ciphertext $c \neq c^*$, and obtains decryption shares. At the end of the game, $\mathcal{A}$ outputs $b' \in \{0, 1\}$ as the guess of b. Let $X_{0'}$ denote the event that $b' = b$, apparently we have $\mathbf{Adv}_{\mathcal{THE},\mathcal{A}}^{the\text{-}cca}(\lambda) = \left|\Pr[X_{0'}] - \frac{1}{2}\right|$.

Game$_0$: This Game is identical to Game$_{0'}$, except that when $\mathcal{A}$ queries $c = (c_1, ..., c_n, vk, \sigma)$, if $vk = vk^*$, then the simulator outputs $\perp$. Otherwise outputs the decryption share. Let X_0 denote the event $b' = b$ in this game. Let F denote the event that $\mathcal{A}$ queries $c = (c_1, ..., c_n, vk, \sigma)$, $c \neq c^*$, $vk = vk^*$ and $\texttt{Ver}(vk, c_1, ..., c_n, \sigma) = 1$. Actually, if this event occurs with non-negligible probability, we can construct a adversary $\mathcal{B}$ to attack the one-time signature scheme. Obviously X_0 and $X_{0'}$ are identical unless F occurs. We claim:

Lemma 2. $\Pr[F]$ *is negligible.*

The proof of this lemma is given in section 3.1. According to Lemma 1 and Lemma 2, we have $\left|\Pr[X_0] - \Pr[X_{0'}]\right|$ is negligible.

We then define a series of games Game$_1$,...,Game$_{n-t_p}$ with gradual changes as follows: In general, for $1 \leq i \leq n-t_p$, Game$_i$ is identical to Game$_0$ except for one step in the computation of the challenge ciphertext c^*. Recall, in Game$_0$ we have $c_j^* \leftarrow \texttt{TEnc}(pk_j, vk^*, s_j^*)$, where s_j^* is the j-th share of the secret sharing scheme. In Game$_i$ we instead do this only for $j > i$, but set $c_j^* \leftarrow \texttt{TEnc}(pk_j, vk^*, 0)$ for $j \leq i$. In other words, for $1 \leq i \leq n-t_p$, Game$_{i-1}$ and Game$_i$ are identical except Game$_{i-1}$ sets $c_i^* \leftarrow \texttt{TEnc}(pk_i, vk^*, s_i^*)$, while Game$_i$ sets $c_i^* \leftarrow \texttt{TEnc}(pk_i, vk^*, 0)$. Denote by X_i for $1 \leq i \leq n - t_p$ the event $b = b'$ in Game$_i$. We claim:

Lemma 3. *For every $1 \leq i \leq n - t_p$, we have $\left|\Pr[X_i] - \Pr[X_{i-1}]\right|$ is negligible.*

The proof of this lemma is given in section 3.2.

Let us now think about the game Game$_{n-t_p}$. When encrypting the challenge m_b, only t_p shares are used in creating the challenge ciphertext. Then the privacy of the secret sharing scheme implies that $|\Pr\,[X_{n-t_p}] - \frac{1}{2}|$ is negligible.

We obtain the following result from the games and lemmas above.

$$\mathbf{Adv}_{\mathcal{THE},\mathcal{A}}^{the-cca}(\lambda) = \left|\Pr\,[X_{0'}] - \tfrac{1}{2}\right|$$

$$= \left|\Pr\,[X_{0'}] - \Pr\,[X_0] + \textstyle\sum_{i=1}^{n-t_p} \left(\Pr\,[X_{i-1}] - \Pr\,[X_i]\right) + \Pr\,[X_{n-t_p}] - \tfrac{1}{2}\right|$$

$$\leq \left|\Pr\,[X_{0'}] - \Pr\,[X_0]\right| + \textstyle\sum_{i=1}^{n-t_p} \left|\Pr\,[X_{i-1}] - \Pr\,[X_i]\right| + \left|\Pr\,[X_{n-t_p}] - \tfrac{1}{2}\right|$$

since $(n-t_p)$ is polynomial in λ, we get $\mathbf{Adv}_{\mathcal{THE},\mathcal{A}}^{the\text{-}cca}(\lambda)$ is negligible in λ, Theorem 1 is proven. □

3.1 Proof of Lemma 2

Proof. We construct a adversary $\mathcal{B}$ to attack the one-time signature scheme as follows: On receiving vk generated by $\texttt{Gen}$, $\mathcal{B}$ sets $vk^* = vk$, and generate $(PK, \overrightarrow{SK}) \leftarrow \texttt{ThGen}$, and does the same as in Game$_{0'}$. Upon any decryption query of the form $c = (c_1, ..., c_n, vk = vk^*, \sigma)$ such that $\texttt{Ver}(vk, c_1, ..., c_n, \sigma) = 1$, $\mathcal{B}$ immediately outputs $(c_1, ..., c_n, \sigma)$ as a forgery and return $\perp$, otherwise $\mathcal{B}$ returns the decryption share.

When $\mathcal{A}$ submits challenge plaintexts m_0, m_1, $\mathcal{B}$ creates the challenge ciphertext $c^* = (c_1^*, ..., c_n^*, vk^*, \sigma^*)$ by running $\texttt{ThEnc}(PK, m_b)$ except that signature σ^* is generated by querying $\mathcal{B}$'s signing oracle on message $c_1^*, ..., c_n^*$.

It is clear by construction that $\mathcal{B}$ simulates Game$_{0'}$ to $\mathcal{A}$. We now show that if event F happens then $\mathcal{B}$ outputs a valid forgery. If F happens before $\mathcal{A}$ is challenged on c^*, then $\mathcal{B}$ outputs a valid signature without making any queries, which is a forgery. If F happens after $\mathcal{A}$ receives c^* via a query $c = (c_1, ..., c_n, vk^*, \sigma)$, then because $c \neq c^*$ we must have $(c_1, ...c_n, \sigma) \neq (c_1^*, ..., c_n^*, \sigma^*)$. In either case, $\mathcal{B}$'s output $(c_1, ..., c_n, \sigma)$ differs from its single signature query, and hence is a forgery. Because the signature scheme is one-time strongly unforgeable, we conclude that event F happens with negligible probability, as desired. □

3.2 Proof of Lemma 3

Proof. Assume existing $1 \leq i \leq n - t_p$ such that $\left|\Pr[X_i] - \Pr[X_{i-1}]\right|$ is non-negligible. We construct an adversary $\mathcal{C}_i$ who succeeds in breaking selective-tag CCA of $\mathcal{TBE}$ by using $\mathcal{A}$ as a subroutine.

Just as the "global" setting of the games sequence, $\mathcal{C}_i$ first chooses a one-time signature key pair $(vk^*, sk^*) \leftarrow \mathtt{Gen}(\lambda)$, then outputs the challenge tag $\tau^* = vk^*$. $\mathcal{C}_i$ gets a public key pk for $\mathcal{TBE}$, sets $pk_i = pk$, and generates the remaining $(n-1)$ public/secret keys by himself. These public keys, as well as the last t_p secret keys, are given to $\mathcal{A}$. Adversary $\mathcal{C}_i$ honestly simulates the running of Game$_{i-1}$/Game$_i$ until $\mathcal{A}$ submits the challenge (m_0, m_1). At this point $\mathcal{C}_i$ chooses a random bit b, computes the shares $(s_1^*, ..., s_n^*) \leftarrow Share(m_b)$, and prepares c_j^* for $j \neq i$ just as in Game$_{i-1}$ and Game$_i$. Furthermore, $\mathcal{C}_i$ outputs the challenge $(s_i^*, 0)$ in its own CCA experiment. Upon receiving ciphertext $\widetilde{c}^*$, it sets $c_i^* = \widetilde{c}^*$, signs whatever is needed, and give the challenge ciphertext c^* to $\mathcal{A}$.

We specify how $\mathcal{C}_i$ deals with oracle queries of $\mathcal{A}$. Notice that $\mathcal{C}_i$ can decrypt all ciphertexts c_j for $j \neq i$ by himself, since the corresponding decryption keys are known. As for c_i, by Lemma 2 $\mathcal{A}$ does not reuse the challenge value vk^*, this means that $\mathcal{C}_i$ can always submit c_i to its own decryption oracle using the tag $\tau = vk \neq vk^*$. Finally $\mathcal{C}_i$ outputs 1 iff $\mathcal{A}$ correctly predicts b. This completes the description of $\mathcal{C}_i$, and it is not hard to see that $\mathcal{C}_i$ gives a perfect simulation of either game Game$_{i-1}$ or Game$_i$ depending on which of s_i^* or 0 was encrypted. It is easy to know that the advantage of $\mathcal{C}_i$ to attack selective-tag CCA of $\mathcal{TBE}$ is $\mathbf{Adv}_{\mathcal{TBE},\mathcal{C}_i}^{tbe\text{-}stag\text{-}cca}(\lambda) = \frac{1}{2}\left|\Pr[X_i] - \Pr[X_{i-1}]\right|$. According to the assumption at the beginning of the proof, we obtain that $\mathbf{Adv}_{\mathcal{TBE},\mathcal{C}_i}^{tbe\text{-}stag\text{-}cca}(\lambda)$ is non-negligible, this is a contradiction to the security of $\mathcal{TBE}$ scheme. □

4 Selective-Tag CCA Secure TBE

In this section, we demonstrate a public key encryption which is secure against selective-tag and adaptive chosen ciphertext attack, from lossy trapdoor function [23]. Peikert and Waters proved that the scheme is passively chosen ciphertext secure (CCA1). However, we show that it also give adaptive chosen ciphertext security under a selective-tag setting. We first review some useful definitions.

Lossy Trapdoor Functions. A collection of (n, ℓ)-lossy trapdoor functions is a triplet of PPT algorithms (S_{ltf}, F, F^{-1}) such that:

1. $S_{ltf}(\lambda, \mathtt{injective})$ outputs a pair $(s, td) \in \{0,1\}^n \times \{0,1\}^n$. The algorithm $F(s, \cdot)$ computes an injective function $f_s(\cdot)$ over $\{0,1\}^n$, and $F^{-1}(td, \cdot)$ computes $f_s^{-1}(\cdot)$.
2. $S_{ltf}(\lambda, \mathtt{lossy})$ outputs $s \in \{0,1\}^n$. The algorithm $F(s, \cdot)$ computes a function $f_s(\cdot)$ over $\{0,1\}^n$ whose image has size at most $2^{n-\ell}$.
3. The description of functions sampled using $S_{ltf}(\lambda, \mathtt{injective})$ and $S_{ltf}(\lambda, \mathtt{lossy})$ are computationally indistinguishable.

All-But-One Trapdoor Functions. In an ABO collection, each function has an extra input called its *branch*. All of the branches are injective trapdoor functions (having the same trapdoor value), except for one branch which is lossy. The lossy branch is specified as a parameter to the function sampler, and its value is hidden by the resulting function description.

Let $\mathcal{B} = \{B_\lambda\}_{\lambda \in \mathbb{N}}$ be a collection of sets whose elements represent the branches. A collection of (n, ℓ)-all-but-one trapdoor functions with branch collection $\mathcal{B}$ is a triplet of PPT algorithms (S_{abo}, G, G^{-1}) such that:

1. For any $b^* \in B_\lambda$, $S_{abo}(\lambda, b^*)$ outputs $(s, td) \in \{0,1\}^n \times \{0,1\}^n$. For any $b \neq b^*$, $G(s, b, \cdot)$ computes an injective function $g_{s,b}(\cdot)$ over $\{0,1\}^n$, and $G^{-1}(td, b, \cdot)$ computes $g_{s,b}^{-1}(\cdot)$. Additionally, $G(s, b^*, \cdot)$ computes a function $g_{s,b^*}(\cdot)$ over $\{0,1\}^n$ whose image has size at most $2^{n-\ell}$.
2. For any $b_0^*, b_1^* \in B_\lambda$, the first output s_0 of $S_{abo}(\lambda, b_0^*)$ and the first output s_1 of $S_{abo}(\lambda, b_1^*)$ are computationally indistinguishable.

Hash Functions. A family of functions $\mathcal{H} = \{h : D \to R\}$ is called pairwise independent [30] if, for every distinct $x, x' \in D$ and every $y, y' \in R$, $\Pr_{h \leftarrow \mathcal{H}}[h(x) = y \wedge h(x') = y'] = 1/|R|^2$.

Extracting Randomness. The min-entropy of a random variable X over a domain S is defined as $H_\infty(X) = -\lg(\max_{s \in S} \Pr[X = s])$. In many natural settings, the variable X is correlated with another variable Y whose value is known to a an adversary. We use the notion of average min-entropy [11], which captures the remaining unpredictability of X conditioned on the value of Y:

$$\widetilde{H}_\infty(X|Y) = -\lg\left(E_{y \leftarrow Y}\left[2^{H_\infty(X|Y=y)}\right]\right) = -\lg\left(E_{y \leftarrow Y}\left[\max_{s \in S} \Pr[X = s]\right]\right).$$

The average min-entropy is the negative logarithm of the average predictability of X conditioned on the random choice of Y; that is, the average maximum probability of predicting X given Y. We review the following useful lemmas.

Lemma 4. ([11], Lemma 2.2). *If Y takes at most 2^r possible values and Z is any random variable, then $\widetilde{H}_\infty(X|(Y,Z)) \geq H_\infty(X|Z) - r$.*

Lemma 5. ([11], Lemma 2.4). *Let X,Y be random variables such that $X \in \{0,1\}^n$ and $\widetilde{H}_\infty(X|Y) \geq k$. Let $\mathcal{H}$ be a family of pairwise independent hash function from $\{0,1\}^n$ to $\{0,1\}^\rho$. Then for $h \leftarrow \mathcal{H}$, we have $\Delta\Big((Y, h, h(x)), (Y, h, U_\rho)\Big) \leq \epsilon$ as long as $\rho \leq k - 2\lg(1/\epsilon)$.*

We show the scheme presented in [23] in the manner of TBE as follows: Let (S_{ltf}, F, F^{-1}) give a collection of (n, ℓ)-lossy trapdoor functions, and let (S_{abo}, G, G^{-1}) give a collection of (n, ℓ')-ABO trapdoor functions having branches $B_\lambda = \{0,1\}^v$. We use the branch space as the tag space in the tag-based encryption scheme.

Just as [23] described, we require that the total residual leakage over the lossy and ABO collections is

$$n - \ell + n - \ell' \leq n - \kappa \tag{1}$$

for some $\kappa = \kappa(\lambda) = \omega(\log \lambda)$, which guarantees the one-wayness of injective functions as in [23]. Let $\mathcal{H}$ be a family of pairwise independent hash functions from $\{0,1\}^n$ to $\{0,1\}^\rho$, where $0 \leq \rho \leq \kappa - 2\lg(1/\epsilon)$ for some negligible $\epsilon = \text{negl}(\lambda)$. The cryptosystem has message space $\{0,1\}^\rho$.

`TGen`: This algorithm generates an injective trapdoor function via $(s, td) \leftarrow S_{ltf}(\lambda, \texttt{injective})$, an ABO trapdoor function having lossy branch 0^v via $(s', td') \leftarrow S_{abo}(\lambda, 0^v)$, and a hash function $h \leftarrow \mathcal{H}$. The public key consists of the two function indices and the hash function: $pk = (s, s', h)$. The secret decryption key consists of td (td' is for the proof), along with the public key: $dk = (td, pk)$. It Outputs (pk, dk).

`TEnc`: This algorithm takes as input (pk, m) where $pk = (s, s', h)$ is a public key and $m \in \{0,1\}^\rho$ is the message. It chooses a branch $\tau \in B_\lambda$ at random, and use it as a tag. Then it chooses $x \leftarrow \{0,1\}^n$ uniformly at random. It computes $c_1 = F(s, x), c_2 = G(s', \tau, x), c_3 = m \oplus h(x)$. The ciphertext is output as $(c, \tau) = \Big((c_1, c_2, c_3), \tau\Big)$.

`TDec`: This algorithm takes as input $\Big(dk, (c, \tau)\Big)$ where $dk = (td, pk)$ is the secret key, and $(c, \tau) = \Big((c_1, c_2, c_3), \tau\Big)$ is a ciphertext. It first computes $x = F^{-1}(td, c_1)$, and checks that $c_1 = F(s, x)$ and $c_2 = G(s', \tau, x)$; if not, it outputs $\perp$. Otherwise, it outputs $m = c_3 \oplus h(x)$.

Theorem 2. *The algorithms* (`TGen`, `TEnc`, `TDec`) *described above is a selective-tag CCA secure tag-based encryption scheme.*

We prove the theorem by defining a sequence of experiments $\text{Game}_0, \ldots, \text{Game}_4$, where Game_0 is identical to the selective-tag CCA experiment. Then for $1 \leq i \leq 3$, the adversary's view in Game_i and Game_{i+1} are statistically or computationally indistinguishable. Finally it follows immediately from the definition of Game_4 that the adversary's view is identical for either value of $b \in \{0,1\}$. By the transitivity we get that Game_0 and Game_4 are computationally indistinguishable, hence the theorem is proved. We describe the games between a simulator and an adversary $\mathcal{A}$ as follows:

Game$_0$: This game is identical to the selective-tag CCA game for TBE. Specifically, at the beginning $\mathcal{A}$ outputs a challenge tag τ^*. Then the simulator runs `TGen` to get (pk, dk), and gives pk to $\mathcal{A}$. After obtaining pk, $\mathcal{A}$ can query a oracle $\texttt{DEC}(dk, \cdot)$ with ciphertext c, tag τ, and receives the plaintext. $\mathcal{A}$ submits two challenge plaintext (m_0, m_1) to the simulator, and the simulator chooses $b \in \{0,1\}$ uniformly at random, and return $c^* \leftarrow \texttt{TEnc}(pk, \tau^*, m_b)$. Upon receiving the ciphertext $\mathcal{A}$ can still query the decryption oracle $\texttt{DEC}(c, \tau)$. The only restriction of $\mathcal{A}$ to query decryption oracle is $\mathcal{A}$ is not allowed to make query with tag τ^*. Finally, $\mathcal{A}$ outputs a b' for the guess of b. If $b' = b$ then $\mathcal{A}$ wins the game. Just as the selective-tag CCA experiment, our goal is to prove that $\mathbf{Adv}_{\mathcal{TBE},\mathcal{A}}^{tbe\text{-}stag\text{-}cca}(\lambda)$ is negligible.

Game$_1$: This game is identical to Game$_0$, except that when the simulator generates the (pk, dk), the simulator replace $(s', td') \leftarrow S_{abo}(\lambda, 0^v)$ with $(s', td') \leftarrow S_{abo}(\lambda, \tau^*)$. Notice that in this game, the simulator still decrypts queries using the injective function trapdoor td, and the ABO function trapdoor td' is never used.

Game$_2$: This game is identical to Game$_1$, except that when the simulator decrypts queries, it replaces $x = F^{-1}(td, c_1)$ with $x = G^{-1}(td', \tau, c_2)$. Note that the simulator performs all the consistency checks, and $\mathcal{A}$ is not allowed to make queries with tag τ^*, the simulator can always answer the query. Also note that the injective function trapdoor td is never used in this game.

Game$_3$: This game is identical to Game$_2$, except that when the simulator generates the (pk, dk), the simulator replaces $(s, td) \leftarrow S_{ltf}(\lambda, \texttt{injective})$ with $(s, \perp) \leftarrow S_{ltf}(\lambda, \texttt{lossy})$.

Game$_4$: This game is identical to Game$_3$, except that when the simulator generates the challenge ciphertext $c^* = (c_1^*, c_2^*, c_3^*)$, it replaces $c_3^* = m_b \oplus h(x)$ with $c_3^* = r \leftarrow \{0,1\}^\rho$.

Observe that the adversary's views in Game$_4$ are identical for either choice of $b \in \{0, 1\}$, because b is never used in the game. We show the following results.

Lemma 6. *The adversary's views in Game$_0$ and Game$_1$ are computationally indistinguishable, assuming the hidden lossy branch property of the ABO trapdoor functions.*

Proof. We show that the adversary's views in Game$_0$ and Game$_1$, conditioned on any fixed value of τ^*, are computationally indistinguishable.
For any fixed τ^*, assume the adversary's views in Game$_0$ and Game$_1$ are distinguishable, we construct a PPT adversary $\mathcal{B}$ to distinguish lossy branches 0^v and τ^* of ABO trapdoor functions. Given an ABO function index s' which is generated as either $(s', td') \leftarrow S_{abo}(\lambda, 0^v)$ or $(s', td') \leftarrow S_{abo}(\lambda, \tau^*)$, $\mathcal{B}$ generates $(s, t) \leftarrow S_{ltf}(\lambda, \texttt{injective})$, and $h \leftarrow \mathcal{H}$, and outputs $pk = (s, s', h)$. $\mathcal{B}$ implements decryption oracle and generates challenge ciphertexts exactly as in Game$_0$ and Game$_1$ which are identical. Note that $\mathcal{B}$ can do so because it generates the injective function trapdoor td itself.
By the construction the view generated by $\mathcal{B}$ is exactly Game$_0$ when s' is generated by $S_{abo}(\lambda, 0^v)$, and is exactly Game$_1$ when s' is generated by $S_{abo}(\lambda, \tau^*)$. Proof is completed. □

Lemma 7. *The adversary's views in* $Game_1$ *and* $Game_2$ *are identical.*

Proof. Notice that Game_1 and Game $_2$ are identical except the decryption operations. These two operations both check that $c_1 = F(s, x)$ and $c_2 = G(s', \tau, x)$ for some x that they compute, and output $\perp$ if not. Thus, it suffices to show that this x is unique, and both decryption manners find it. In both games, (s, td) is generated by $S_{ltf}(\lambda, \texttt{injective})$, and (s', td') is generated by $S_{abo}(\lambda, \tau^*)$. $F(s, \cdot)$ and $G(s', \tau, \cdot)$ for $(\tau \neq \tau^*)$ are both injective. Thus, there is a unique x such that $(c_1, c_2) = (F(s, x), G(s', \tau, x))$. In Game_1 x is found by computing $F^{-1}(td, c_1)$, while in Game_2 by computing $G^{-1}(td', \tau, c_2)$. □

Lemma 8. *The adversary's views in* $Game_2$ *and* $Game_3$ *are computationally indistinguishable, assuming the indistinguishability of injective and lossy functions of lossy trapdoor functions.*

Proof. Assume the adversary's views in Game_2 and Game_3 are distinguishable, then, we can construct a PPT adversary $\mathcal{C}$ to distinguish lossy and injective functions. Given an index s which was generated as $(s, td) \leftarrow S_{ltf}(\lambda, \texttt{injective})$ or $(s, \perp) \leftarrow S_{ltf}(\lambda, \texttt{lossy})$, $\mathcal{C}$ generates $(s', td') \leftarrow S_{abo}(\lambda, \tau^*)$, and $h \leftarrow \mathcal{H}$. $\mathcal{C}$ implements decryption oracle and generates challenge ciphertexts exactly as in Game_2 and Game_3. Notice that $\mathcal{C}$ generates the ABO trapdoor td' itself, but does not know the trapdoor t corresponding to s even if it exists. The views generated by $\mathcal{C}$ is exactly Game_2 when s is generated by $S_{ltf}(\lambda, \texttt{injective})$, and is exactly Game_3 when s is generated by $S_{ltf}(\lambda, \texttt{lossy})$. This concludes the lemma. □

Lemma 9. *The adversary's views in* $Game_3$ *and* $Game_4$ *are statistically indistinguishable.*

Proof. Observe that in Game_3 and Game_4 we have $F(s, \cdot)$ and $G(s', \tau^*, \cdot)$ are lossy functions with image sizes at most 2^{n-l} and $2^{n-l'}$. Therefore, the random variables $(c_1^*, c_2^*) = (F(s, x), G(s', \tau^*, x))$ can take at most $2^{n-l+n-l'} \leq 2^{n-\kappa}$ values by our hypothesis in (1). By Lemma 4, and the independence of x from s, s', we have $\widetilde{H}_\infty(x|c_1^*, c_2^*, s, s') \geq H_\infty(x|s, s') - (n - \kappa) = n - (n - \kappa) = \kappa$. Since $\rho \leq \kappa - 2\lg(1/\epsilon)$ and Lemma 5, we have $\Delta\Big((c_1^*, c_2^*, h, h(x)),\ (c_1^*, c_2^*, h, r')\Big) \leq \epsilon = \text{negl}(\lambda)$, where $r' \leftarrow \{0,1\}^\rho$ is uniform and independent of all other variables. In Game_3 we have $c_3^* = m_b \oplus h(x)$, while in Game_4 we have $c_3^* = r \leftarrow \{0,1\}^\rho$, which is identically distributed to $m_b \oplus r'$. Thus, the two games are statistically indistinguishable, and this completes the proof. □

5 Discussions and Comparisons

Discussions. Since the scheme proposed in section 4 is a selective-tag CCA secure TBE, It can be used as a component for our general threshold scheme. Remark that the scheme proposed above actually is not full-tag CCA secure. We can easily break the full-tag CCA security by simply XOR c_3^* with a message we know and query to the decryption oracle. We also point out that our construction

Table 1. Comparisons among schemes

Schemes	PK Size	SK Size of Each Sever	Ciphertext Size	Assumption	RO Free	Quantum Attack Resistance
SG98	$(n+2)\|\mathbb{G}\|$	$\|\mathbb{Z}_q\|$	$5\|\mathbb{G}\|+2\|\mathbb{Z}_q\|$	CDH	×	×
CG99	$5\|\mathbb{G}\|$	$(L+5)\|\mathbb{Z}_q\|$	$4\|\mathbb{G}\|$	DDH	√	×
BBH06	$(n+4)\|\mathbb{G}\|$	$\|\mathbb{G}\|$	$2\|\mathbb{G}\|+\|\mathbb{G}_T\|+\|\mathtt{SIGN}\|$	DBDH	√	×
AT09	$(n+4)\|\mathbb{G}\|$	$\|\mathbb{Z}_q\|$	$2\|\mathbb{G}\|+\|\mathbb{G}_T\|+\|\mathtt{SIGN}\|$	DBDH	√	×
BD10	γ^5	$(2n-1)\gamma$	γ^2	SIVP_{γ^4}	√	√
Ours	$2n\gamma^3\log\gamma$	$\gamma^2\log\gamma$	$2n\gamma^2\log\gamma$	$\mathrm{SIVP}_{\delta^\gamma}$	√	√

† n denotes the number of the server, γ denotes the dimension of the lattice, $\delta > 1$ denotes some constant (e.g. $\delta = 1.01$) such that an approximation factor δ^γ for SIVP cannot be achieved in polynomial time, and L denotes the number of decryption performed before the public and secret keys need to be refreshed.

fits the definition of threshold tag-based encryption schemes [1], TTBE for short, and achieves the selective-tag CCA security without using the one-time signature. Briefly, a threshold tag-based encryption scheme is very similar to a threshold encryption scheme, except that the encryption, decryption and combination algorithms take additionally a "tag" as input. The security definition of selective-tag CCA of TTBE is very similar to the traditional tag-based encryption.

Comparisons. A few constructions of lossy trapdoor functions have been proposed. Peikert and Waters [23] presented constructions based on the decisional Diffie-Hellman (DDH) and learn with errors (LWE) assumption. Then, Rosen and Segev [26] proposed composite residuosity assumption based constructions. Recently Freeman et al. [14] proposed constructions based on quadratic residuosity assumption and d-Linear assumption with slight lossiness. We focus on the construction proposed by Peikert and Waters [23] which is based on the LWE assumption which was shown by Regev [25] and Peikert [22] to be as *worst-case* instances of shortest independent vector problem (SIVP) and gap shortest vector problem (GapSVP) in integer lattices. In [20] they show how to construct a strongly unforgeable one-time signature from chameleon hash function. A chameleon hash function can be obtained based on SIS assumption [5] which is related to worst-case lattice-based assumptions such as the hardness of SIVP problem. There is also a method [18], which is analogous to the construction of chameleon hash, to construct commitment schemes based on SIS assumption. Therefore, we obtain lattice based *robust* secret sharing scheme.

Table 1 shows comparisons among our scheme and other threshold schemes [29,4,3,1,2], We take attention to the efficiency between the BD10 [2] scheme and ours. In practice, The number of servers n will be small, such as 7 or 10. According to the table, when using the same large lattice dimension, the public key size of our scheme is less than BD10 [2] by a factor γ^2, but with the expense of larger secret size and ciphertext size by a factor γ. Overall the efficiency of our scheme is slightly better than the BD10 [2] scheme.

References

1. Arita, S., Tsurudome, K.: Construction of Threshold Public-Key Encryptions Through Tag-Based Encryptions. In: Abdalla, M., Pointcheval, D., Fouque, P.-A., Vergnaud, D. (eds.) ACNS 2009. LNCS, vol. 5536, pp. 186–200. Springer, Heidelberg (2009)
2. Bendlin, R., Damgård, I.: Threshold Decryption and Zero-Knowledge Proofs for Lattice-Based Cryptosystems. In: Micciancio, D. (ed.) TCC 2010. LNCS, vol. 5978, pp. 201–218. Springer, Heidelberg (2010)
3. Boneh, D., Boyen, X., Halevi, S.: Chosen Ciphertext Secure Public Key Threshold Encryption without Random Oracles. In: Pointcheval, D. (ed.) CT-RSA 2006. LNCS, vol. 3860, pp. 226–243. Springer, Heidelberg (2006)
4. Canetti, R., Goldwasser, S.: An Efficient threshold Public Key Cryptosystem Secure Against Adaptive Chosen Ciphertext Attack (Extended Abstract). In: Stern, J. (ed.) EUROCRYPT 1999. LNCS, vol. 1592, pp. 90–106. Springer, Heidelberg (1999)
5. Cash, D., Hofheinz, D., Kiltz, E., Peikert, C.: Bonsai Trees, or How to Delegate a Lattice Basis. In: Gilbert, H. (ed.) EUROCRYPT 2010. LNCS, vol. 6110, pp. 523–552. Springer, Heidelberg (2010)
6. Chaum, D.L.: Untraceable electronic mail, return addresses, and digital pseudonyms. Communications of the ACM 24(2), 84–90 (1981)
7. Damgård, I., Hofheinz, D., Kiltz, E., Thorbek, R.: Public-Key Encryption with Non-Interactive Opening. In: Malkin, T. (ed.) CT-RSA 2008. LNCS, vol. 4964, pp. 239–255. Springer, Heidelberg (2008)
8. Desmedt, Y.G.: Society and Group Oriented Cryptography: A New Concept. In: Pomerance, C. (ed.) CRYPTO 1987. LNCS, vol. 293, pp. 120–127. Springer, Heidelberg (1988)
9. Desmedt, Y., Frankel, Y.: Threshold Cryptosystems. In: Brassard, G. (ed.) CRYPTO 1989. LNCS, vol. 435, pp. 307–315. Springer, Heidelberg (1990)
10. Dodis, Y., Katz, J.: Chosen-Ciphertext Security of Multiple Encryption. In: Kilian, J. (ed.) TCC 2005. LNCS, vol. 3378, pp. 188–209. Springer, Heidelberg (2005)
11. Dodis, Y., Reyzin, L., Smith, A.: Fuzzy Extractors: How to Generate Strong Keys from Biometrics and other Noisy Data. In: Cachin, C., Camenisch, J.L. (eds.) EUROCRYPT 2004. LNCS, vol. 3027, pp. 523–540. Springer, Heidelberg (2004)
12. Dolev, D., Dwork, C., Naor, M.: Non-malleable cryptography. In: Proceedings of the Twenty-Third Annual ACM Symposium on Theory of Computing, pp. 542–552. ACM (1991)
13. Frankel, Y.: A Practical Protocol for Large Group Oriented Networks. In: Quisquater, J.-J., Vandewalle, J. (eds.) EUROCRYPT 1989. LNCS, vol. 434, pp. 56–61. Springer, Heidelberg (1990)
14. Freeman, D.M., Goldreich, O., Kiltz, E., Rosen, A., Segev, G.: More Constructions of Lossy and Correlation-Secure Trapdoor Functions. In: Nguyen, P.Q., Pointcheval, D. (eds.) PKC 2010. LNCS, vol. 6056, pp. 279–295. Springer, Heidelberg (2010)
15. Galindo, D.: Breaking and Repairing Damgård et al. Public Key Encryption Scheme with Non-interactive Opening. In: Fischlin, M. (ed.) CT-RSA 2009. LNCS, vol. 5473, pp. 389–398. Springer, Heidelberg (2009)
16. Galindo, D., Libert, B., Fischlin, M., Fuchsbauer, G., Lehmann, A., Manulis, M., Schröder, D.: Public-Key Encryption with Non-Interactive Opening: New Constructions and Stronger Definitions. In: Bernstein, D.J., Lange, T. (eds.) AFRICACRYPT 2010. LNCS, vol. 6055, pp. 333–350. Springer, Heidelberg (2010)

17. Gemmell, P.: An introduction to threshold cryptography. RSA CryptoBytes 2(3), 7–12 (1997)
18. Kawachi, A., Tanaka, K., Xagawa, K.: Concurrently Secure Identification Schemes Based on the Worst-Case Hardness of Lattice Problems. In: Pieprzyk, J. (ed.) ASIACRYPT 2008. LNCS, vol. 5350, pp. 372–389. Springer, Heidelberg (2008)
19. Kiltz, E.: Chosen-Ciphertext Security from Tag-Based Encryption. In: Halevi, S., Rabin, T. (eds.) TCC 2006. LNCS, vol. 3876, pp. 581–600. Springer, Heidelberg (2006)
20. Mohassel, P.: One-Time Signatures and Chameleon Hash Functions. In: Biryukov, A., Gong, G., Stinson, D.R. (eds.) SAC 2010. LNCS, vol. 6544, pp. 302–319. Springer, Heidelberg (2011)
21. Naor, M., Yung, M.: Public-key cryptosystems provably secure against chosen ciphertext attacks. In: Proceedings of the Twenty-Second Annual ACM Symposium on Theory of Computing, pp. 427–437. ACM (1990)
22. Peikert, C.: Public-key cryptosystems from the worst-case shortest vector problem: extended abstract. In: Proceedings of the 41st Annual ACM Symposium on Theory of Computing, pp. 333–342. ACM (2009)
23. Peikert, C., Waters, B.: Lossy trapdoor functions and their applications. In: Proceedings of the 40th Annual ACM Symposium on Theory of Computing, pp. 187–196. ACM (2008)
24. Rackoff, C., Simon, D.: Non-Interactive Zero-Knowledge Proof of Knowledge and Chosen Ciphertext Attack. In: Feigenbaum, J. (ed.) CRYPTO 1991. LNCS, vol. 576, pp. 433–444. Springer, Heidelberg (1992)
25. Regev, O.: On lattices, learning with errors, random linear codes, and cryptography. In: Proceedings of the Thirty-Seventh Annual ACM Symposium on Theory of Computing, pp. 84–93. ACM (2005)
26. Rosen, A., Segev, G.: Efficient lossy trapdoor functions based on the composite residuosity assumption. IACR ePrint Archive (2008)
27. Shamir, A.: How to share a secret. Communications of the ACM 22(11), 612–613 (1979)
28. Shoup, V.: A proposal for an ISO standard for public key encryption. Manuscript 100, 101–103 (2001)
29. Shoup, V., Gennaro, R.: Securing Threshold Cryptosystems Against Chosen Ciphertext Attack. In: Nyberg, K. (ed.) EUROCRYPT 1998. LNCS, vol. 1403, pp. 1–16. Springer, Heidelberg (1998)
30. Wegman, M.N., Carter, J.L.: New hash functions and their use in authentication and set equality. Journal of Computer and System Sciences 22(3), 265–279 (1981)
31. Zhang, R., Hanaoka, G., Shikata, J., Imai, H.: On the Security of Multiple Encryption or CCA-Security+ CCA-Security= CCA-Security? In: Bao, F., Deng, R., Zhou, J. (eds.) PKC 2004. LNCS, vol. 2947, pp. 360–374. Springer, Heidelberg (2004)

Monoidic Codes in Cryptography

Paulo S.L.M. Barreto[1,⋆], Richard Lindner[2], and Rafael Misoczki[3]

[1] Departmento de Engenharia de Computação e Sistemas Digitais (PCS)
Escola Politécnica, Universidade de São Paulo, Brasil
pbarreto@larc.usp.br
[2] Technische Universität Darmstadt, Department of Computer Science
Hochschulstraße 10, 64289 Darmstadt, Germany
rlindner@cdc.informatik.tu-darmstadt.de
[3] Project SECRET, INRIA-Rocquencourt
Domaine de Voluceau, 78150 Rocquencourt, France
rafael.misoczki@inria.fr

Abstract. At SAC 2009, Misoczki and Barreto proposed a new class of codes, which have parity-check matrices that are quasi-dyadic. A special subclass of these codes were shown to coincide with Goppa codes and those were recommended for cryptosystems based on error-correcting codes. Quasi-dyadic codes have both very compact representations and allow for efficient processing, resulting in fast cryptosystems with small key sizes. In this paper, we generalize these results and introduce quasi-monoidic codes, which retain all desirable properties of quasi-dyadic codes. We show that, as before, a subclass of our codes contains only Goppa codes or, for a slightly bigger subclass, only Generalized Srivastava codes. Unlike before, we also capture codes over fields of odd characteristic. These include wild Goppa codes that were proposed at SAC 2010 by Bernstein, Lange, and Peters for their exceptional error-correction capabilities. We show how to instantiate standard code-based encryption and signature schemes with our codes and give some preliminary parameters.

Keywords: post-quantum cryptography, codes, efficient algorithms.

1 Introduction

In 1996, conventional public-key cryptography deployed in practice was shown to be susceptible to feasible attacks, if sufficiently large quantum computers were ever built. In order to counter such attacks preemptively, several computational problems resistant to quantum computer attacks have been studied for their usage as foundation of cryptographic security [BBD08].

One promising candidate of such computational problems is the syndrome decoding problem. McEliece showed in 1978 how to construct a public-key encryption scheme based on the problem of decoding binary Goppa codes to their

⋆ Supported by the Brazilian National Council for Scientific and Technological Development (CNPq) under research productivity grant 303163/2009-7.

B.-Y. Yang (Ed.): PQCrypto 2011, LNCS 7071, pp. 179–199, 2011.

full error-correction capability when given their generator matrix in a disguised form [McE78]. At ASIACRYPT 2001, Courtois, Finiasz, and Sendrier showed that a signature scheme can be based on the same problem [CFS01].

So far, no algorithm is capable of decoding Goppa codes, or the closely related Generalized Srivastava (GS) codes, better than completely random linear codes. And the problem of decoding random linear codes is widely believed to be very hard. The main drawback of cryptographic schemes which use Goppa/GS codes is that their keys are several orders of magnitude bigger than those of classical schemes with comparable practical security. This issue of big key sizes is directly related to the size of the code description. This is the main problem which we will address.

Related Work. The problem of finding Goppa/GS codes with small descriptions is not new.

In [BLP10], Bernstein, Lange, and Peters find that Goppa codes over $\mathbb{F}_q$, where the Goppa polynomial has t roots of multiplicity $r-1$ and r divides q, have the capability of correcting $\lfloor rq/2 \rfloor$ errors instead of the usual $\lfloor (r-1)q/2 \rfloor$ errors they can correct with an alternant decoder. These codes are called wild Goppa codes and due to their increased correction capability, one can use codes with smaller descriptions for the same level of practical security.

Another major breakthrough in saving description size has been achieved in [MB09] by Barreto and Misoczki. They define a new class of quasi-dyadic codes, which have very compact descriptions, and show that this has a non-empty intersection with the class of binary Goppa codes. They also show how to generate codes in this intersection efficiently and give some preliminary parameters. Later, in [BCMN10], they are joined by Cayrel and Niebuhr and go on to show that quasi-dyadic Goppa codes can be generated in such a way that they are dense enough to be usable with the CFS signature scheme.

More generally, other proposals aimed at key reduction not restricted to Goppa codes were proposed [BC07, Gab05, MRS00] but subsequently broken [OTD10]; in special, [FOPT10a] and [GL10] presented structural attacks against McEliece variants with compact keys, being effective against quasi-cyclic codes [BCGO09]. With respect to the binary quasi-dyadic Goppa codes, this attack was not successful and, focused on increasing the effort of this attack, Persichetti proposed a construction using quasi-dyadic Srivastava codes [Per11], instead of Goppa ones, providing keys with similar size to the keys presented in [MB09].

Most attempts at decreasing key sizes deal with codes in characteristic 2, in spite of evidence [Pet10] that odd characteristics may offer security advantages.

Our Contribution. In this paper we introduce a new class of codes which allow for an extremely small representation and efficient processing. Our so called quasi-monoidic codes are a generalization of quasi-dyadic codes to finite fields of odd characteristics.

Using quasi-monoidic Goppa codes for the McEliece cryptosystems and CFS signature scheme, one can potentially obtain smaller key sizes than before, as exemplified by Tables 1 and 3 in Section 6. For example, we find that many wild Goppa codes are in fact quasi-monoidic.

Organization. In Section 2, we introduce the basic concepts of coding theory, which are relevant to our proposal. In Section 3, we introduce our new class of quasi-monoidic codes and show how to construct Goppa/GS codes that are quasi-monoidic. Next, we describe how to instantiate the standard code-based encryption and signature schemes with this family in Section 4. Afterwards, in Section 5 we assess the security properties of our proposal, and in Section 6 we suggest a few actual parameters to encourage further analysis. Finally, in Section 7 we briefly argue why the matrix-vector products for quasi-monoidic matrices can be computed efficiently using a discrete Fourier transform.

2 Coding Theory

Basic concepts. We will start with some matrix descriptions. For both descriptions, t is an integer greater than zero. Given a sequence $L = (L_0, \ldots, L_{n-1}) \in \mathbb{F}_q^n$, the *Vandermonde matrix* $\mathrm{vdm}(t, L)$ is the $t \times n$ matrix with elements $V_{ij} = L_j^i$. Given a polynomial g with coefficients $(g_1, \ldots, g_t) \in \mathbb{F}_q^n$, the *Toeplitz matrix* $\mathrm{toep}(g_1, \ldots, g_t)$ is the $t \times t$ matrix with elements $T_{ij} := g_{t-i+j}$ for $j \leqslant i$ and $T_{ij} := 0$ otherwise. The following are the *GRS*, *alternant* and *Goppa codes* definitions.

Definition 1. *Given a sequence $L = (L_0, \ldots, L_{n-1}) \in \mathbb{F}_q^n$ of distinct elements and a sequence $D = (D_0, \ldots, D_{n-1}) \in \mathbb{F}_q^n$ of nonzero elements, the* Generalized Reed-Solomon code *$GRS_r(L, D)$ is the $[n, k, r]$ linear error-correcting code defined by the parity-check matrix*

$$H = \mathrm{vdm}(r-1, L) \cdot \mathrm{diag}(D).$$

An alternant code *is a subfield subcode of a Generalized Reed-Solomon code.*

Definition 2. *Given a prime power p, $q = p^d$ for some d, a sequence $L = (L_0, \ldots, L_{n-1}) \in \mathbb{F}_q^n$ of distinct elements and a polynomial $g(x) \in \mathbb{F}_q[x]$ of degree t such that $g(L_i) \neq 0$ for $0 \leqslant i < n$, the* Goppa code *$\Gamma(L, g)$ over $\mathbb{F}_p$ is defined by the parity-check matrix*

$$\begin{aligned} H = {} & \mathrm{toep}(g_1, \ldots, g_t) \\ & \cdot \mathrm{vdm}(t, L_0, \ldots L_{n-1}) \\ & \cdot \mathrm{diag}(g(L_0)^{-1}, \ldots, g(L_{n-1})^{-1}) \end{aligned} \tag{1}$$

By [MS77][Ch. 12, §3], *we can omit* $\mathrm{toep}(g_1, \ldots, g_t)$ *from this construction, making it easy to see that* Goppa codes *are also* alternant codes *over $\mathbb{F}_p$ corresponding to $GRS_t(L, D)$ where $D = (g(L_0)^{-1}, \ldots, g(L_{n-1})^{-1})$.*

Goppa codes have minimum distance at least $t+1$. Binary Goppa codes improve this to at least $2t+1$. It turns out that, although this improvement does not hold in general for larger characteristics, codewords that differ by vectors whose components are all equal are on average much more sparsely distributed. Thus, while the unambiguous correction of general errors in odd characteristics can in general not proceed beyond about $t/2$ errors, correction of error patterns of homogeneous (all-equal) error magnitudes can probabilistically reach as much as t errors [BLM10].

Goppa codes in Tzeng-Zimmermann form. It was shown by Tzeng and Zimmermann [TZ75], that all Goppa codes with Goppa polynomial $g(x) = h(x)^r$, for some square-free $h(x)$ and number $r > 0$, admit a parity-check matrix consisting solely of Cauchy power matrices over the splitting field of $g(x)$.

Definition 3. *Let $\mathbb{F}$ be a finite field, and $\beta = (\beta_0, \beta_1 \ldots, \beta_{t-1}), \gamma = (\gamma_0, \gamma_1, \ldots, \gamma_{n-1})$ be two disjoint sequences of distinct elements in $\mathbb{F}$. The* Cauchy matrix *$C(\beta, \gamma)$ associated with these sequences is one where $C_{i,j} = (\beta_i - \gamma_j)^{-1}$, i.e.,*

$$C = \begin{pmatrix} (\beta_0 - \gamma_0)^{-1} & \cdots & (\beta_0 - \gamma_{n-1})^{-1} \\ \vdots & & \vdots \\ (\beta_{t-1} - \gamma_0)^{-1} & \cdots & (\beta_{t-1} - \gamma_{n-1})^{-1} \end{pmatrix}.$$

For any additional integer $r > 0$, the associated Cauchy power matrix *$C(\beta, \gamma, r)$ is a Cauchy matrix, where each coordinate is raised to the r-th power, i.e., $C_{i,j} = (\beta_i - \gamma_j)^{-r}$.*

Finally, the Cauchy layered matrix *$CL(\beta, \gamma, r)$ consists of all Cauchy power matrices with exponents up to r, i.e.,*

$$CL(\beta, \gamma, r) = \begin{pmatrix} C(\beta, \gamma) \\ C(\beta, \gamma, 2) \\ \vdots \\ C(\beta, \gamma, r) \end{pmatrix}.$$

There is an ambivalence in this definition, i.e., there is no bijection from all sequences β and γ to all Cauchy matrices. Specifically, for any $\omega \in \mathbb{F}$, we have $C(\beta, \gamma) = C(\beta + \omega, \gamma + \omega)$.

In terms of properties, Cauchy matrices are very similar to Vandermonde matrices. For example, there are efficient algorithms to compute matrix-vector products, submatrices of Cauchy matrices are again Cauchy, all Cauchy matrices have full-rank, and there are closed formulas for computing their determinant.

As mentioned before, Tzeng and Zimmermann showed that all Goppa codes, where the Goppa polynomial is the r-th power of an square-free polynomial, admit a parity-check matrix which is a Cauchy layered matrix. This parity-check matrix is in TZ form. Specifically, the parity-check matrix H in TZ form of the Goppa code with support $L = \{\gamma_0, \ldots, \gamma_{n-1}\}$ and Goppa polynomial $g(x) = \prod_{i=0}^{t-1}(x - \beta_i)^r$ is $H = CL(\beta, \gamma, r)$.

This is particularly interesting for the case of wild Goppa codes as introduced by Bernstein, Lange, and Peters [BLP10]. They show that if r divides the field characteristic, then the rows of this TZ parity-check matrix are not linearly independent, but the rows of $H' = CL(\beta, \gamma, r - 1)$, where we omit the last Cauchy block, are already a parity-check matrix of the full code. This allows wild Goppa codes to achieve error-correcting capabilities surpassing general alternant codes and make them particularly interesting for various application including cryptography.

Generalized Srivastava codes.

Definition 4. *Let* $(\alpha_1, \ldots, \alpha_n)$, $(\omega_1, \ldots, \omega_s)$ *are* $n+s$ *distinct elements of* $\mathbb{F}_{q^m}$, *and* $(z_1, \ldots, z_n)$ *are nonzero elements of* $\mathbb{F}_{q^m}$. *The* Generalized Srivastava code *is an* $[n, k \geq n - mst, d \geq st + 1]$ *code over* $\mathbb{F}_q$, *is also an alternant code, and is defined by the parity-check matrix*

$$H = \begin{pmatrix} H_1 \\ H_2 \\ \vdots \\ H_s \end{pmatrix}$$

where

$$H_l = \begin{pmatrix} \frac{z_1}{\alpha_1 - \omega_l} & \frac{z_2}{\alpha_2 - \omega_l} & \cdots & \frac{z_n}{\alpha_n - \omega_l} \\ \frac{z_1}{(\alpha_1 - \omega_l)^2} & \frac{z_2}{(\alpha_2 - \omega_l)^2} & \cdots & \frac{z_n}{(\alpha_n - \omega_l)^2} \\ \cdots & \cdots & \cdots & \cdots \\ \frac{z_1}{(\alpha_1 - \omega_l)^t} & \frac{z_2}{(\alpha_2 - \omega_l)^t} & \cdots & \frac{z_n}{(\alpha_n - \omega_l)^t} \end{pmatrix}$$

for $l = 1, \ldots, s$. *The original Srivastava codes are the case* $t = 1$, $z_i = \alpha_i^{\mu}$ *for some* μ.

For more details about Generalized Srivastava codes, see [MS77][Ch. 12, §6].

3 Quasi-monoidic Codes

Monoidic matrices.

Definition 5. *Let* R *be a commutative ring,* $A = \{a_0, \cdots, a_{N-1}\}$ *a finite abelian group of size* $|A| = N$ *with neutral element* $a_0 = 0$, *and* $h \colon A \longrightarrow R$ *a sequence indexed by* A. *The* A*-adic matrix* $M(h)$ *associated with this sequence is one for which* $M_{i,j} = h(a_i - a_j)$ *holds, i.e.,*

$$M = \begin{pmatrix} h(0) & h(-a_1) & \cdots & h(-a_{N-1}) \\ h(a_1) & h(0) & \cdots & h(a_1 - a_{N-1}) \\ \vdots & \vdots & \ddots & \vdots \\ h(a_{N-1}) & h(a_{N-1} - a_1) & \cdots & h(0) \end{pmatrix}.$$

All A-adic matrices form a ring that is isomorphic to the monoid ring $R[A]$, which is studied in abstract algebra [Lan02]. We use the additive notation for the finite abelian group A here for practical purposes, but the definition can be generalized to all groups, in which case one might prefer the multiplicative notation.

Some A-adic matrices have special names, for example the $\mathbb{Z}_2^d$-adic matrices are dyadic and the $\mathbb{Z}_3^d$-adic matrices are triadic. If we do not want to specify the group A explicitly, we will say the matrix is monoidic. So, to identify all Goppa codes with a monoidic representation, we continue by giving necessary and sufficient conditions for Cauchy matrices to be monoidic and show that the case for Cauchy power matrices follows from that.

Conditions for which monoidic implies Cauchy.

Theorem 1. *Let $M(h)$ be A-adic for a sequence h of length N over $\mathbb{F}$. Then M is Cauchy iff*

(1) $h(a_i)$ *are distinct and invertible in* $\mathbb{F}$ *for all* $0 \le i < N$, *and*
(2) $(h(a_i - a_j))^{-1} = (h(a_i))^{-1} + (h(-a_j))^{-1} - (h(0))^{-1}$ *for all* $0 \le i, j < N$.

In this case $M(h) = C(\beta, \gamma)$, *where* $\beta(a_i) = (h(a_i))^{-1}$ *and* $\gamma(a_i) = (h(0))^{-1} - (h(-a_i))^{-1}$.

Proof. We start by showing that our conditions indeed imply that M is Cauchy. For the disjointness, assume that there are indices i and j, such that $\beta(a_i) = \gamma(a_j)$. In this case we get $0 = \beta(a_i) - \gamma(a_j) = 1/h(a_i - a_j)$, which is a contradiction. Finally we compare the matrices $M(h)$ and $C(\beta, \gamma)$ resulting in the equality

$$M_{i,j} = h(a_i - a_j) = 1/(1/h(a_i) + 1/h(-a_j) - 1/h(0)) = 1/(\beta(a_i) - \gamma(a_j)) = C_{i,j}.$$

We continue by showing that if M is Cauchy, i.e., $M(h) = C(\beta', \gamma')$, then indeed our conditions must hold. Since $C(\beta', \gamma') = C(\beta' + \omega, \gamma' + \omega)$ for any $\omega \in \mathbb{F}$, we can choose the sequences in such a way that $\gamma'(0) = 0$. Now, $M_{i,0} = C_{i,0}$ for all i, which means $h(a_i) = 1/\beta'(a_i)$. By the properties of β' this gives us condition (1), i.e., that all $h(a_i)$ are distinct and invertible, as well as $\beta' = \beta$. We use similarly that $M_{0,i} = C_{0,i}$ which implies $h(-a_i) = 1/(\beta(0) - \gamma'(a_i))$. Solving for γ' reveals that it equals γ. Since $\beta = \beta'$ and $\gamma = \gamma'$, we get that $M(h) = C(\beta, \gamma)$ implying condition (2). □

Note that if the A-adic matrix of a sequence h is also Cauchy, then the sequence of r-th powers, i.e., $h^r = (h_0^r, h_{a_1}^r, \ldots, h_{a_{n-1}}^r)$ yields the corresponding Cauchy power matrix. In other words, for any number $r > 0$ we have $M(h) = C(\beta, \gamma) \implies M(h^r) = C(\beta, \gamma, r)$.

Now, we will show how to construct random monoidic Cauchy matrices.

Construction of monoidic Cauchy matrices.

Corollary 1. *Let A be a finite, abelian group with set of generators $b_1, \ldots, b_d$ and $M(h)$ be A-adic and Cauchy for a sequence h over $\mathbb{F}$, then for all $c_1, \ldots, c_d \in \mathbb{Z}$,*

$$(h(c_1 b_1 + \cdots + c_d b_d))^{-1} = c_1 (h(b_1))^{-1} + \cdots + c_d (h(b_d))^{-1} - (c_1 + \cdots + c_d - 1)(h(0))^{-1}.$$

Furthermore, the field characteristic $\mathrm{char}(\mathbb{F})$ *divides the order of any element in* $A \setminus \{0\}$.

Proof. By Theorem 1, we know that for all $a, a' \in A$ the following holds

$$(h(a + a'))^{-1} = (h(a))^{-1} + (h(a'))^{-1} - (h(0))^{-1}.$$

By repeatedly using this equation, we prove the first claim.

$$
\begin{aligned}
(h(c_1b_1+\cdots+c_db_d))^{-1} &= (h(\underbrace{b_1+\cdots+b_1}_{c_1 \text{ times}}+\cdots+\underbrace{b_d+\cdots+b_d}_{c_d \text{ times}}))^{-1} \\
&= (h(b_1))^{-1} + (h(\underbrace{b_1+\cdots+b_1}_{(c_1-1) \text{ times}}+\cdots+\underbrace{b_d+\cdots+b_d}_{c_d \text{ times}}))^{-1} - (h(0))^{-1} \\
&= c_1(h(b_1))^{-1} + (h(\underbrace{b_2+\cdots+b_2}_{(c_2) \text{ times}}+\cdots+\underbrace{b_d+\cdots+b_d}_{c_d \text{ times}}))^{-1} - c_1(h(0))^{-1} \\
&= c_1(h(b_1))^{-1}+\cdots+c_d(h(b_d))^{-1} - (c_1+\cdots+c_d-1)(h(0))^{-1}.
\end{aligned}
$$

For the second claim, let $a \in A \setminus \{0\}$ be a non-neutral group element and $k = \mathrm{ord}(a)$, i.e., $ka = 0$. By the equation we have just shown, we know that

$$
\begin{aligned}
h(0)^{-1} = h(ka)^{-1} &= kh(a)^{-1} - (k-1)h(0)^{-1} \\
k(h(0)^{-1} - h(a)^{-1}) &= 0
\end{aligned}
$$

Since a is not the neutral element, all elements of h are distinct, and the field characteristic is prime, the second claim follows. □

Since the field characteristic p divides the order of any element, only groups of size $N = p^d$ can be used. Conversely, let $b_1, \ldots, b_d$ be group elements that form an $\mathbb{F}_p$ set of generators, then the sequence elements $h(0), h(b_1), \ldots, h(b_d)$ completely determine the sequence. We call these values the essence of the sequence h.

For example, if $A = \mathbb{F}_p^d$, then such a set of generators $b_1, \ldots, b_d$ is given by the generators of the d distinct copies of $\mathbb{F}_p$ in A. For a given set of generators, we can sample a monoidic sequence uniformly at random with the algorithm in Figure 2.

We will briefly argue why the algorithm in Figure 2 is correct. Assume that it is not. The only situation resulting in an error is in line 7, if the computed quantity is not invertible, so let us assume this to be the case. Since only zero is not invertible, we have

$$0 = c_1h(b_1)+\cdots+c_dh(b_d) - (c_1+\cdots+c_d-1)h(0).$$

Now, not all coefficients of $h(0), h(b_1), \ldots, h(b_d)$ can be zero simultaneously, so there is an $\mathbb{F}_p$-linear dependency among them. However, by our choice of F in line 4, from which all $h(b_i)$ are chosen, no such dependency can exist.

As a consequence of our algorithm, the total number of possible sequences is

$$|\{h\colon \mathbb{F}_p^d \to \mathbb{F}_Q \mid M(h) \text{ is monoidic and Cauchy}\}| = (Q-1)\cdots(Q-p^d).$$

Quasi-monoidic Generalized Srivastava codes. Our final goal is to describe a way of disguising the Cauchy block structures of the code that is used for error-correction, while simultaneously keeping much of the monoidic structure intact in the form of small monoidic blocks. This will allow us to obtain code-based public-key schemes with small keys.

Description	Parameter	Restriction
Field char	p	prime
Base field	q	p^s
Extension field	Q	q^m
Group order	N	$p^d \le Q/p$

Description	Parameter	Restriction
Goppa roots	t	$< n/m$
Goppa multiplicity	r	$< n/(tm)$
Blocksize	b	$\gcd(t, N)$
Code length	n	$\mathsf{b}\ell < N$

Fig. 1. Parameters for quasi-monoidic GS codes. Let $s, m, \ell > 0$. For brevity, we will focus on the case where $s = 1$ (smallest base field size), $r = p - 1$ (wild case).

MonoidCauchy(p, Q, d):

1. $F \longleftarrow \mathbb{F}_Q \setminus \{0\}$
2. $h(0) \longleftarrow U(F)$
3. **For** $i = 1, \ldots, d$:
4. $\quad F \longleftarrow \mathbb{F}_Q \setminus (\mathbb{F}_p\, h(0) + \mathbb{F}_p\, h(b_1) + \cdots + \mathbb{F}_p\, h(b_{i-1}))$
5. $\quad h(b_i) \longleftarrow U(F)$
6. **For** $c_1, \ldots, c_d \in \mathbb{F}_p$:
7. $\quad h(c_1 b_1 + \cdots + c_d b_d) \longleftarrow c_1 h(b_1) + \cdots + c_d h(b_d) - (c_1 + \cdots + c_d - 1)h(0)$
8. **Output** $(h(0)^{-1}, h(a_1)^{-1}, \ldots, h(a_{p^d-1})^{-1})$

Fig. 2. Choosing A-adic Cauchy sequences, where $A = \{0, a_1, \ldots, a_{p^d-1}\}$ has set of generators $b_1, \ldots, b_d$.

QuasiMonoidic$(\ldots)$:

1. $h \longleftarrow$ MonoidCauchy(p, Q, d); $\omega \longleftarrow U(\mathbb{F}_Q)$
2. **For** $i = 0, \ldots, t-1$: $\beta_i \longleftarrow (h(a_i))^{-1} + \omega$ "Goppa roots"
3. **For** $i = 0, \ldots, N-1$: $\gamma_i \longleftarrow (h(0))^{-1} - (h(-a_i))^{-1} + \omega$ "Goppa support"
4. $\tau \longleftarrow U(S_{N/\mathsf{b}})$ "Block permutation"
5. $\pi_0, \ldots, \pi_{\ell-1} \longleftarrow U(\{0, \ldots, \mathsf{b}-1\})$ "Support permutations"
6. $\sigma_0, \ldots, \sigma_{\ell-1} \longleftarrow U(\mathbb{F}_q^*)$ "Scaling"
7. **For** $i = 0, \ldots, \ell-1$: $\widehat{\gamma}_i \longleftarrow (\gamma_{\tau(i)\mathsf{b}}, \ldots, \gamma_{\tau(i)\mathsf{b}+\mathsf{b}-1})$ "Select blocks"
8. **For** $i = 0, \ldots, \ell-1$: $\widehat{\gamma}_i \longleftarrow \widehat{\gamma}_i M(\chi_{a_{\pi_i}})$ "Permute support"
9. $H \longleftarrow [CL(\beta, \widehat{\gamma}_0, r)\sigma_0 \mid \cdots \mid CL(\beta, \widehat{\gamma}_{\ell-1}, r)\sigma_{\ell-1}]$ "Parity-check matrix"
10. $H \longleftarrow$ QMTrace(q, b, H)
11. $H \longleftarrow$ QMGauss(b, H)
12. $H \longleftarrow$ QMSignature(b, H)
13. **Output** private $\beta, \widehat{\gamma}_0, \ldots, \widehat{\gamma}_{\ell-1}, \sigma_0, \ldots, \sigma_{\ell-1}$; public H

Fig. 3. Choosing quasi-monoidic GS codes with private and public description. Here, $S_{N/\mathsf{b}}$ is the group of permutations on $\{0, \ldots, N/\mathsf{b} - 1\}$ and $\chi_{a_{\pi_i}}$ is the characteristic function of the group element a_{π_i}.

The relevant parameters used for this process are described in Figure 1 and the corresponding algorithm is presented in Figure 3. We will continue by explaining some details including the QM-subroutines used therein and conclude with a clarifying example.

We start the generation process by choosing a random fully monoidic Goppa code of length N. Then we split the support in blocks of length $\mathtt{b}$ and select ℓ such blocks at random to comprise the support of our quasi-monoidic code. To each chosen support block, we apply a random monoidic permutation, i.e., we multiply with the matrix $M(\chi_{a_\pi})$, where χ_{a_π} is the characteristic function of the group element a_π, for a randomly chosen π. Since χ is a characteristic function, this matrix will have a single non-zero coefficient being 1 per row and column, so it is a permutation matrix. Furthermore, this transformation preserves the monoidic structure of the block and indeed all monoidic permutations have this form.

We continue by creating the parity-check matrix H of our code consisting of the ℓ scaled Cauchy layered matrices corresponding to each block. The resulting matrix consists of $tr/\mathtt{b} \times \ell$ monoidic blocks of size $\mathtt{b}$ and we will keep this quasi-monoidic structure intact for the remainder. Note that if $q > 2$, i.e., we have non-trivial scaling factors, then the code defined via our parity-check matrix need not be Goppa anymore, but it is always a Generalized Srivastava code.

The first subroutine QMTrace will generate a parity-check matrix for the corresponding subfield subcode over the base field $\mathbb{F}_q$. Recall that $\mathbb{F}_Q = \mathbb{F}_q[x]/\langle f \rangle$ for some irreducible polynomial f of degree m. We can identify each matrix coefficient $h_{i,j}$ with its representative polynomial $h_{i,j,0} + h_{i,j,1}x + \cdots + h_{i,j,m-1}x^{m-1}$ of smallest degree. We expand the matrix rows by a factor of m and distribute the entries as follows $h^{\text{new}}_{kt+i,j} \longleftarrow h_{i,j,k}$,i.e., in order to keep the block structure intact, we first take all constant terms of coefficients in a block then all linear terms and so on.

The second subroutine QMGauss will compute the quasi-monoidic systematic form of the parity-check matrix. It does so by identifying each monoidic block with an element of the corresponding ring of monoidic matrices and performing the usual Gauss algorithm on those elements. Since this ring is not necessarily an integral domain, the algorithm may find that a pivot element is not invertible. In this case, the systematic form we seek does not exist and the algorithm has to loop back to the "Block permutation" step. Fortunately, the chance of this is small. The probability that the matrix is nonsingular is $\prod_{j=0}^{k-1} 1 - 1/p^{k-j}$, which approaches a constant (to be determined numerically) for large k. This constant is different for each p but tends to 1 for large p. In order to avoid redundancy, this subroutine omits those columns of the systematic form, which we know to be the identity matrix.

The third and final subroutine QMSignature will simply extract the monoidic signature of each block, i.e., its first column. For the whole quasi-monoidic matrix, this simply amounts to extracting each $\mathtt{b}$-th column. This concludes our description of the algorithm.

In Appendix A, we give a detailed example of the generation process that illustrates some subtleties of the algorithm.

Decoding quasi-monoidic Goppa codes. In [BLM10], an efficient decoding algorithm for square-free (irreducible or otherwise) Goppa codes over $\mathbb{F}_p$ for any prime p is presented. Since it fits perfectly to decode quasi-monoidic Goppa codes, we will provide a brief description of this method in Appendix B.

Decoding GS codes is less studied than the case of Goppa codes, though both are closely related. Let D be a diagonal matrix containing the scaling factors, then adding an error pattern e to a GS codeword c amounts to adding the pattern eD to the codeword cD of the associated unscaled Goppa code. So, if the Goppa decoder capability depends only on the weight of the error pattern (like the wild decoder [BLP10]), then it can be used equally well for GS codes and scaling could be used. On the other hand, if the Goppa decoder capability is best if all error magnitudes coincide (like the "equal magnitude" decoder [BLM10]), then scaling must not be used. It turns out that, in the latter case, keys also get potentially smaller due to the larger number of correctable errors.

4 Monoidic Encryption and Signatures

In this section we provide the basic description about the McEliece encryption scheme [McE78] and the Parallel-CFS signature scheme [Fin10]. Both of them can be instantiated with our monoidic codes.

4.1 McEliece Encryption Scheme

Let the security level be λ. The parameters for the code below are assumed to be chosen so that the cost of the best attack against it is at least 2^λ (see [Pet11] for a recent survey) takes at least 2^λ operations.

Key Generation: Choose a prime p, a finite field $\mathbb{F}_q$ with $q = p^m$ for some $m > 0$ and a Quasi-Monoidic code $\Gamma(L, g)$ with support $L = (L_0, \ldots, L_{n-1}) \in (\mathbb{F}_q)^n$ of distinct elements and a square-free generator polynomial $g \in \mathbb{F}_q[x]$ of degree t, satisfying $g(L_j) \neq 0$, $0 \leqslant j < n$, both provided by the algorithm of Figure 3. Let $k = n - mt$. Compute a systematic generator matrix $G \in \mathbb{F}_p^{k \times n}$ for $\Gamma(L, g)$, i.e. $G = [I_k \mid -M^T]$ for some matrix $M \in \mathbb{F}_p^{mt \times k}$ and I_k an identity matrix of size k. The private key is $sk := (L, g)$ and the public key is $pk := (M, t)$.

Encryption: To encrypt a plain text $d \in \mathbb{F}_p^k$, choose an error-vector $e \in \{0, 1\}^n \subseteq \mathbb{F}_p^n$ with weight $\mathsf{wt}(e) \leqslant t$, and compute the cipher text $c \leftarrow dG + e \in \mathbb{F}_p^n$.

Decryption: To decrypt a cipher text $c \in \mathbb{F}_p^n$ knowing L and g, compute the decodable syndrome of c, apply a decoder to determine the error-vector e, and recover the plain text d from the first k columns of $c - e$.

4.2 Parallel-CFS

To sign a document with the standard CFS signature schemes [CFS01] we should hash the document into a syndrome and then decode it to an error vector of certain weight t. Since not all syndromes are decodable, a counter is hashed with the message, and the signer tries successive counter values until a decodable syndrome is found. The signature consists of both the error pattern of weight t corresponding to the syndrome and the counter value yielding this syndrome. In [FS09] is described an unpublished attack by D. Bleichenbacher showing that the usual parameters are insecure and the improved parameters result in a signature scheme with excessive cost of signing time or key length.

To address this problems, M. Finiasz proposed in [Fin10] the Parallel-CFS, which can be described as follows: instead of producing one hash (using a function $\mathcal{H}$) from a document D and signing it, one can produce i hashes (using i different functions $\mathcal{H}_1, \ldots, \mathcal{H}_i$) and sign all $\mathcal{H}_1(D), \ldots, \mathcal{H}_i(D)$ in parallel. Then Parallel-CFS can be described by the following algorithms.

Key Generation: Choose parameters m, t and let $n = 2^m$. Select δ such that $\binom{2^m}{t+\delta} > 2^{mt}$. Choose a Quasi-Monoidic code $\Gamma(g, L)$, where g is a polynomial of degree t in $\mathbb{F}_{2^m}[X]$ and a support $L = (L_0, \ldots, L_{n-1}) \in \mathbb{F}_{2^m}^n$. Let H be a $mt \times n$ systematic parity-check matrix of Γ. H is the public verification key and $\Gamma(g, L)$ represents the private signature key.

Signature: For i signatures in parallel (see Table 2 column "sigs", based on [Fin10], for this estimation), the signer tries to guess δ errors, searching all error patterns $\phi_\delta(j)$ of weight δ, and then applies the decoding algorithm to the resulting syndrome $s_{j,i} = \mathcal{H}_i(D) + H \cdot \phi_\delta(j)^T$. Once a decodable syndrome is found for an $j_{0,i}$, then there exists a plain text $p'_{j_0,i}$, such that $H \cdot \phi_t(p'_{j_0,i})^T = s_{j_0,i} = \mathcal{H}_i(D) + H \cdot \phi_\delta(j_0)^T$.
With the error patterns $e_i = \phi_t(p'_{j_0,i}) + \phi_\delta(j_0)$ of weight at most $t+\delta$, it holds that $H \cdot e_i^T = \mathcal{H}_i(D)$, for i signatures. The signature is $(\phi_{t+\delta}^{-1}(e_1) \| \ldots \| \phi_{t+\delta}^{-1}(e_i))$.

Verification: Given a signature $(p_1 \| \ldots \| p_i)$ for a document D, the verification step consists of checking the i equalities $H \cdot \phi_{t+\delta}(p_i)^T \stackrel{?}{=} \mathcal{H}_i(D)$.

CFS-friendly quasi-monoidic Goppa codes. There is a simple extension to the construction of quasi-dyadic codes that applies to our quasi-monoidic codes as well. These CFS-friendly codes were proposed in [BCMN10] and we will briefly describe the idea. Recall that MonoidCauchy constructs a full monoidic $N \times N$ parity-check matrix of which, after some scaling and permuting, we will use only a $t \times n$ submatrix. The idea is to relax the construction of the full matrix, allowing for some undefined entries, as long as they do not end up in the submatrix we actually use.

This relaxation is realized by omitting line 4 of MonoidCauchy (Fig. 2), i.e., the condition of linear independence of the essential entries in the inverted

Table 1. Encryption quasi-monoidic codes

level	p	m	n	k	t	key(bits)	syndrome(bits)
80	2	12	3840	768	256	9216	3072
80	3	8	2430	486	243	6163	3082
80	5	5	1000	375	125	4354	1452
80	167	3	668	167	167	3700	3700
112	2	12	2944	1408	128	16896	1536
112	3	8	2673	729	243	9244	3082
112	11	5	1089	484	121	8372	2093
112	241	3	964	241	241	5722	5722
128	2	12	3200	1664	128	19968	1536
128	3	9	3159	972	243	13866	3467
128	5	5	5000	625	625	10159	10159
128	373	3	1492	373	373	9560	9560
192	2	14	6144	2560	256	35840	3584
192	3	10	4131	1701	243	26961	3852
192	29	6	5887	841	841	24514	24514
192	547	4	2735	547	547	19901	19901
256	2	15	11264	3584	512	53760	7680
256	7	9	5145	2058	343	51998	8667
256	37	6	9583	1369	1369	42791	42791
256	907	4	4535	907	907	35645	35645

Table 2. Encryption quasi-monoidic codes yielding short syndromes

level	p	m	n	k	t	key(bits)	syndrome(bits)
80	2	11	1792	1088	64	11968	704
80	7	5	735	490	49	6879	688
80	41	3	451	328	41	5272	659
128	2	12	3200	1664	128	19968	1536
128	3	9	2106	1377	81	19643	1156
128	7	6	1813	1519	49	25587	826
192	2	14	5376	3584	128	50176	1792
192	3	11	4536	3645	81	63550	1413

monoidic sequence. This may cause some entries in the sequence to be 0, so we cannot invert them in the final step of the algorithm and just leave them at 0, since no legal entry can have that value. Now, after selecting the submatrix in QuasiMonoidic (Fig. 3), i.e., after line 7, we need to check that all matrix coefficients are non-zero and restart if there are any. This is unlikely since the submatrix is usually small.

The whole relaxation allows us to work with smaller extension fields $\mathbb{F}_Q$, because we now need only $t+n$ distinct elements in $\mathbb{F}_Q^*$, where before we needed $2N$. So the codes we produce will be denser and thus more suited for the CFS signature scheme.

Table 3. Parallel CFS quasi-monoidic codes

level	p	m	n	k	t	key (bits / KiB)	sigs	δ	sigbits
80	2	15	32580	32400	12	1458000 / 178	2	4	326
80	3	11	177048	176949	9	3085033 / 377	3	2	375
80	13	4	28509	28457	13	421214 / 52	2	4	342
112	2	20	1048332	1048092	12	62885520 / 7677	3	3	636
112	11	6	1771495	1771429	11	36768825 / 4489	3	2	558
112	13	5	371228	371163	13	6867332 / 839	3	3	624
128	2	23	8388324	8388048	12	578775312 / 70652	3	2	684
128	5	8	390495	390375	15	21754145 / 2656	3	4	759
128	13	6	4826731	4826653	13	107164431 / 13082	2	3	514

5 Security Assessment

Decoding attacks. In estimating *concrete* security (rather than asymptotic behavior only), we adopt the following criteria, which were discussed and analyzed by Finiasz and Sendrier [FS09] and by Peters [Pet11, Observation 6.9] (see also [BLP11]), whereby directly decoding a code of length n, dimension k, and generic error patterns of weight w over $\mathbb{F}_q$, without using the trapdoor, has a *workfactor* at least WF_q measured in bit operations. Typically $\wp \approx w/2$ and $\ell \gtrsim \log_q \binom{k/2}{\wp} + \wp \log_q(q-1)$:

$$\mathrm{WF}_2 = \min_{\wp,\ell} \left\{ \begin{array}{c} \frac{1}{2}(n-k)^2(n+k) + \left(k/2 - \wp + 1 + \binom{\lfloor k/2 \rfloor}{\wp} + \binom{\lceil k/2 \rceil}{\wp}\right)\ell \\ +(w - 2\wp + 1)4\wp\binom{\lfloor k/2 \rfloor}{\wp}\binom{\lceil k/2 \rceil}{\wp} \end{array} \right\} \quad (2)$$

$$\mathrm{WF}_q = \min_{\wp,\ell} \left\{ \begin{array}{c} (n-k)^2(n+k) + \left(k/2 - \wp + 1 + \left(\binom{\lfloor k/2 \rfloor}{\wp} + \binom{\lceil k/2 \rceil}{\wp}\right)(q-1)^{\wp}\right)\ell \\ +\frac{q}{q-1}(w - 2\wp + 1)2\wp\left(1 + \frac{q-2}{q-1}\right)\frac{\binom{\lfloor k/2 \rfloor}{\wp}\binom{\lceil k/2 \rceil}{\wp}(q-1)^{2\wp}}{q^{\ell}} \end{array} \right\} \quad (3)$$

When it is known beforehand that all errors have equal magnitude and $q > 2$, we simplify Equation 3 accordingly:

$$\mathrm{WF}'_q = \min_{\wp,\ell} \left\{ \begin{array}{c} (n-k)^2(n+k) + \left(k/2 - \wp + 1 + \left(\binom{\lfloor k/2 \rfloor}{\wp} + \binom{\lceil k/2 \rceil}{\wp}\right)(q-1)\right)\ell \\ +\frac{q}{q-1}(w - 2\wp + 1)2\wp\left(1 + \frac{q-2}{q-1}\right)\frac{\binom{\lfloor k/2 \rfloor}{\wp}\binom{\lceil k/2 \rceil}{\wp}(q-1)}{q^{\ell}} \end{array} \right\} \quad (4)$$

The results of this estimations are provided in Tables 1, 2, and 3 and discussed in Section 6.

Structural attacks. Structural attacks against families of codes that yield compact keys McEliece have also been proposed. In [FOPT10a], the idea is to convert the public code into a multivariate nonlinear system and then trying to solve it with Gröbner basis techniques. A related technique inspired by the Sidelnikov-Shestakov attack [SS92] is described in [GL10].

The former attack recovers variables x_i and y_i which denote respectively the diagonal and the support of the code, i.e. the x_i define the Vandermonde matrix V, and the y_i define the diagonal matrix D, which compose the parity-check matrix $H = VD$ in the alternant case. In the Goppa case, these variables are coupled by a more complex relationship, namely $y_i = g(x_i)^{-1}$. In both cases, the result is a multivariate system, with equations of degree up to t, namely, $H_{ij} = x_j^i y_j$ for $0 \leq i < t - 1$ and $0 \leq j < n - 1$.

For generic codes, this system is too complex to be feasibly solved with Gröbner bases. However in the dyadic case (and, by extension, the monoidic case), many equations are redundant, due to relation (2) of Theorem 1. Furthermore, for subcodes defined over extension fields (but not over the base field itself), it turns out that only linear and quadratic equations are enough to specify variables x_i and y_i. This feature yields a simpler multivariate system that can be tackled with, and in fact the corresponding quasi-dyadic codes over extension fields in [MB09] can be broken this way.

However, for the case of a subcode defined over the base field, the associated Gröbner basis is trivial if only linear and quadratic equations are used to define the x_i and the y_i variables, and the attack fails [FOPT10b]. Although those results were obtained for characteristic 2, at the time of writing there does not appear to be any way to take advantage of larger characteristics to improve this attack. Exploring this line of attack is thus left as an open problem for followup research.

Regarding the attack in [GL10], a 'small' extension degree m could lead to a successful break, but it is unclear how small m must be so that such an attack would become feasible. Just how small m should be for the attack to be successful in each characteristic $p > 2$ is unclear, though. In this sense, the parameters listed in this paper deliberately use relatively small values of m, in the hope that they stimulate further cryptanalysis research. While we stress that these parameters are not designed for effective deployment, the indicated security levels correspond to the best known generic decoding attacks so as to give a realistic impression of what practical might look like.

6 Parameters of Cryptographic Interest

We now assess the efficiency of the proposed codes in possible practical cryptographic scenarios.

Tables 1, 2, and 3 compare some of the best quasi-monoidic codes achievable for each characteristic at several security levels. These figures only assume the ability to correct t errors of equal magnitude, already taking into account that this choice of introduced errors decreases the WF to break it. Correcting such error patterns is possible using e.g. the decoding method for square-free Goppa codes proposed in [BLM10]. The design minimum distance is at least $t+1$ when differences between codewords are allowed to assume any pattern, but codewords that differ by patterns where all magnitudes are equal are much more sparse than that; the distribution is much more similar to what holds for binary codes, since the difference patterns only fail to be binary because of the overall magnitude. The error correcting strategy described in [BLM10] (algorithm 1) benefits from this observation, which allows for the correction of t errors with high probability as long as all error magnitudes are equal. The entries on Table 1 describe codes suitable for McEliece or Niederreiter encryption [Nie86].

One can argue that minimizing keys may not be the best way to reduce bandwidth occupation. After all, usually one expects to exchange encrypted messages considerably more often than certified keys, so it pays to minimize the encryption overhead per message instead. This is particularly easy to achieve using the Niederreiter cryptosystem, as long as the adopted codes yield short syndromes. Table 2 lists suggestions for codes that satisfy these requirements (including protection against structural interpolation attacks), without incurring unduly long keys. One sees that the choice for short syndromes often implies longer codes for larger characteristics.

Table 3 describes codes suitable for parallel CFS digital signatures [Fin10, BCMN10]. The signature size is slightly smaller than the product of the the syndrome size by the number of parallel signatures, and signing times are $O(t!)$. Quasi-monoidic codes in larger characteristics yield either shorter keys and signatures than in the binary case, or else considerably shorter signing times due to smaller values of t.

7 Efficiency

We will show that for all groups A relevant to cryptography, the matrix-vector products involving A-adic matrices can be computed in $\widetilde{O}(N)$ operations with a multidimensional discrete Fourier transform. As we have seen in Corollary 1, all relevant groups have the form $A = \mathbb{Z}_p^d$. Recall that the ring of A-adic matrices over R is isomorphic to the monoid ring $R[A]$ (hence the name, 'monoidic' matrices and codes). In the following lemma, we show that this has the structure of a multivariate polynomial quotient ring.

Lemma 1. *Let R be a commutative ring, then* $R[\mathbb{Z}_p^d] \cong R[x_1, \ldots, x_d] / \langle x_1^p - 1, \ldots, x_d^p - 1\rangle$.

Proof. Let $A = \mathbb{Z}_p^d$. Consider the following R-bases for the left and right ring respectively, left we have $[\chi_{(a_1,\ldots,a_d)}]_{a\in A}$ and right $[x_1^{a_1} \cdots x_k^{a_d}]_{a\in A}$, where a ranges through all d-tuples in A for each ring.

We define ψ for all basis elements of the left ring to be $\psi(\chi_{(a_1,\ldots,a_k)}) = x_1^{a_1} \cdots x_k^{a_k}$. This can be extended canonically to an R-module isomorphism on the whole ring. It only remains to check that ψ respects multiplication. It suffices to check this for the generators, so let $a, b \in A$ then

$$\begin{aligned}
\psi(\chi_a) \cdot \psi(\chi_b) &= (x_1^{a_1} \cdots x_k^{a_d}) \cdot (x_1^{b_1} \cdots x_k^{b_d}) \bmod x_1^p - 1, \ldots, x_d^p - 1 \\
&= x_1^{a_1+b_1 \bmod p} \cdots x_d^{a_d+b_k \bmod p} \\
&= \psi(\chi_{(a_1+b_1 \bmod p, \ldots, a_d+b_d \bmod p)}) = \psi(\chi_a \cdot \chi_b)
\end{aligned}$$

□

We propose to compute the polynomial products by means of the several size-p fast discrete Fourier transform (DFT). This requires that the ring we work over has an element ω of order p and its characteristic is not p. One way to achieve this, is to lift our field $\mathbb{F}_q$ into a ring R of characteristic 0 that has been extended with a primitive p-th root of unity. Now, we can perform the operation in R, and project the results back.

The DFT itself works like the Walsh-Hadamard transform in [MB09], except that the matrices describing the transformation and its inverse are H_d and H_d^{-1}, which are recursively defined as

$$H_1 = \begin{pmatrix} 1 & 1 & 1 & \cdots & 1 \\ 1 & \omega^1 & \omega^2 & \cdots & \omega^{p-1} \\ 1 & \omega^2 & \omega^4 & \cdots & \omega^{2(p-1)} \\ \vdots & \vdots & \vdots & \ddots & \vdots \\ 1 & \omega^{p-1} & \omega^{2(p-1)} & \cdots & \omega^{(p-1)(p-1)} \end{pmatrix}, \quad H_1^{-1} = \frac{1}{p} \begin{pmatrix} 1 & 1 & 1 & \cdots & 1 \\ 1 & \omega^{-1} & \omega^{-2} & \cdots & \omega^{-(p-1)} \\ 1 & \omega^{-2} & \omega^{-4} & \cdots & \omega^{-2(p-1)} \\ \vdots & \vdots & \vdots & \ddots & \vdots \\ 1 & \omega^{-(p-1)} & \omega^{-2(p-1)} & \cdots & \omega^{-(p-1)(p-1)} \end{pmatrix},$$

$$H_k = H_1 \otimes H_{k-1}, \qquad H_k^{-1} = H_1^{-1} \otimes H_{k-1}^{-1},$$

where $\otimes$ is the Kronecker product.

Acknowledgements. We thank Rafael Dahmen for his invaluable advice; Christiane Peters for her always patient comments, and the anonymous reviewers for helping in the improvement of this paper.

References

[BBD08] Bernstein, D.J., Buchmann, J., Dahmen, E. (eds.): PQCrypto 2008. LNCS, vol. 5299. Springer, Heidelberg (2008)

[BC07] Baldi, M., Chiaraluce, F.: Cryptanalysis of a new instance of McEliece cryptosystem based on QC-LDPC code. In: IEEE International Symposium on Information Theory – ISIT 2007, pp. 2591–2595. IEEE (2007)

[BCGO09] Berger, T.P., Cayrel, P.-L., Gaborit, P., Otmani, A.: Reducing Key Length of the McEliece Cryptosystem. In: Preneel, B. (ed.) AFRICACRYPT 2009. LNCS, vol. 5580, pp. 77–97. Springer, Heidelberg (2009)

[BCMN10] Barreto, P.S.L.M., Cayrel, P.-L., Misoczki, R., Niebuhr, R.: Quasi-Dyadic CFS Signatures. In: Lai, X., Yung, M., Lin, D. (eds.) Inscrypt 2010. LNCS, vol. 6584, pp. 336–349. Springer, Heidelberg (2011)

[BLM10] Barreto, P.S.L.M., Lindner, R., Misoczki, R.: Decoding square-free Goppa codes over $\mathbb{F}_p$. Cryptology ePrint Archive, Report 2010/372 (2010), `http://eprint.iacr.org/2010/372.pdf`

[BLP10] Bernstein, D.J., Lange, T., Peters, C.: Wild McEliece. In: Biryukov, A., Gong, G., Stinson, D.R. (eds.) SAC 2010. LNCS, vol. 6544, pp. 143–158. Springer, Heidelberg (2011)

[BLP11] Bernstein, D.J., Lange, T., Peters, C.: Smaller Decoding Exponents: Ball-Collision Decoding. In: Rogaway, P. (ed.) CRYPTO 2011. LNCS, vol. 6841, pp. 743–760. Springer, Heidelberg (2011)

[CFS01] Courtois, N., Finiasz, M., Sendrier, N.: How to Achieve a McEliece-Based Digital Signature Scheme. In: Boyd, C. (ed.) ASIACRYPT 2001. LNCS, vol. 2248, pp. 157–174. Springer, Heidelberg (2001)

[Fin10] Finiasz, M.: Parallel-CFS. In: Biryukov, A., Gong, G., Stinson, D.R. (eds.) SAC 2010. LNCS, vol. 6544, pp. 159–170. Springer, Heidelberg (2011)

[FOPT10a] Faugère, J.-C., Otmani, A., Perret, L., Tillich, J.-P.: Algebraic Cryptanalysis of McEliece Variants with Compact Keys. In: Gilbert, H. (ed.) EUROCRYPT 2010. LNCS, vol. 6110, pp. 279–298. Springer, Heidelberg (2010)

[FOPT10b] Faugère, J.-C., Otmani, A., Perret, L., Tilllich, J.-P.: Algebraic cryptanalysis of compact McEliece's variants – toward a complexity analysis. In: International Conference on Symbolic Computation and Cryptography – SCC 2010, pp. 45–56 (2010)

[FS09] Finiasz, M., Sendrier, N.: Security Bounds for the Design of Code-Based Cryptosystems. In: Matsui, M. (ed.) ASIACRYPT 2009. LNCS, vol. 5912, pp. 88–105. Springer, Heidelberg (2009)

[Gab05] Gaborit, P.: Shorter keys for code based cryptography. In: International Workshop on Coding and Cryptography – WCC 2005, pp. 81–91. ACM Press, Bergen (2005)

[GL10] Gauthier Umaña, V., Leander, G.: Practical key recovery attacks on two McEliece variants. In: Cid, C., Faugère, J.-C. (eds.) International Conference on Symbolic Computation and Cryptography – SCC 2010, pp. 27–44 (2010)

[Lan02] Lang, S.: Algebra, revised 3rd edn. Springer, Heidelberg (2002)

[MB09] Misoczki, R., Barreto, P.S.L.M.: Compact McEliece Keys from Goppa Codes. In: Jacobson Jr., M.J., Rijmen, V., Safavi-Naini, R. (eds.) SAC 2009. LNCS, vol. 5867, pp. 376–392. Springer, Heidelberg (2009)

[McE78] McEliece, R.J.: A public-key cryptosystem based on algebraic coding theory. Deep Space Network Progress Report 44, 114–116 (1978)

[MRS00] Monico, C., Rosenthal, J., Shokrollahi, A.: Using low density parity check codes in the McEliece cryptosystem. In: IEEE International Symposium on Information Theory – ISIT 2000, p. 215. IEEE, Sorrento (2000)

[MS77] MacWilliams, F.J., Sloane, N.J.A.: The theory of error-correcting codes. North-Holland Mathematical Library, vol. 16 (1977)

[Nie86] Niederreiter, H.: Knapsack-type cryptosystems and algebraic coding theory. Problems of Control and Information Theory 15(2), 159–166 (1986)

[OTD10] Otmani, A., Tillich, J.-P., Dallot, L.: Cryptanalysis of two McEliece cryptosystems based on quasi-cyclic codes. Mathematics in Computer Science 3(2), 129–140 (2010)

[Per11] Persichetti, E.: Compact McEliece keys based on quasi-dyadic Srivastava codes. Cryptology ePrint Archive, Report 2011/179 (2011), `http://eprint.iacr.org/2011/179.pdf`

[Pet10] Peters, C.: Information-set Decoding For Linear Codes over $\mathbb{F}_q$. In: Sendrier, N. (ed.) PQCrypto 2010. LNCS, vol. 6061, pp. 81–94. Springer, Heidelberg (2010)

[Pet11] Peters, C.: Curves, Codes, and Cryptography. Ph.D. thesis, Technische Universiteit Eindhoven, the Netherlands (2011), `http://alexandria.tue.nl/extra2/711052.pdf`

[SS92] Sidelnikov, V., Shestakov, S.: On the insecurity of cryptosystems based on generalized Reed-Solomon codes. Discrete Mathematics and Applications 1(4), 439–444 (1992)

[TZ75] Tzeng, K.K., Zimmermann, K.: On extending Goppa codes to cyclic codes. IEEE Transactions on Information Theory 21, 712–716 (1975)

A An Exemplary Quasi-Monoidic Srivastava Code

For our example, let $p = 3, s = 1, m = 4, d = 3, t = 3$. We use the extension field $\mathbb{F}_{3^4} = \mathbb{F}_3[u]/\langle u^4 + 2u^3 + 2\rangle$, the group $A = \mathbb{Z}_3^3$ of size $N = p^d = 27$, with set of generators $b_1 = (1,0,0)$, $b_2 = (0,1,0)$, $b_3 = (0,0,1)$.

We randomly select the images of the $\mathbb{F}_3$-linearly dependent

$$h(0)^{-1} = u^3 + u^2 + u + 2, \qquad h(b_1)^{-1} = u^2 + 2u + 1,$$
$$h(b_2)^{-1} = u^3 + 2u^2 + u + 1, \qquad h(b_3)^{-1} = u^2 + 1.$$

We also select a shift $\omega = u^3 + 2u + 2$, compute $\beta = (2u^3 + u^2 + 1, u^3 + u^2 + u, u^2 + 2u + 2)$, and $\gamma = (\gamma_0, \ldots, \gamma_8)$ with

$$\gamma_0 = (u^3 + 2u + 2, 1, 2u^3 + u), \qquad \gamma_1 = (u^3 + u^2 + 2u + 1, u^2, 2u^3 + u^2 + u + 2),$$
$$\gamma_2 = (u^3 + 2u^2 + 2u, 2u^2 + 2, 2u^3 + 2u^2 + u + 1), \qquad \gamma_3 = (u + 1, 2u^3 + 2u, u^3 + 2),$$
$$\gamma_4 = (u^2 + u, 2u^3 + u^2 + 2u + 2, u^3 + u^2 + 1), \qquad \gamma_5 = (2u^2 + u + 2, 2u^3 + 2u^2 + 2u + 1, u^3 + 2u^2),$$
$$\gamma_6 = (2u^3, u^3 + u + 2, 2u + 1), \qquad \gamma_7 = (2u^3 + u^2 + 2, u^3 + u^2 + u + 1, u^2 + 2u),$$
$$\gamma_8 = (2u^3 + 2u^2 + 1, u^3 + 2u^2 + u, 2u^2 + 2u + 2).$$

where the group indices are ordered $0 = (0,0,0), a_1 = (1,0,0), a_2 = (2,0,0), \ldots, a_{p^d-1} = (2,2,2)$. Our blocksize is $\mathsf{b} = \gcd(t, N) = 3$ and we randomly choose the permutation $\tau = \binom{012345678}{567834012}$. We use only the first $\ell = 6$ blocks chosen by the permutation, i.e., blocks $5, 6, 7, 8, 3, 4$, resulting in a code of length $n = \mathsf{b}\ell = 18$.

We continue and select the support permutations

$$\pi_0 = 0, \qquad \pi_1 = 2, \qquad \pi_2 = 1, \qquad \pi_3 = 2, \qquad \pi_4 = 0, \qquad \pi_5 = 1.$$

corresponding to the monoidic permutation matrices $M(\chi_{a_{\pi_i}})$, where

$$M(\chi_{a_0}) = \begin{pmatrix} 1 & 0 & 0 \\ 0 & 1 & 0 \\ 0 & 0 & 1 \end{pmatrix}, \qquad M(\chi_{a_1}) = \begin{pmatrix} 0 & 0 & 1 \\ 1 & 0 & 0 \\ 0 & 1 & 0 \end{pmatrix}, \qquad M(\chi_{a_2}) = \begin{pmatrix} 0 & 1 & 0 \\ 0 & 0 & 1 \\ 1 & 0 & 0 \end{pmatrix}.$$

We compute

$$\widehat{\gamma}_0 = (2u^2 + u + 2, 2u^3 + 2u^2 + 2u + 1, u^3 + 2u^2), \qquad \widehat{\gamma}_1 = (u^3 + u + 2, 2u + 1, 2u^3),$$
$$\widehat{\gamma}_2 = (u^2 + 2u, 2u^3 + u^2 + 2, u^3 + u^2 + u + 1), \qquad \widehat{\gamma}_3 = (u^3 + 2u^2 + u, 2u^2 + 2u + 2, 2u^3 + 2u^2 + 1),$$
$$\widehat{\gamma}_4 = (u + 1, 2u^3 + 2u, u^3 + 2), \qquad \widehat{\gamma}_5 = (u^3 + u^2 + 1, u^2 + u, 2u^3 + u^2 + 2u + 2).$$

Afterwards, we have to set the scaling factors σ and to compute the layered parity-check matrix from the sequence β and $\widehat{\gamma}$. Since we would like to end up with a Goppa code (to be able to use the superior error-correction capabilities of "equal magnitude" decoding described in Appendix B), we will set all $\sigma_i = 1$ and the Cauchy layered exponent to be $r = 1$.

Finally, using the QMTRACE step, we can produce the subfield subcode

$$H=\left(\begin{array}{ccc|ccc|ccc|ccc|ccc|ccc}
2&0&2&1&1&1&0&2&1&1&1&0&2&1&0&2&0&0\\
2&2&0&1&1&1&1&0&2&0&1&1&0&2&1&0&2&0\\
0&2&2&1&1&1&2&1&0&1&0&1&1&0&2&0&0&2\\ \hline
2&0&2&0&2&2&2&0&2&0&1&1&2&1&0&1&1&0\\
2&2&0&2&0&2&2&2&0&1&0&1&0&2&1&0&1&1\\
0&2&2&2&2&0&0&2&2&1&1&0&1&0&2&1&0&1\\ \hline
2&2&1&2&1&2&0&0&2&1&0&2&1&1&1&0&1&0\\
1&2&2&2&2&1&2&0&0&2&1&0&1&1&1&0&0&1\\
2&1&2&1&2&2&0&2&0&0&2&1&1&1&1&1&0&0\\ \hline
0&0&1&1&0&1&2&0&2&0&0&0&0&1&1&0&0&1\\
1&0&0&1&1&0&2&2&0&0&0&0&1&0&1&1&0&0\\
0&1&0&0&1&1&0&2&2&0&0&0&1&1&0&0&1&0
\end{array}\right),\quad H_{\mathrm{sys}}=\left(\begin{array}{ccc|ccc|ccc|ccc|ccc|ccc}
\mathbf{2}&1&0&\mathbf{2}&1&1&1&0&0&0&0&0&0&0&0&0&0&0\\
\mathbf{0}&2&1&\mathbf{1}&2&1&0&1&0&0&0&0&0&0&0&0&0&0\\
\mathbf{1}&0&2&\mathbf{1}&1&2&0&0&1&0&0&0&0&0&0&0&0&0\\ \hline
\mathbf{0}&0&0&\mathbf{1}&2&0&0&0&0&1&0&0&0&0&0&0&0&0\\
\mathbf{0}&0&0&\mathbf{0}&1&2&0&0&0&0&1&0&0&0&0&0&0&0\\
\mathbf{0}&0&0&\mathbf{2}&0&1&0&0&0&0&0&1&0&0&0&0&0&0\\ \hline
\mathbf{0}&2&2&\mathbf{1}&2&2&0&0&0&0&0&0&1&0&0&0&0&0\\
\mathbf{2}&0&2&\mathbf{2}&1&2&0&0&0&0&0&0&0&1&0&0&0&0\\
\mathbf{2}&2&0&\mathbf{2}&2&1&0&0&0&0&0&0&0&0&1&0&0&0\\ \hline
\mathbf{1}&2&2&\mathbf{1}&0&2&0&0&0&0&0&0&0&0&0&1&0&0\\
\mathbf{2}&1&2&\mathbf{2}&1&0&0&0&0&0&0&0&0&0&0&0&1&0\\
\mathbf{2}&2&1&\mathbf{0}&2&1&0&0&0&0&0&0&0&0&0&0&0&1
\end{array}\right).$$

The systematic form H_{sys} can be computed by inverting the matrix consisting of the last trm columns. From the systematic parity-check matrix, QMSIGNATURE extracts those entries marked in boldface, which are sufficient to describe the public generator matrix

$$G_{\mathrm{sys}}=\left(\begin{array}{ccc|ccc|ccc|ccc|ccc|ccc}
1&0&0&0&0&0&\mathbf{2}&\mathbf{0}&\mathbf{1}&\mathbf{0}&\mathbf{0}&\mathbf{0}&\mathbf{0}&\mathbf{2}&\mathbf{2}&\mathbf{1}&\mathbf{2}&\mathbf{2}\\
0&1&0&0&0&0&1&2&0&0&0&0&2&0&2&2&1&2\\
0&0&1&0&0&0&0&1&2&0&0&0&2&2&0&2&2&1\\ \hline
0&0&0&1&0&0&\mathbf{2}&\mathbf{1}&\mathbf{1}&\mathbf{1}&\mathbf{0}&\mathbf{2}&\mathbf{1}&\mathbf{2}&\mathbf{2}&\mathbf{1}&\mathbf{2}&\mathbf{0}\\
0&0&0&0&1&0&1&2&1&2&1&0&2&1&2&0&1&2\\
0&0&0&0&0&1&1&1&2&0&2&1&2&2&1&2&0&1
\end{array}\right).$$

Note that for the parity-check matrix, the signature of each monoidic block is its first column, but for the generator, which contains transposed monoidic blocks, the signature is the first row.

B Decoding Square-Free Goppa Codes

For codes with degree t and its average distance at least $(4/p)t+1$, the proposed decoder can uniquely correct $(2/p)t$ errors, with high probability. The correction capability is higher if the distribution of error magnitudes is not uniform, approaching or reaching t errors when any particular error value occurs much more often than others or exclusively. The parity-check matrix used by this algorithm is in the form (1).

At some point of this algorithm, we will call the WEAKPOPOVFORM algorithm (also present in [BLM10] and described below) to find the short vectors in the lattice spanned by the rows of

$$A=\begin{bmatrix} g & 0\,0\ldots\,0 \\ -v_1 & 1\,0\ldots\,0 \\ -v_2 & 0\,1\ldots\,0 \\ \vdots & \vdots\;\vdots\;\ddots\;\vdots \\ -v_{p-1} & 0\,0\ldots\,1 \end{bmatrix}, \tag{5}$$

Where g is the Goppa polynomial and the v_i's values will be computed through the execution of Algorithm 1.

Algorithm 1. Decoding p-ary square-free Goppa codes

INPUT: $\Gamma(L, g)$, a Goppa code over $\mathbb{F}_p$ where g is square-free.
INPUT: $H \in \mathbb{F}_q^{r \times n}$, a parity-check matrix in the form of Equation 1.
INPUT: $c' = c + e \in \mathbb{F}_p^n$, the received codeword with errors.
OUTPUT: set of corrected codeword $c \in \Gamma(L, g)$ ($\varnothing$ upon failure).

```
s^T ← Hc'^T ∈ F_q^n, s_e(x) ← Σ_i s_i x^i. ▷ N.B. Hc'^T = He^T.
if ∄ s_e^{-1}(x) mod g(x) then
    return ∅ ▷ g(x) is composite
end if
S ← ∅
for φ ← 1 to p − 1 do ▷ guess the correct scale factor φ
    for k ← 1 to p − 1 do
        u_k(x) ← x^k + φ k x^{k−1} / s_e(x)  mod g(x)
        if ∄ (u_k(x))^{1/p} mod g(x) then
            try next φ ▷ g(x) is composite
        end if
        v_k(x) ← (u_k(x))^{1/p} mod g(x)
    end for
    Build the lattice basis A defined by Equation 5.
    Apply WEAKPOPOVFORM (Algorithm 2) to reduce the basis of Λ(A).
    for i ← 1 to p do
        a ← A_i ▷ with a_j indices in range 0...p − 1
        for j ← 0 to p − 1 do
            if deg(a_j) > ⌊(t − j)/p⌋ then
                try next i ▷ not a solution
            end if
        end for
        σ(x) ← Σ_j x^j a_j(x)^p
        Compute the set J such that σ(L_j) = 0, ∀j ∈ J.
        for j ∈ J do
            Compute the multiplicity μ_j of L_j.
            e_j ← φ μ_j
        end for
        if He^T = s^T then
            S ← S ∪ {c' − e}
        end if
    end for
    return S
end for
```

Algorithm 2. (WEAKPOPOVFORM) Computing the weak Popov form

INPUT: $A \in \mathbb{F}_q[x]^{p \times p}$ in the form of Equation 5.
OUTPUT: weak Popov form of A.

▷ Compute I^A:
for $j \leftarrow 1$ **to** p **do**
    $I^A_j \leftarrow$ **if** $\deg(A_{j,1}) > 0$ **then** 1 **else** j
end for
▷ Put A in weak Popov form:
while $\mathrm{rep}(I^A) > 1$ **do**
    ▷ Find suitable k and ℓ to apply simple transform of first kind:
    for $k \leftarrow 1$ **to** p **such that** $I^A_k \neq 0$ **do**
        for $\ell \leftarrow 1$ **to** p **such that** $\ell \neq k$ **do**
            while $\deg(A_{\ell, I^A_k}) \geqslant \deg(A_{k, I^A_k})$ **do**
                $c \leftarrow \mathrm{lead}(A_{\ell, I^A_k}) / \mathrm{lead}(A_{k, I^A_k})$
                $e \leftarrow \deg(A_{\ell, I^A_k}) - \deg(A_{k, I^A_k})$
                $A_\ell \leftarrow A_\ell - c x^e A_k$
            end while
            ▷ Update I^A_ℓ and hence $\mathrm{rep}(I^A)$ if necessary:
            $d \leftarrow \max\{\deg(A_{\ell,j}) \mid j = 1, \ldots, p\}$
            $I^A_\ell \leftarrow \max\{j \mid \deg(A_{\ell,j}) = d\}$
        end for
    end for
end while
return A

Simplified High-Speed High-Distance List Decoding for Alternant Codes

Daniel J. Bernstein

Department of Computer Science
University of Illinois at Chicago, Chicago, IL 60607–7045, USA
djb@cr.yp.to

Abstract. This paper presents a simplified list-decoding algorithm to correct any number w of errors in any alternant code of any length n with any designed distance $t+1$ over any finite field $\mathbf{F}_q$; in particular, in the classical Goppa codes used in the McEliece and Niederreiter public-key cryptosystems. The algorithm is efficient for w close to, and in many cases slightly beyond, the $\mathbf{F}_q$ Johnson bound $J' = n' - \sqrt{n'(n'-t-1)}$ where $n' = n(q-1)/q$, assuming $t+1 \le n'$. In the typical case that $qn/t \in (\lg n)^{O(1)}$ and that the parent field has $(\lg n)^{O(1)}$ bits, the algorithm uses $n(\lg n)^{O(1)}$ bit operations for $w \le J' - n/(\lg n)^{O(1)}$; $O(n^{4.5})$ bit operations for $w \le J' + o((\lg n)/\lg\lg n)$; and $n^{O(1)}$ bit operations for $w \le J' + O((\lg n)/\lg\lg n)$.

1 Introduction

Take any prime power q; integer $m \ge 1$; integer $n \ge m$ with $n \le q^m$; integer $t \ge 1$ with $t \le n/m$; distinct $\alpha_1, \ldots, \alpha_n \in \mathbf{F}_{q^m}$; and nonzero $\beta_1, \ldots, \beta_n \in \mathbf{F}_{q^m}$. Define

$$C = \big\{(\beta_1 f(\alpha_1), \ldots, \beta_n f(\alpha_n)) : \\ f \in \mathbf{F}_{q^m}[x]; \quad \deg f < n - t; \quad \beta_i f(\alpha_i) \in \mathbf{F}_q \text{ for each } i\big\}.$$

This set C is an $[n, \ge n-mt, \ge t+1]$ linear code over $\mathbf{F}_q$. In other words: it is a subspace of the $\mathbf{F}_q$-vector space $\mathbf{F}_q^n$; it has dimension at least $n-mt$, i.e., at least q^{n-mt} elements; and any two distinct elements of it have Hamming distance at least $t+1$, i.e., differ in at least $t+1$ coordinates.

Any code C defined as above is called an **alternant code**. This class of codes was introduced by Helgert in [**38**], independently by Chien and Choy in [**22**], and independently by Delsarte in [**28**]. The class includes binary Reed–Solomon codes, which had been introduced by Reed and Solomon in [**47**]; BCH codes, which had been introduced by Hocquenghem in [**39**] and independently by Bose and Ray-Chaudhuri in [**16**]; various odd-characteristic generalizations

---

This work was supported in part by NIST grant 60NANB10D263 and in part by the Cisco University Research Program. Permanent ID of this document: f529c2ab14c4ec22244b0a3f0190089b. Date: 2011.09.10.

B.-Y. Yang (Ed.): PQCrypto 2011, LNCS 7071, pp. 200–216, 2011.

introduced by Gorenstein and Zierler in [**33**]; and classical Goppa codes, which had been introduced by Goppa in [**31**] and [**32**].

The w-error-correction problem for C is the problem of finding $c \in C$, given a vector at distance w from c. For $w \leq \lfloor t/2 \rfloor$ the vector dictates a unique possibility for c, but this does not mean that c is easy to find. There are at least q^{n-mt} codewords, and the cost of enumerating them all is exponential in n, except in the (rarely used) case that t is very close to n/m. Fortunately, early research produced much better algorithms for the $\lfloor t/2 \rfloor$-error-correction problem:

- Peterson in [**46**] introduced an algorithm using $n^{O(1)}$ arithmetic operations in $\mathbf{F}_{q^m}$. Each of those operations uses a polynomial number of bit operations, under the extremely weak assumption that q^m has $n^{O(1)}$ bits. Applications typically choose q^m to have only $O(\lg n)$ bits.
- Berlekamp in [**7**] introduced an algorithm using only $O(n^2)$ operations in $\mathbf{F}_{q^m}$. If q^m has $(\lg n)^{O(1)}$ bits then each operation in $\mathbf{F}_{q^m}$ uses $(\lg n)^{O(1)}$ bit operations, so Berlekamp's algorithm uses $n^2(\lg n)^{O(1)}$ bit operations.
- Justesen in [**42**], and independently Sarwate as reported in [**48**], introduced an algorithm using only $n(\lg n)^{2+o(1)}$ operations in $\mathbf{F}_{q^m}$. If q^m has only $(\lg n)^{O(1)}$ bits then this algorithm uses only $n(\lg n)^{O(1)}$ bit operations.

What about $w > \lfloor t/2 \rfloor$? The big-field Johnson bound states that there are only polynomially many possibilities for c if $w < n - \sqrt{n(n-t-1)}$, assuming $t+1 \leq n$. Guruswami and Sudan, in a famous 1998 paper [**35**], introduced a polynomial-time algorithm to compute the list of possibilities for c if $w < n - \sqrt{n(n-t-1)}$. An intermediate range of w was already covered by an algorithm of Sudan in [**50**], but [**35**] introduced "multiplicities" to push w much higher.

Even better, the $\mathbf{F}_q$ Johnson bound states that there are only polynomially many possibilities for c if $w < n' - \sqrt{n'(n'-t-1)}$ where $n' = n(q-1)/q$, assuming $t+1 \leq n'$ and $q \in n^{O(1)}$. In 2000 Koetter and Vardy introduced a polynomial-time algorithm to compute the list of possibilities; see [**34**, Section 6.3.8]. Compared to the unique-decoding case $w = \lfloor t/2 \rfloor$, the big-field Johnson bound extends the distance by approximately $t^2/8n$, and the $\mathbf{F}_q$ Johnson bound further extends the distance by approximately $t^2/8n(q-1)$; this improvement is particularly impressive for $q = 2$.

Unfortunately, "polynomial time" does not mean fast. Several subsequent papers have improved the complexity of list decoding, but each paper fails at least one, if not all, of the following desiderata:

- Speed. For example, the recent paper [**4**] reports list-decoding cost "quadratic in the blocklength n" (counting the number of operations in $\mathbf{F}_{q^m}$); but this is asymptotically much larger than the $n(\lg n)^{2+o(1)}$ that had been achieved decades earlier for $w = \lfloor t/2 \rfloor$.
- Effectiveness (how many errors are decoded). For example, the recent paper [**52**] is limited to the big-field Johnson distance, significantly below the $\mathbf{F}_q$ Johnson distance if q is small.
- Simplicity. For example, [**3**]—one of the few papers reporting essentially-linear-time list decoding—is sufficiently general to handle arbitrary weights,

such as the optimized Koetter–Vardy weights; but the user is required to trace the desired weights (after scaling and rounding to integers) through a thicket of degree computations.

It seems that every implementation of code-based cryptography avoids list decoding, even though [**11**, Section 7] pointed out years ago that list decoding improves the tradeoff between key size and security level against all known attacks. One can blame the non-use of list decoding on the lack of simple high-speed high-distance decoding algorithms. Some list-decoding papers try to compensate by adding generality, for example studying higher-genus algebraic-geometry codes, but if list decoding is not usable even for the most basic constructions of alternant codes then obviously it will also not be usable for higher-genus codes!

This paper presents a list-decoding algorithm that is simultaneously (1) fast, (2) effective, and (3) simple. The algorithm continues to work for arbitrarily large values of w, although its speed degrades as w approaches and passes the $\mathbf{F}_q$ Johnson bound. Specifically, in the typical case that n/t, q, and $\lg q^m$ are all in $(\lg n)^{O(1)}$, the algorithm uses

- $n(\lg n)^{O(1)}$ bit operations for $w \le n' - \sqrt{n'(n'-t-1)} - n/(\lg n)^{O(1)}$;
- $O(n^{4.5})$ bit operations for $w \le n' - \sqrt{n'(n'-t-1)} + o((\lg n)/\lg\lg n)$; and
- $n^{O(1)}$ bit operations for $w \le n' - \sqrt{n'(n'-t-1)} + O((\lg n)/\lg\lg n)$.

Note that the $n(\lg n)^{O(1)}$ bound does not imply competitive speed with other $n(\lg n)^{O(1)}$ algorithms; it merely implies that the speed ratio is bounded by $(\lg n)^{O(1)}$. However, the $O(n^{4.5})$ bound allows easy comparisons to, e.g., the $n^{7+o(1)}$ achieved in [**4**, Corollary 5.8] for w slightly below $n' - \sqrt{n'(n'-t-1)}$, or the $n^{6+o(1)}$ achieved in [**6**] for w slightly below $n - \sqrt{n(n-t-1)}$.

The word "simplified" in the title might suggest that I obtained this algorithm by starting from an existing acceleration of the Koetter–Vardy algorithm and simplifying it. I actually obtained the algorithm in a completely different way. I started with a very simple algorithm by Howgrave-Graham that was published in 1997 and that was subsequently understood to have the same decoding capability as the Guruswami–Sudan algorithm. I then tweaked the Howgrave-Graham algorithm to match the Koetter–Vardy results. The Howgrave-Graham algorithm does not seem to be widely known among coding theorists, including those working on code-based cryptography; see Section 3 and [**9**] for further discussion of the history.

2 Review of Fast Arithmetic

This section reviews several standard subroutines for fast multiplication, fast lattice-basis reduction, etc.

All of the algorithms here are $\mathbf{F}_{q^m}$-algebraic algorithms, i.e., sequences of additions, subtractions, multiplications, divisions, and comparisons of elements of $\mathbf{F}_{q^m}$. For a formal definition of this model of computation see, e.g., [**18**]. **Cost** here refers to total algebraic complexity over $\mathbf{F}_{q^m}$, i.e., the number of arithmetic operations performed in $\mathbf{F}_{q^m}$.

The weak assumption $\lg q^m \in (\lg n)^{O(1)}$ implies that each of these operations in $\mathbf{F}_{q^m}$ can be carried out using $(\lg n)^{O(1)}$ bit operations. The weaker assumption $\lg q^m \in n^{O(1)}$ implies that each of these operations in $\mathbf{F}_{q^m}$ can be carried out using $n^{O(1)}$ bit operations.

Fast multiplication. Multiplying two d-coefficient polynomials in $\mathbf{F}_{q^m}[x]$—i.e., two polynomials of degree below d—costs $d(\lg d)^{1+o(1)}$. See, e.g., my online survey paper [**8**, Section 4] for algorithmic details and credits.

Fast multiplication of many inputs. Computing a product of d linear polynomials costs $d(\lg d)^{2+o(1)}$. See, e.g., [**8**, Section 12].

Fast evaluation. Computing $Y(x_1), Y(x_2), \ldots, Y(x_d)$, given $x_1, \ldots, x_d \in \mathbf{F}_{q^m}$ and a d-coefficient polynomial $Y \in \mathbf{F}_{q^m}[x]$, costs $d(\lg d)^{2+o(1)}$. See, e.g., [**8**, Section 18].

Fast interpolation. For any distinct $x_1, \ldots, x_d \in \mathbf{F}_{q^m}$ and any $y_1, \ldots, y_d \in \mathbf{F}_{q^m}$ there is a unique polynomial $Y \in \mathbf{F}_{q^m}[x]$ of degree below d having $Y(x_1) = y_1$, $Y(x_2) = y_2$, and so on through $Y(x_d) = y_d$. Computing this polynomial Y from $x_1, \ldots, x_d, y_1, \ldots, y_d$ costs $d(\lg d)^{2+o(1)}$. See, e.g., [**8**, Section 23].

Fast lattice-basis reduction. If an $\ell \times \ell$ matrix over $\mathbf{F}_{q^m}[x]$ has nonzero determinant D then there is a nonzero linear combination Q of the matrix columns such that $\deg Q \leq (\deg D)/\ell$. Here $\deg Q$ means the maximum degree of the entries of Q.

If each of the matrix entries is a d-coefficient polynomial then computing such a Q costs $\ell^{\Omega} d(\lg \ell d)^{O(1)}$ by [**30**, Theorem 3.8]. Here Ω is any positive real number such that $\ell \times \ell$ matrix multiplication costs $O(\ell^{\Omega})$. One can trivially take $\Omega = 3$, but state-of-the-art matrix-multiplication techniques have pushed Ω below 2.5.

There is an error in the proof of [**30**, Theorem 3.8]: the authors assume, without justification, that they can quickly find $x_0 \in \mathbf{F}_{q^m}$ such that $D(x_0) \neq 0$. Unfortunately, it is entirely possible that *every* $x_0 \in \mathbf{F}_{q^m}$ will have $D(x_0) = 0$; in such cases, the algorithm stated in [**30**, Section 3] will fail. The simplest workaround is to replace $\mathbf{F}_{q^m}$ by an extension having significantly more than $\deg D$ elements; extension degree $(\lg \ell d)^{O(1)}$ always suffices, leaving the cost bound $\ell^{\Omega} d(\lg \ell d)^{O(1)}$ unaffected. (Extension degree 2 suffices for the matrix shape used later in this paper, since D visibly splits into linear factors in $\mathbf{F}_{q^m}[x]$.)

A closer look at the algorithm in [**30**] shows that the cost is $d(\lg d)^{2+o(1)}$ if ℓ and the required extension degree are bounded by $(\lg d)^{o(1)}$. The same complexity also appeared later in [**3**]. As ℓ increases, the algorithm in [**3**] scales as $\ell^{3+o(1)}$ rather than $\ell^{\Omega+o(1)}$.

Fast root-finding. The traditional factorization method for a polynomial in $\mathbf{Q}[y]$, introduced by Zassenhaus in [**55**] four decades ago, begins with a factorization of the polynomial modulo a small prime number p, and then uses Newton iteration ("Hensel's lemma") to lift the factorization to factorizations modulo p^2, p^4, etc. A few Newton steps produce enough p-adic precision to determine the factorization in $\mathbf{Q}[y]$; see, e.g., [**29**, Theorem 15.20]. This procedure relies on a preliminary "squarefree factorization" of the polynomial, but that factorization has essentially linear cost; see [**29**, Theorem 14.23].

In the case of linear factors (i.e., roots) the entire factorization procedure uses $\ell^{2+o(1)}d(\lg d)^{2+o(1)}$ bit operations for ℓ-coefficient polynomials with d-bit integer coefficients; see [**29**, Theorem 15.21]. There has been a tremendous amount of research on algorithms for the first step, factoring in $(\mathbf{Z}/p)[y]$, but rather naive algorithms are adequate if ℓ is much smaller than d and if one allows randomization. There has also been a tremendous amount of research on algorithms to handle higher-degree factors, but for this paper linear factors are adequate.

One can obtain essentially the same speed by computing approximate roots in $\mathbf{R}$ with an analogous Newton iteration, but working with the p-adic numbers $\mathbf{Q}_p$ is simpler because it avoids roundoff error. There are still a few technical details that require attention: one must avoid primes p that divide denominators of the original coefficients; one must also avoid primes p that create new squared factors. There are not many bad choices of p; see [**29**, Lemma 15.1].

Zassenhaus's method is not limited to the rational number field $\mathbf{Q}$. Replacing $\mathbf{Q}$ by the rational function field $\mathbf{F}_{q^m}(x)$, and replacing the small prime p of $\mathbf{Z}$ by a small irreducible element p of $\mathbf{F}_{q^m}[x]$, produces a factorization method for $\mathbf{F}_{q^m}(x)[y]$; see, e.g., [**29**, Theorem 15.23]. Squarefree factorization becomes slightly more complicated, as discussed in [**29**, page 447], but is still fast. The cost for the initial factorization modulo p is $\ell^{2+o(1)}(\lg q^m)^{1+o(1)}$ by [**29**, Theorem 14.14]. There are subquadratic factorization algorithms in the literature, but this refinement is not necessary for this paper.

The root-finding conclusion that matters for this paper—the polynomial analogue of [**29**, Theorem 15.21]—is the following. There is a standard algorithm that, given a nonzero polynomial $Q \in \mathbf{F}_{q^m}(x)[y]$, finds all y-roots of Q. If Q is an ℓ-coefficient polynomial (in y), each coefficient in turn being a d-coefficient polynomial (in x), then the entire procedure costs $\ell^{2+o(1)}((\lg q^m)^{1+o(1)}+d(\lg d)^{2+o(1)})$. Note that this cost bound is influenced by $\lg q^m$, the number of bits of the parent field $\mathbf{F}_{q^m}$; one needs to put limits on q^m not merely to control the translation from cost into bit operations, but also to control the cost of factorization.

3 Correcting Nearly $n - \sqrt{n(n-t-1)}$ Errors

This section states a simple high-speed list-decoding algorithm that corrects errors up to the big-field Johnson bound. The algorithm in the next section is more general and more powerful, correcting more errors; but the algorithm in this section is slightly simpler, and the reader is encouraged to read it first.

Parameters. This algorithm has three parameters: a positive integer $w \le n$, the number of errors to be corrected; an integer $k \ge 0$; and an integer $\ell \ge k$. The algorithm assumes that $t+1 \le n$ and that these parameters satisfy

$$n\frac{k(k+1)}{2} + (n-t-1)\frac{\ell(\ell-1)}{2} < \ell k(n-w),$$

i.e., $(1-(t+1)/n)(1-1/\ell) < (1-w/n)^2 - (1-w/n-k/\ell)^2 - k/\ell^2$.

One can take ℓ in $O(n^2)$ for any w smaller than the big-field Johnson bound. My main interest is in the case $\ell \in (\lg n)^{O(1)}$, achievable when there is a noticeable gap between w and the big-field Johnson bound. Further notes on parameter selection appear below. The total cost of the algorithm will turn out to be bounded by

- $n(\lg n)^{O(1)}$ if $\ell \in (\lg n)^{O(1)}$ and $\lg q^m \in (\lg n)^{O(1)}$; and by
- $n^{\Omega+2+o(1)}$ if $\ell \in O(n)$ and $\lg q^m \in O(n^{\Omega})$; and by
- $n^{2\Omega+3+o(1)}$ if $\ell \in O(n^2)$ and $\lg q^m \in O(n^{2\Omega-1})$; and by
- $n^{O(1)}$ if $\ell \in O(n^2)$ and $\lg q^m \in n^{O(1)}$.

For example, Step 2 below costs $\ell^3 n(\lg \ell n)^{1+o(1)}$, which is visibly within each of these bounds. I will state the cost of each step as a function of ℓ, n, and (when relevant) q^m.

Input and output. The algorithm input is a vector $v \in \mathbf{F}_q^n$. The algorithm output is the set of $c \in C$ of Hamming distance at most w from v.

Step 1: initial interpolation. Compute the polynomial $A = (x - \alpha_1)(x - \alpha_2)\cdots(x - \alpha_n) \in \mathbf{F}_{q^m}[x]$. Also compute the unique polynomial $V \in \mathbf{F}_{q^m}[x]$ with $\deg V < n$ satisfying $V(\alpha_1) = v_1/\beta_1$, $V(\alpha_2) = v_2/\beta_2$, and so on through $V(\alpha_n) = v_n/\beta_n$. This costs $n(\lg n)^{2+o(1)}$.

Step 2: lattice-basis construction. Define $X = x^{n-t-1}$ and $F = Xy - V \in \mathbf{F}_{q^m}[x, y]$. Compute the ℓ polynomials

$$
\begin{aligned}
M_0 &= A^k;\\
M_1 &= A^{k-1}F = A^{k-1}Xy - A^{k-1}V;\\
M_2 &= A^{k-2}F^2 = A^{k-2}X^2y^2 - 2A^{k-2}XVy + A^{k-2}V^2;\\
&\vdots\\
M_{k-1} &= AF^{k-1} = AX^{k-1}y^{k-1} - \cdots;\\
M_k &= F^k = X^k y^k - \cdots;\\
M_{k+1} &= F^{k+1} = X^{k+1}y^{k+1} - \cdots;\\
&\vdots\\
M_{\ell-1} &= F^{\ell-1} = X^{\ell-1}y^{\ell-1} - \cdots
\end{aligned}
$$

in $\mathbf{F}_{q^m}[x, y]$. If $\ell = k$ then $M_{\ell-1}$ is defined as AF^{k-1}, not $F^{\ell-1}$. (One can save time by replacing $F^k, F^{k+1}, \ldots, F^{\ell-1}$ with $F^k, XyF^k, \ldots, (Xy)^{\ell-1-k}F^k$, but the speedup is not visible at the level of detail of the analysis below.)

The coefficients of powers of y here form an $\ell \times \ell$ triangular matrix. There are several straightforward ways to compute all of the matrix entries with a total of $O(\ell^2)$ multiplications in $\mathbf{F}_{q^m}[x]$, each multiplication involving polynomials of degree $O(\ell n)$. The total cost is just $\ell^3 n(\lg \ell n)^{1+o(1)}$.

Step 3: lattice-basis reduction. The determinant of the aforementioned $\ell \times \ell$ matrix of coefficients of $M_0, \ldots, M_{\ell-1}$ is the product of the diagonal entries of the matrix (since the matrix is triangular), i.e., the product of the leading coefficients of $M_0, \ldots, M_{\ell-1}$, namely

$$A^k \cdot A^{k-1}X \cdot A^{k-2}X^2 \cdots X^k \cdot X^{k+1} \cdots X^{\ell-1} = A^{k(k+1)/2}X^{\ell(\ell-1)/2},$$

of degree $nk(k+1)/2 + (n-t-1)\ell(\ell-1)/2$. Inside the lattice $\mathbf{F}_{q^m}[x]M_0 + \cdots + \mathbf{F}_{q^m}[x]M_{\ell-1} \subseteq \mathbf{F}_{q^m}[x,y]$ find a nonzero polynomial Q having x-degree at most $(nk(k+1)/2 + (n-t-1)\ell(\ell-1)/2)/\ell$, and therefore x-degree below $k(n-w)$. This costs $\ell^\Omega n\ell(\lg \ell^2 n)^{O(1)} = \ell^{\Omega+1}n(\lg \ell n)^{O(1)}$.

Step 4: factorization. Compute all $f \in \mathbf{F}_{q^m}[x]$ such that $Q(x, f/X) = 0$; i.e., compute all factors of Q having the form $y - f/X$ with $f \in \mathbf{F}_{q^m}[x]$. Note that there are at most $\ell - 1$ such factors, since Q has y-degree at most $\ell - 1$. This costs $\ell^{2+o(1)}((\lg q^m)^{1+o(1)} + n\ell(\lg \ell n)^{2+o(1)})$.

For each polynomial $f \in \mathbf{F}_{q^m}[x]$ such that $Q(x, f/X) = 0$ and $\deg f < n - t$: Compute $c = (\beta_1 f(\alpha_1), \ldots, \beta_n f(\alpha_n)) \in \mathbf{F}_{q^m}^n$. Output c if $c \in \mathbf{F}_q^n$ and $|c - v| \le w$, where $|c - v|$ means the Hamming weight of $c - v$. This costs $n(\lg n)^{2+o(1)}$.

Why the algorithm works. Each output c from the algorithm is checked, in Step 4, to be an element of C with $|c - v| \le w$.

Conversely, consider any $c \in C$ with $|c - v| \le w$. There is a polynomial $f \in \mathbf{F}_{q^m}[x]$ with $\deg f < n - t$ such that $c = (\beta_1 f(\alpha_1), \ldots, \beta_n f(\alpha_n))$. The goal is to show that the algorithm outputs c; equivalently, that f is found in Step 4 of the algorithm.

The hypothesis $|c - v| \le w$ means that there are at least $n - w$ indices i for which $c_i = v_i$; i.e., for which $\beta_i f(\alpha_i) = \beta_i V(\alpha_i)$; i.e., for which α_i is a root of $f - V$. In other words, $\gcd\{A, f - V\}$ has degree at least $n - w$.

Consider the map $y \mapsto f/X$ from $\mathbf{F}_{q^m}[x, Xy]$ to $\mathbf{F}_{q^m}[x]$. The image of $F = Xy - V$ is $f - V$, so the images of $M_0, M_1, \ldots, M_{\ell-1}$ are $A^k, A^{k-1}(f-V), \ldots, (f-V)^k, \ldots, (f-V)^\ell$. Each of these polynomials is divisible by $\gcd\{A, f - V\}^k$. The image of Q, namely $Q(x, f/X)$, is therefore also divisible by $\gcd\{A, f - V\}^k$.

Write Q as $Q_0 + Q_1 y + \cdots + Q_{\ell-1}y^{\ell-1}$. Then $Q(x, f/X) = Q_0 + Q_1(f/X) + \cdots + Q_{\ell-1}(f/X)^{\ell-1}$. Each Q_i has degree below $k(n-w)$, and f/X has degree at most 0, so $Q(x, f/X)$ has degree below $k(n-w)$; but $Q(x, f/X)$ is divisible by $\gcd\{A, f - V\}^k$, which has degree at least $k(n-w)$. Consequently $Q(x, f/X) = 0$ as claimed.

Notes on parameter selection. Suitable k, ℓ exist with $\ell \in O(nt)$ whenever w is smaller than the big-field Johnson bound. For example, the integers $k = (n-w)(t+1) \ge 0$ and $\ell = n(t+1) > k$ have $(1 - (t+1)/n)(1 - 1/\ell) = 1 - (t+1)/n - 1/\ell + 1/n^2$ and $(1 - w/n - k/\ell)^2 + k/\ell^2 = (1 - w/n)/\ell < 1/\ell$; so w, k, ℓ are in the parameter space if $(1 - w/n)^2 \ge 1 - (t+1)/n + 1/n^2$, i.e., if $(n-w)^2 \ge n(n-t-1) + 1$. Both $(n-w)^2$ and $n(n-t-1)$ are integers, so this condition is equivalent to $(n-w)^2 > n(n-t-1)$, i.e., $w < n - \sqrt{n(n-t-1)}$.

This choice of ℓ is simpler and smaller than the choice made in [**36**, Lemma 7 and Proposition 9]. Here is an absurdly large numerical example to illustrate the worst-case asymptotics: for $n = 1000007$ and $t = 67774$ and $w = 34482$, one can take $k = 65438456875$ and $\ell = 67775474425$, while [**36**, Lemma 7] chooses $k = 932238525625$.

My main interest is in much smaller values of ℓ. Choosing k as $\lfloor(1 - w/n)\ell\rfloor$ guarantees $0 \le k < \ell$ since $w > 0$, and guarantees $(1 - w/n - k/\ell)^2 + k/\ell^2 < 1/\ell^2 + 1/\ell$, so w, k, ℓ are in the parameter space if $(1 - w/n)^2 \ge (1 - (t + 1)/n)(1 - 1/\ell) + 1/\ell^2 + 1/\ell$; i.e., $(1 - w/n)^2 \ge 1 - (t + 1)/n + (t + 1)/n\ell + 1/\ell^2$; i.e., $(1 - w/n)^2 - (1 - J/n)^2 \ge (t + 1)/n\ell + 1/\ell^2$ where J is the big-field Johnson bound; i.e., $J - w \ge ((t + 1)/\ell + n/\ell^2)/(2 - w/n - J/n)$. One can achieve this with $\ell \in (\lg n)^{O(1)}$ if $J - w$ is at least $n/(\lg n)^{O(1)}$.

There are limits to how far this idea can be pushed. For example, it is tempting to take k, ℓ as constants, so that cost factors such as ℓ^2 can be replaced by $O(1)$. The same replacement was used to justify, e.g., the statement "quadratic in the blocklength n" in [**4**, Abstract]. Apparently it is not instantly obvious that—at least for small q, such as the case $q = 2$ highlighted in [**4**]—this replacement is fundamentally flawed!

The difficulty is the following. If q is constant, or more generally $n^{o(1)}$, then $t \in o(n)$, so $J - t/2 \in o(t)$. Choosing $k, \ell \in O(1)$ then forces w to be smaller than $\lfloor t/2 \rfloor$ for all sufficiently large n: in other words, the algorithm cannot correct more errors than Berlekamp's algorithm once n is sufficiently large. For the same reason, the "quadratic" claim in [**4**] is content-free: it might be true that taking constants k, ℓ limits the algorithm in [**4**] to cost $O(n^2)$, but then the algorithm cannot correct more errors than a trivial combination of brute-force list decoding for small n and Berlekamp's algorithm for large n, which also costs $O(n^2)$.

Of course, this criticism does not apply to bounds that treat ϵ, k, ℓ as variables, such as the bound $O(n^2/\epsilon^5)$ in [**4**, Corollary 5.7]. Furthermore, the "rational" algorithms of [**54**] and [**10**] allow a better tradeoff between k, ℓ, w and can meaningfully take $k, \ell \in O(1)$.

History. Håstad showed in 1988 that one could find all small roots of a polynomial modulo a large integer N by applying the famous LLL lattice-basis reduction algorithm. The same result was found independently by Vallée, Girault, and Toffin in 1989. See [**37**] and [**53**].

Coppersmith, in a famous 1996 paper, incorporated multiplicities into the Vallée–Girault–Toffin algorithm, drastically increasing the range of roots that could be found. Coppersmith also showed that similar lattices could be used to find not merely polynomial values that are *multiples* of N but also polynomial values that are *divisors* of N. See [**24**] and [**25**].

The next year Howgrave-Graham in [**40**] introduced a critical simplification in Coppersmith's algorithm. Coppersmith had identified the relevant lattice by linear constraints; Howgrave-Graham directly wrote down generators for the lattice. For example, for the problem of finding a divisor of N within X of V, Howgrave-Graham chose parameters k, ℓ, wrote down the lattice generated by $N^k, N^{k-1}(Xy + V), \ldots, (Xy + V)^k, \ldots, (Xy + V)^k(Xy)^{\ell-k-1}$, found a short vector Q in the lattice, and found small roots of Q. See [**41**, page 101] (with "p_0" for V, "u" for k, "h" for ℓ, "b_1" for Q, "N" for N, and "X" for X).

The same algorithm finds any integer within X of V that has a sufficiently large common divisor with N. One does not need the integer to *be* the divisor. This generalized perspective did not appear in [**24**], [**25**], [**40**], or [**41**], but did appear in papers a few years later, as discussed below.

For comparison, the problem of decoding Reed–Solomon codes is the problem of finding a polynomial f no larger than $X = x^{n-t-1}$ sharing many values with a received polynomial V (interpolated from the received word); i.e., the problem of finding a polynomial (namely $V - f$) that is within X of V and that has a large common divisor with $(x - \alpha_1) \cdots (x - \alpha_n)$. Except for a trivial replacement of integers with polynomials, this problem is a special case of the problem stated in the previous paragraph, and the decoding algorithm displayed in this section—correcting approximately $n - \sqrt{n(n-t-1)}$ errors—is a special case of the Howgrave-Graham algorithm.

The first announcement of this decoding effectiveness was by Guruswami and Sudan in [**35**] in 1998. With hindsight it is easy to see that [**35**] constructs the same lattice as Howgrave-Graham, finds the same short vector Q in the lattice, and finds the same roots of Q. Like Coppersmith, and unlike Howgrave-Graham, [**35**] identifies the lattice through linear constraints. Unlike Coppersmith, [**35**] states these constraints locally: the lattice is exactly the set of polynomials of degree below ℓ that vanish to multiplicity at least k at various points. This local perspective allowed Guruswami and Sudan to generalize, varying multiplicities separately at each point; but this generalization is not necessary for any of the decoding problems that I am considering, and it makes the algorithm very slow. [**35**] uses linear algebra to solve a large two-dimensional interpolation problem, finding a short vector Q in the specified lattice; it is much more efficient to first solve a simpler one-dimensional interpolation problem (computing V), and then write down basis vectors for the same lattice (namely A^k, $A^{k-1}F$, etc.).

Boneh in [**14**], motivated by the Guruswami–Sudan results, stated a CRT list-decoding algorithm with quantitatively analogous error-correcting capabilities. Boneh also stated an algorithm for the more general problem of finding any polynomial value having a large gcd with N; this obviously includes the multiple-of-N problems and the divisor-of-N problems. The algorithm in [**14**] constructs the same lattice as the Howgrave-Graham algorithm (in the same way), finds the same Q, and finds the same roots; the only difference is that the Howgrave-Graham algorithm throws away more of the outputs. The very large overlap between the algorithms was not pointed out in [**14**].

In 2003 I posted the first draft of a survey paper [**9**] giving a unified algorithm statement for univariate polynomials over $\mathbf{Q}$. I showed that a unified parameter optimization produced, as special cases, the quantitative results that had been obtained by Coppersmith, Howgrave-Graham, Boneh, et al. for various applications. I took a slightly broader perspective, allowing a large gcd for polynomial values on *rational* inputs, although at the time I did not see any way to use this extra generality; subsequent applications include [**54**], [**10**], and [**20**].

I discussed CRT decoding in [**9**, Section 7], and said that replacing $\mathbf{Q}$ with a rational function field in the same algorithm would decode Reed–Solomon codes as effectively as the Guruswami–Sudan algorithm. I had not actually read the Guruswami–Sudan paper at that point, and I did not realize that Guruswami and Sudan were missing the Howgrave-Graham simplification. I also had no idea that Koetter and Vardy had quantitatively improved the Guruswami–Sudan results, moving from the big-field Johnson bound to the $\mathbf{F}_q$ Johnson bound; I learned this much later when Augot kindly sent me a copy of [4]. I do not see any way to use the algorithm stated in [9] to obtain the Koetter–Vardy results: an extra tweak is required, and is the main content of Section 4 of this paper. The advantages of this tweaked algorithm over the Koetter–Vardy algorithm are analogous to the advantages of the Howgrave-Graham algorithm over the Guruswami–Sudan algorithm: most importantly, the local specification of the lattice is eliminated in favor of directly writing down lattice generators starting from V.

Cohn and Heninger in [23] presented an explicit function-field version of the Howgrave-Graham algorithm, including a generalization from the rational function field $\mathbf{F}_{q^m}(x)$ to arbitrary function fields; this generalization includes list decoding for algebraic-geometry codes. In the case of Reed–Solomon codes, [23, Section 6] reaches the big-field Johnson bound with cost only $n^{2\Omega+3+o(1)}$. Cost bounds for other choices of ℓ can also be extracted straightforwardly from the analysis in [23] and match the cost bounds shown in this section. However, this generalization still does not cover the Koetter–Vardy results.

4 Correcting Nearly $n' - \sqrt{n'(n'-t-1)}$ Errors

This section states a simple high-speed list-decoding algorithm that corrects errors up to the $\mathbf{F}_q$ Johnson bound.

Parameters. The algorithm has four parameters: a positive integer $w \le n$, the number of errors to be corrected; an integer $j \ge 0$; an integer $k \ge j$; and an integer $\ell \ge (q-1)j + k$. The algorithm assumes that $t + 1 \le n$ and that these parameters satisfy

$$n\frac{k(k+1)}{2} + n(q-1)\frac{j(j+1)}{2} + (n-t-1)\frac{\ell(\ell-1)}{2} < \ell(k(n-w)+jw),$$

i.e., $(1-(t+1)/n)(1-1/\ell) < (1-w/n)^2 + (w/n)^2/(q-1) - (1-w/n-k/\ell)^2 - (w/n-(q-1)j/\ell)^2/(q-1) - k/\ell^2 - (q-1)j/\ell^2$.

Suitable j, k, ℓ exist whenever w is smaller than the $\mathbf{F}_q$ Johnson bound, as discussed below. The special case $j = 0$ of this algorithm (with the computations of B and E straightforwardly eliminated) is exactly the algorithm of the previous section, and is usable only when w is smaller than the big-field Johnson bound.

The asymptotic cost bounds for this algorithm, as functions of n, ℓ, q^m, are exactly as in the previous section: for example, the cost is bounded by $n(\lg n)^{O(1)}$ if $\ell \in (\lg n)^{O(1)}$ and $\lg q^m \in (\lg n)^{O(1)}$, and is bounded by $n^{\Omega+2+o(1)}$ if $\ell \in O(n)$ and $\lg q^m \in O(n^\Omega)$.

Input and output. The algorithm input is a vector $v \in \mathbf{F}_q^n$. The algorithm output is the set of $c \in C$ of Hamming distance at most w from v.

Step 1: initial interpolation. Compute the polynomial $A = (x - \alpha_1)(x - \alpha_2) \cdots (x - \alpha_n) \in \mathbf{F}_{q^m}[x]$; the unique polynomial $V \in \mathbf{F}_{q^m}[x]$ with $\deg V < n$ satisfying $V(\alpha_1) = v_1/\beta_1$, $V(\alpha_2) = v_2/\beta_2$, and so on through $V(\alpha_n) = v_n/\beta_n$; and the unique polynomial $B \in \mathbf{F}_{q^m}[x]$ with $\deg B < n$ satisfying $B(\alpha_1) = 1/\beta_1^{q-1}$, $B(\alpha_2) = 1/\beta_2^{q-1}$, and so on through $B(\alpha_n) = 1/\beta_n^{q-1}$.

Step 2: lattice-basis construction. Define $X = x^{n-t-1}$; $F = Xy - V \in \mathbf{F}_{q^m}[x, y]$; and $E = F^q - FB$. Compute the ℓ polynomials $M_0, M_1, \ldots, M_{\ell-1} \in \mathbf{F}_{q^m}[x, y]$ shown in Figure 4.1. Observe that each of $M_0, M_1, \ldots, M_{\ell-1}$ includes A, E, and F to a total power of at least k; that each of $M_0, M_1, \ldots, M_{\ell-1}$ includes A and E to a total power of at least j; and that M_i has y-degree i.

The simplest strategy is to begin by computing $E, E^2, \ldots, E^j$; $A, A^2, \ldots, A^k$; and $F, F^2, \ldots, F^{\max\{k-j+q-1, \ell-qj-1\}}$. Each M_i is then a product of three known polynomials. Overall this procedure uses $O(\ell)$ polynomial products in $\mathbf{F}_{q^m}[x, y]$, each of product degree $\leq \ell - 1$ in y and $O(\ell n)$ in x. Kronecker substitution $x \mapsto y^\ell$ reduces these products to $O(\ell^2 n)$-coefficient products in $\mathbf{F}_{q^m}[y]$, each of which costs $\ell^2 n(\lg \ell^2 n)^{1+o(1)}$, for a total cost of $\ell^3 n(\lg \ell n)^{1+o(1)}$.

Step 3: lattice-basis reduction. The matrix of coefficients of $M_0, \ldots, M_{\ell-1}$ has determinant

$$A^{(k-j)(k+j+1)/2+qj(j+1)/2} X^{\ell(\ell-1)/2} = A^{k(k+1)/2+(q-1)j(j+1)/2} X^{\ell(\ell-1)/2}$$

of degree $nk(k+1)/2 + n(q-1)j(j+1)/2 + (n-t-1)\ell(\ell-1)/2$. Inside the lattice $\mathbf{F}_{q^m}[x]M_0 + \cdots + \mathbf{F}_{q^m}[x]M_{\ell-1} \subseteq \mathbf{F}_{q^m}[x, y]$ find a nonzero polynomial Q having x-degree at most $(nk(k+1)/2 + n(q-1)j(j+1)/2 + (n-t-1)\ell(\ell-1)/2)/\ell$, and therefore x-degree below $k(n-w) + jw$.

Step 4: factorization. Compute all $f \in \mathbf{F}_{q^m}[x]$ such that $Q(x, f/X) = 0$; i.e., compute all factors of Q having the form $y - f/X$ with $f \in \mathbf{F}_{q^m}[x]$. For each polynomial $f \in \mathbf{F}_{q^m}[x]$ such that $Q(x, f/X) = 0$ and $\deg f < n - t$: Compute $c = (\beta_1 f(\alpha_1), \ldots, \beta_n f(\alpha_n)) \in \mathbf{F}_{q^m}^n$. Output c if $c \in \mathbf{F}_q^n$ and $|c - v| \leq w$.

Why the algorithm works. Consider any $c \in C$ with $|c - v| \leq w$. There is a polynomial $f \in \mathbf{F}_{q^m}[x]$ with $\deg f < n-t$ such that $c = (\beta_1 f(\alpha_1), \ldots, \beta_n f(\alpha_n))$. The goal, as in the previous section, is to show that the algorithm finds f in Step 4.

As before consider the map $y \mapsto f/X$ from $\mathbf{F}_{q^m}[x, Xy]$ to $\mathbf{F}_{q^m}[x]$. This map takes A, F, E to $A, f - V, (f - V)^q - (f - V)B$ respectively.

There are exactly $n - |c - v|$ indices i for which $c_i = v_i$, i.e., for which $f(\alpha_i) = V(\alpha_i)$. Each of these indices has $x - \alpha_i$ dividing $f - V$, A, and $(f-V)^q - (f-V)B$, so $(x - \alpha_i)^k$ divides the images of $M_0, M_1, \ldots, M_{\ell-1}$.

There are also exactly $|c - v|$ indices i for which $c_i \neq v_i$, i.e., for which $\beta_i f(\alpha_i) \neq \beta_i V(\alpha_i)$. Both $\beta_i f(\alpha_i)$ and $\beta_i V(\alpha_i)$ are in $\mathbf{F}_q$, so the difference $\beta_i f(\alpha_i) - \beta_i V(\alpha_i)$ is a nonzero element of $\mathbf{F}_q$; i.e., $\beta_i^{q-1}(f(\alpha_i) - V(\alpha_i))^{q-1} = 1$; i.e., $(f(\alpha_i) - V(\alpha_i))^{q-1} = B(\alpha_i)$. Each of these indices has $x - \alpha_i$ dividing both $(f-V)^q - (f-V)B$ and A, so $(x - \alpha_i)^j$ divides the images of $M_0, M_1, \ldots, M_{\ell-1}$.

The image of Q is thus divisible by $\prod_{i:c_i=v_i}(x - \alpha_i)^k \cdot \prod_{i:c_i \neq v_i}(x - \alpha_i)^j$, which has degree $k(n - |c - v|) + j|c - v| = kn - (k - j)|c - v| \geq kn - (k - j)w =$

$$
\begin{aligned}
M_0 &= A^k F^0; && \text{(start of initial batch)}\\
M_1 &= A^{k-1} F^1;\\
&\vdots\\
M_{k-j-1} &= A^{j+1} F^{k-j-1};\\
M_{k-j} &= A^j F^{k-j}; && \text{(start of intermediate batch 0)}\\
M_{k-j+1} &= A^j F^{k-j+1};\\
&\vdots\\
M_{k-j+q-1} &= A^j F^{k-j+q-1};\\
M_{k-j+q} &= A^{j-1} E F^{k-j}; && \text{(start of intermediate batch 1)}\\
M_{k-j+q+1} &= A^{j-1} E F^{k-j+1};\\
&\vdots\\
M_{k-j+2q-1} &= A^{j-1} E F^{k-j+q-1};\\
&\vdots && \vdots\\
M_{k-j+(j-1)q} &= A E^{j-1} F^{k-j}; && \text{(start of intermediate batch } j-1)\\
M_{k-j+(j-1)q+1} &= A E^{j-1} F^{k-j+1};\\
&\vdots\\
M_{k-j+jq-1} &= A E^{j-1} F^{k-j+q-1};\\
M_{k-j+jq} &= E^j F^{k-j}; && \text{(start of final batch)}\\
M_{k-j+jq+1} &= E^j F^{k-j+1};\\
&\vdots\\
M_{\ell-1} &= E^j F^{\ell-qj-1}
\end{aligned}
$$

Fig. 4.1. Polynomials constructed in the new algorithm. There is an initial batch of length $k-j$; j intermediate batches, each of length q; and a final batch of length $\ell-(q-1)j-k$. If $\ell=(q-1)j+k$ and $j>0$ then the last polynomial is $AE^{j-1}F^{k-j+q-1}$; if $\ell=(q-1)j+k$ and $j=0$ then the last polynomial is AF^{k-1}.

$k(n-w)+jw$; but the image of Q has degree below $k(n-w)+jw$, so it must be 0 as desired.

Notes on parameter selection. Assume that $t+1 \le n'$ where $n'=n(q-1)/q$. As before write $J'=n'-\sqrt{n'(n'-t-1)}$.

Suitable j,k,ℓ exist with $\ell \in O(qnt)$ for each positive integer $w<J'$. For example, the integers $j=2w(t+1)$, $k=2(q-1)(n-w)(t+1)$, and $\ell=2(q-1)n(t+1)$ have $(1-(t+1)/n)(1-1/\ell)=1-(t+1)/n-1/\ell+1/2(q-1)n^2$

and $(1-w/n-k/\ell)^2+(w/n-(q-1)j/\ell)^2/(q-1)+k/\ell^2+(q-1)j/\ell^2=1/\ell$; so w,j,k,ℓ are in the parameter space if $1-(t+1)/n+1/2(q-1)n^2<(1-w/n)^2+(w/n)^2/(q-1)$, i.e., $(q-1)n(n-t-1)+1/2<(q-1)(n-w)^2+w^2$. Both $(q-1)n(n-t-1)$ and $(q-1)(n-w)^2+w^2$ are integers, so this inequality holds if and only if $(q-1)n(n-t-1)<(q-1)(n-w)^2+w^2$, which is equivalent to $(n'-w)^2>n'(n'-t-1)$, i.e., $w<n'-\sqrt{n'(n'-t-1)}$.

These parameters have $\ell\in O(n^2)$ if $q\in O(1)$; and $\ell\le n^2(\lg n)^{O(1)}$ if $q\in(\lg n)^{O(1)}$; and $\ell\in n^{O(1)}$ if $q\in n^{O(1)}$. If q grows superpolynomially with n then this algorithm obviously cannot run in polynomial time, except in the special case $j=0$ covered in the previous section. Such a large q would also force the $\mathbf{F}_q$ Johnson bound to be extremely close to the big-field Johnson bound; if there is an integer w between the two bounds then correcting w errors in polynomial time is, as far as I know, an open problem.

My main interest is in small q and, as in the previous section, small ℓ. It seems reasonable, although not always exactly optimal, to choose k as $\lfloor(1-w/n)\ell\rfloor$ and j as $\lfloor(w/n)\ell/(q-1)\rfloor$. Then $0\le j\le k$ since $(w/n)/(q-1)\le 1-w/n$, and $\ell\ge(q-1)j+k$. These choices also guarantee that $(1-w/n-k/\ell)^2<1/\ell^2$, that $k/\ell^2\le(1-w/n)/\ell$, that $(w/n-(q-1)j/\ell)^2/(q-1)<(q-1)/\ell^2$, and that $(q-1)j/\ell^2\le(w/n)/\ell$, so w,j,k,ℓ are in the parameter space if $(1-(t+1)/n)(1-1/\ell)\le(1-w/n)^2+(w/n)^2/(q-1)-1/\ell-q/\ell^2$; i.e., $1-(t+1)/n+(t+1)/n\ell\le(1-w/n)^2+(w/n)^2/(q-1)-q/\ell^2$; i.e., $(1-J'/n)^2+(J'/n)^2/(q-1)+(t+1)/n\ell+q/\ell^2\le(1-w/n)^2+(w/n)^2/(q-1)$; i.e.,

$$J'-w\ge\frac{(t+1)/\ell+qn/\ell^2}{2-(w+J')/n'}.$$

Assume from now on that $q\in(\lg n)^{O(1)}$. Then $t\le n/m\le(n\lg q)/\lg n\in O((n\lg\lg n)/\lg n)$, so w and J' are both bounded by $O((n\lg\lg n)/\lg n)$, so $2-(w+J')/n'$ is bounded below by 1 for all sufficiently large n. If the gap $J'-w$ is at least 1 then one can push $(t+1)/\ell+qn/\ell^2$ below $J'-w$ by taking ℓ larger than both $2(t+1)$ and $\sqrt{2qn}$; this is achievable with $\ell\in O(n)$. If the gap $J'-w$ is at least $n/(\lg n)^{O(1)}$ then one can take $\ell\in(\lg n)^{O(1)}$.

5 Correcting More Errors

One can trivially build a w-error-correcting algorithm from a $(w-1)$-error-correcting algorithm as follows: guess an error position (probability w/n); guess the error value (probability $1/(q-1)$); correct the error; apply the $(w-1)$-error-correcting algorithm. If the guess does not find the desired $c\in C$, try again.

This procedure takes $(q-1)n/w$ repetitions on average. With more repetitions one can confidently list *all* $c\in C$ at distance w; but I will focus on the effort required to find a *particular* $c\in C$ at distance w. Note that in the previous sections there was no reason to distinguish between these problems: the algorithms in the previous sections find all answers at almost exactly the same cost as finding the first answer.

A consequence of this reduction is that, for small q, there is no point in pushing the algorithms of the previous sections very close to their limits: instead of correcting $J' - 0.001$ errors one can much more cheaply correct $J' - 1.001$ errors and guess the remaining error.

More generally, one can build a w-error-correcting algorithm as follows: guess e distinct error positions (probability $w(w-1)\cdots(w-e+1)/n(n-1)\cdots(n-e+1)$); guess the error values (probability $1/(q-1)^e$); correct the errors; apply a $(w-e)$-error-correcting algorithm. This takes $(q-1)^e n(n-1)\cdots(n-e+1)/w(w-1)\cdots(w-e+1)$ repetitions on average.

Assume that $q \in (\lg n)^{O(1)}$, that $n/t \in (\lg n)^{O(1)}$, and that $w - e \geq \lfloor t/2 \rfloor$. The average number of repetitions is then bounded by $(2(q-1)n/t)^e \in (\lg n)^{O(e)}$; i.e., by $n^{O(1)}$ if $e \in O((\lg n)/\lg\lg n)$, and by $n^{o(1)}$ if $e \in o((\lg n)/\lg\lg n)$. In particular, this algorithm corrects $J'+o((\lg n)/\lg\lg n)$ errors using $n^{\Omega+2+o(1)}$ bit operations, and corrects $J' + O((\lg n)/\lg\lg n)$ errors using $n^{O(1)}$ bit operations.

6 Application to Classical Goppa Codes

The code C is called a **classical Goppa code** if there is a monic degree-t polynomial $g \in \mathbf{F}_{q^m}[x]$ such that each β_i can be expressed as $g(\alpha_i)/A'(\alpha_i)$. Here $A = \prod_i (x - \alpha_i) \in \mathbf{F}_{q^m}[x]$ as in Sections 3 and 4. In this case C is denoted $\Gamma_q(\alpha_1, \ldots, \alpha_n, g)$.

Sugiyama, Kasahara, Hirasawa, and Namekawa showed in [**51**] that

$$\Gamma_q(\alpha_1, \ldots, \alpha_n, \prod_i g_i^{e_i}) = \Gamma_q(\alpha_1, \ldots, \alpha_n, \prod_i g_i^{e_i + [e_i \bmod q = q-1]})$$

when the g_i's are distinct monic irreducible polynomials. Here $[e_i \bmod q = q-1]$ means 1 if $e_i \in \{q-1, 2q-1, \ldots\}$, otherwise 0. For example, $\Gamma_2(\ldots, g) = \Gamma_2(\ldots, g^2)$ if g is squarefree; this had been proven earlier by Goppa in [**31**] using a different technique.

Write $g = \prod_i g_i^{e_i}$ and $\overline{g} = \prod_i g_i^{e_i + [e_i \bmod q = q-1]}$. The Sugiyama–Kasahara–Hirasawa–Namekawa identity $\Gamma_q(\ldots, g) = \Gamma_q(\ldots, \overline{g})$ implies that one can correct w errors in $\Gamma_q(\ldots, g)$ by using any w-error-correcting algorithm for $\Gamma_q(\ldots, \overline{g})$. If some $e_i \bmod q = q-1$ then $\overline{g}$ has larger degree than g, making all of these error-correcting algorithms more effective for $\overline{g}$ than for g.

In particular, combining the SKHN identity with Berlekamp's algorithm corrects $\lfloor qt/2 \rfloor$ errors in "wild Goppa codes" $\Gamma_q(\ldots, g^{q-1})$ with squarefree g. Combining the SKHN identity with the Guruswami–Sudan algorithm corrects nearly $n - \sqrt{n(n-qt-1)}$ errors in the same codes in polynomial time, as discussed in [**12**, Section 5]. Combining the SKHN identity with the Koetter–Vardy algorithm corrects nearly $n' - \sqrt{n'(n'-qt-1)}$ errors in polynomial time, as pointed out in [**4**]. Combining the SKHN identity with the algorithm in this paper corrects even more errors in polynomial time.

See also [**5**] for a different approach that decodes more errors in some cases, particularly for $q = 3$.

References

[1] — (no editor): 39th annual symposium on foundations of computer science, FOCS '98, November 8-11, 1998, Palo Alto, California, USA. IEEE Computer Society (1998); See [35]

[2] — (no editor): Proceedings of the 32nd annual ACM symposium on theory of computing. Association for Computing Machinery, New York (2000); See [14]

[3] Alekhnovich, M.: Linear diophantine equations over polynomials and soft decoding of Reed-Solomon codes. IEEE Transactions on Information Theory 51, 2257–2265 (2005); Cited from §1, §2, §2

[4] Augot, D., Barbier, M., Couvreur, A.: List-decoding of binary Goppa codes up to the binary Johnson bound (2010); Cited from §1, §1, §3, §3, §3, §3, §3, §3, §6

[5] Barreto, P.S.L.M., Lindner, R., Misoczki, R.: Decoding square-free Goppa codes over $\mathbf{F}_p$ (2010); Cited from §6

[6] Beelen, P., Brander, K.: Key equations for list decoding of Reed–Solomon codes and how to solve them. Journal of Symbolic Computation 45, 773–786 (2010); Cited from §1

[7] Berlekamp, E.R.: Algebraic coding theory. McGraw-Hill, New York (1968); Cited from §1

[8] Bernstein, D.J.: Fast multiplication and its applications. In: [**19**], pp. 325–384 (2008); Cited from §2, §2, §2, §2

[9] Bernstein, D.J.: Reducing lattice bases to find small-height values of univariate polynomials. In: [**19**], pp. 421–446 (2008); Cited from §1, §3, §3, §3

[10] Bernstein, D.J.: List decoding for binary Goppa codes. In: IWCC 2011 [**21**], pp. 62–80 (2011); Cited from §3, §3

[11] Bernstein, D.J., Lange, T., Peters, C.: Attacking and defending the McEliece cryptosystem. In: PQCrypto 2008 [**17**], pp. 31–46 (2008); Cited from §1

[12] Bernstein, D.J., Lange, T., Peters, C.: Wild McEliece. In: SAC 2010 [**13**], pp. 143–158 (2011); Cited from §6

[13] Biryukov, A., Gong, G., Stinson, D.R. (eds.): Selected areas in cryptography—17th international workshop, SAC 2010, Waterloo, Ontario, Canada, August 12-13, 2010, revised selected papers. Lecture Notes in Computer Science, vol. 6544. Springer (2011); See [12]

[14] Boneh, D.: Finding smooth integers in short intervals using CRT decoding. In: STOC 2000 [**2**], pp. 265–272 (2000); see also newer version [**15**]; Cited from §3, §3, §3

[15] Boneh, D.: Finding smooth integers in short intervals using CRT decoding. Journal of Computer and System Sciences 64, 768–784 (2002); see also older version [**14**]

[16] Bose, R.C., Ray-Chaudhuri, D.K.: On a class of error correcting binary group codes. Information and Control 3, 68–79 (1960); Cited from §1

[17] Buchmann, J., Ding, J. (eds.): Post-quantum cryptography, second international workshop, PQCrypto 2008, Cincinnati, OH, USA, October 17-19, 2008, proceedings. Lecture Notes in Computer Science, vol. 5299. Springer (2008); See [11]

[18] Bürgisser, P., Clausen, M., Shokrollahi, M.A.: Algebraic complexity theory. Springer, Berlin (1997); Cited from §2

[19] Buhler, J.P., Stevenhagen, P. (eds.): Surveys in algorithmic number theory. Mathematical Sciences Research Institute Publications, vol. 44. Cambridge University Press, New York (2008); See [8], [9]

[20] Castagnos, G., Joux, A., Laguillaumie, F., Nguyen, P.Q.: Factoring pq^2 with quadratic forms: nice cryptanalyses. In: Asiacrypt 2009 [**43**], pp. 469–486 (2009); Cited from §3
[21] Chee, Y.M., Guo, Z., Ling, S., Shao, F., Tang, Y., Wang, H., Xing, C. (eds.): Coding and cryptology—third international workshop, IWCC 2011, Qingdao, China, May 30-June 3, 2011, proceedings. Lecture Notes in Computer Science, vol. 6639. Springer (2011); See [10]
[22] Chien, R.T., Choy D.M.: Algebraic generalization of BCH-Goppa-Helgert codes. IEEE Transactions on Information Theory 21, 70–79; Cited from §1
[23] Cohn, H., Heninger, N.: Ideal forms of Coppersmith's theorem and Guruswami-Sudan list decoding (2010); Cited from §3, §3, §3
[24] Coppersmith, D.: Finding a small root of a univariate modular equation. In: Eurocrypt 1996 [**44**], pp. 155–165 (1996); see also newer version [**26**]; Cited from §3, §3
[25] Coppersmith, D.: Finding a small root of a bivariate integer equation; factoring with high bits known. In: Eurocrypt 1996 [**44**], pp. 178–189 (1996); see also newer version [**26**]; Cited from §3, §3
[26] Coppersmith, D.: Small solutions to polynomial equations, and low exponent RSA vulnerabilities. Journal of Cryptology 10, 233–260 (1997); see also older version [**24**] and [**25**]
[27] Darnell, M. (ed.): Cryptography and coding: proceedings of the 6th IMA International Conference held at the Royal Agricultural College, Cirencester, December 17-19, 1997. Lecture Notes in Computer Science, vol. 1355. Springer (1997); See [40]
[28] Delsarte, P.: On subfield subcodes of modified Reed-Solomon codes. IEEE Transactions on Information Theory 21, 575–576 (1975); Cited from §1
[29] von zur Gathen, J., Gerhard J.: Modern computer algebra, 2nd edn. Cambridge University Press, Cambridge (2003); Cited from §2, §2, §2, §2, §2, §2, §2, §2
[30] Giorgi, P., Jeannerod, C.-P., Villard, G.: On the complexity of polynomial matrix computations. In: ISSAC 2003 [**49**], pp. 135–142 (2003); Cited from §2, §2, §2, §2
[31] Goppa, V.D.: A new class of linear error correcting codes. Problemy Peredachi Informatsii 6, 24–30 (1970); Cited from §1, §6
[32] Goppa, V.D.: Rational representation of codes and (L, g)-codes. Problemy Peredachi Informatsii 7, 41–49 (1971); Cited from §1
[33] Gorenstein, D., Zierler, N.: A class of error-correcting codes in p^m symbols. Journal of the Society for Industrial and Applied Mathematics 9, 207–214 (1961); Cited from §1
[34] Guruswami, V.: List decoding of error-correcting codes, Ph.D. thesis. Massachusetts Institute of Technology (2001); Cited from §1
[35] Guruswami, V., Sudan, M.: Improved decoding of Reed-Solomon and algebraic-geometry codes. In: FOCS 1998 [**1**], pp. 28–39 (1998); see also newer version [**36**]; Cited from §1, §1, §3, §3, §3, §3, §3
[36] Guruswami, V., Sudan, M.: Improved decoding of Reed-Solomon and algebraic-geometry codes. IEEE Transactions on Information Theory 45, 1757–1767 (1999); see also older version [**36**]; Cited from §3, §3
[37] Håstad, J.: Solving simultaneous modular equations of low degree. SIAM Journal on Computing 17, 336–341 (1988); Cited from §3
[38] Helgert, H.J.: Alternant codes. Information and Control 26, 369–380 (1974); Cited from §1
[39] Hocquenghem, A.: Codes correcteurs d'erreurs. Chiffres 2, 147–156 (1959); Cited from §1

[40] Howgrave-Graham, N.: Finding small roots of univariate modular equations revisited. In: Cirencester 1997 [**27**], pp. 131–142 (1997); Cited from §3, §3
[41] Howgrave-Graham, N.: Computational mathematics inspired by RSA, Ph.D. thesis (1998); Cited from §3, §3
[42] Justesen, J.: On the complexity of decoding Reed–Solomon codes. IEEE Transactions on Information Theory 22, 237–238 (1976); Cited from §1
[43] Matsui, M. (ed.): Advances in cryptology—ASIACRYPT 2009, 15th international conference on the theory and application of cryptology and information security, Tokyo, Japan, December 6-10, 2009, proceedings. Lecture Notes in Computer Science, vol. 5912. Springer (2009); See [20]
[44] Maurer, U.M. (ed.): Advances in cryptology—EUROCRYPT '96: proceedings of the fifteenth international conference on the theory and application of cryptographic techniques held in Saragossa, May 12-16, 1996. Lecture Notes in Computer Science, vol. 1070. Springer, Berlin (1996); See [24], [25]
[45] Mora, T. (ed.): Applied algebra, algebraic algorithms and error-correcting codes: proceedings of the sixth international conference (AAECC-6) held in Rome, July 4-8, 1988. Lecture Notes in Computer Science, vol. 357. Springer, Berlin (1989); See [53]
[46] Wesley Peterson, W.: Encoding and error-correction procedures for the Bose-Chaudhuri codes. Transactions of the Institute of Radio Engineers 6, 459–470 (1960); Cited from §1
[47] Reed, I.S., Solomon, G.: Polynomial codes over certain finite fields. Journal of the Society for Industrial and Applied Mathematics 8, 300–304 (1960); Cited from §1
[48] Sarwate, D.V.: On the complexity of decoding Goppa codes. IEEE Transactions on Information Theory 23, 515–516 (1977); Cited from §1
[49] Rafael Sendra, J. (ed.): Symbolic and algebraic computation, international symposium ISSAC 2003, Drexel University, Philadelphia, Pennsylvania, USA, August 3-6, 2003, proceedings. Association for Computing Machinery (2003); See [30]
[50] Sudan, M.: Decoding of Reed Solomon codes beyond the error-correction bound. Journal of Complexity 13, 180–193 (1997); Cited from §1
[51] Sugiyama, Y., Kasahara, M., Hirasawa, S., Namekawa, T.: Further results on Goppa codes and their applications to constructing efficient binary codes. IEEE Transactions on Information Theory 22, 518–526 (1976); Cited from §6
[52] Trifonov, P.: Efficient interpolation in the Guruswami–Sudan algorithm. IEEE Transactions on Information Theory 56, 4341–4349 (2010); Cited from §1
[53] Vallée, B., Girault, M., Toffin, P.: How to guess ℓth roots modulo n by reducing lattice bases. In: AAECC 1989 [**45**], pp. 427–442 (1989); Cited from §3
[54] Wu, Y.: New list decoding algorithms for Reed–Solomon and BCH codes. IEEE Transactions On Information Theory 54 (2008); Cited from §3, §3
[55] Zassenhaus, H.: On Hensel factorization. I. Journal of Number Theory 1, 291–311 (1969); Cited from §2

Statistical Decoding of Codes over $\mathbb{F}_q$

Robert Niebuhr

Technische Universität Darmstadt
Fachbereich Informatik
Kryptographie und Computeralgebra,
Hochschulstraße 10
64289 Darmstadt
Germany
rniebuhr@cdc.informatik.tu-darmstadt.de

Abstract. In this paper we analyze statistical decoding over a finite field $\mathbb{F}_q$. We generalize Overbeck's binary statistical decoding algorithm to codes over $\mathbb{F}_q$, and analyze the success probability of our algorithm. We provide experimental data for different field sizes. In addition to that, we describe two techniques how knowledge about structure of the code or of the solution can be used in order to speed up the decoding algorithm.

Keywords: Statistical decoding, general decoding, code-based cryptography, public-key cryptography.

1 Introduction

Shor's attack [16] from 1994 allows to solve the integer factoring and the discrete logarithm problem in polynomial time once large enough quantum computers are available. Code-based cryptography is a very promising candidate for post-quantum cryptography, i.e. cryptosystems that are not vulnerable to quantum computer attacks. The first code-based cryptosystem was the McEliece encryption scheme [9], published in 1978. It is as old as RSA and has resisted cryptanalysis to date (except for a parameter adjustment).

Statistical decoding was introduced in 2001 by Al Jabri [7] and improved by Overbeck in 2006 [13]. While Al Jabri claimed that statistical decoding can be used effectively against the McEliece cryptosystem, Overbeck showed that the required precomputation is far greater than expected by Al Jabri, and that therefore the time as well as the memory requirements are much higher compared with other kinds of attacks. However, statistical decoding is quite efficient against short codes (i.e. codes with a small length n) and can even be faster than attacks based on information set decoding (ISD) or the generalized birthday algorithm (see the recent resources [3,11] and [10] for more information on these attacks).

Our contribution. In this paper, we generalize Overbeck's statistical decoding algorithm to codes over non-binary fields $\mathbb{F}_q$. We analyze the success probability of our algorithm theoretically and experimentally, and show that it seems to be independent of the size of the field $\mathbb{F}_q$.

B.-Y. Yang (Ed.): PQCrypto 2011, LNCS 7071, pp. 217–227, 2011.

In addition to that, we briefly describe two techniques how additional structure of the underlying code or the solution can be exploited to increase the algorithm efficiency.

Organization of the paper. We recall some preliminaries and notations from code-based cryptography in Section 2. In the subsequent Section 3 we present our generalized algorithm and the corresponding statistical properties. Experimental results are given in Section 4, and we conclude in Section 5.

2 Preliminaries and Notation

In this section, we recall standard definitions from code-based cryptography.

Definition 1 (Linear code). *A linear code $\mathcal{C}$ of length n and dimension k over a finite field $\mathbb{F}_q$ is defined as a k-dimensional subspace of the n-dimensional vector space $\mathbb{F}_q^n$. $\mathcal{C}$ is denoted a (n,k) code over $\mathbb{F}_q$.*

Definition 2 (Generator and parity check matrix). *A matrix $G \in \mathbb{F}_q^{k\times n}$ of full rank is call generator matrix for an (n,k) code $\mathcal{C}$ if $\mathcal{C} = \{mG : m \in \mathbb{F}_q^k\}$. A parity check matrix for $\mathcal{C}$ is any full rank matrix $H \in \mathbb{F}_q^{(n-k)\times n}$ such that $\mathcal{C} = \{x \in \mathbb{F}_q^n : Hx^T = 0\}$. A parity check matrix H is a generator matrix for the dual code $\mathcal{C}^\perp$.*

Definition 3 (Hamming distance). *The Hamming weight $\mathrm{wt}(x)$ of a vector x is defined as the number of non-zero entries, and the Hamming distance $d(x,y)$ between two vectors x and y is $\mathrm{wt}(x-y)$. The minimum distance d of a code $\mathcal{C}$ is given by $d := \min_{x\in\mathcal{C}\setminus\{0\}} \mathrm{wt}(x)$. Let $t := \lfloor (d-1)/2 \rfloor$, then $\mathcal{C}$ is also denoted an (n,k,t) code.*

Definition 4 (General decoding problem). *The general decoding problem is defined as follows: Given a vector $c \in \mathbb{F}_q^n$ and an (n,k,t) code $\mathcal{C}$ over $\mathbb{F}_q$, find $x \in \mathcal{C}$ such that $d(c,x)$ is minimal. If $d(c,x) \le t$, then this decoding is unique.*

For any vector h, the entry at position i is denoted by h_i. We write $H_{\cdot i}$ for the i-th column of a matrix H. Let $I \subseteq \{1,\ldots,n\}$, then $H_{\cdot I}$ denotes the submatrix of H consisting of the columns indexed by I. Similarly, h_I denotes the vector consisting of the corresponding entries of h.

3 Statistical Decoding

The idea of statistical decoding is as follows: After receiving a codeword with error $c = mG + e$, where m and e are unknown, a precomputed set $H_w \subseteq \mathcal{C}^\perp$ is used as a mask to obtain information about e. Since $GH_w^T = 0$, we have

$$H_w c^T = H_w e^T.$$

Al Jabri showed that if $hc^T = 1$ for some $h \in H_w$ then the non-zero bits of h give some information about the non-zero bits of e. Overbeck has improved this algorithm by also using the vectors h where $hc^T = 0$ to gain information about e.

We will briefly describe Overbeck's algorithm, and then generalize it to codes over non-binary fields $\mathbb{F}_q$.

3.1 Binary Statistical Decoding

Let $\mathcal{C}$ be an (n, k, t) code over $\mathbb{F}_q$, $w < n/2$ be an integer, and $H_w \subseteq \mathcal{C}^\perp$ a sufficiently large subset of the dual space of $\mathcal{C}$, where $\forall h \in H_w : \mathrm{wt}(h) = w$. Given a word $c = x + e$, where $x = mG \in \mathcal{C}$ and $\mathrm{wt}(e)$ is small, the algorithm attempts to find e.

For every $h \in H_w$, we have an *odd error detection* at bit i if $hc^T = 1$ and $h_i = 1$, and an *even error detection* at bit i if $hc^T = 0$ and $h_i = 1$. In each case we can compute the probabilities that e contains an error at bit i. In the case of an odd error detection, the probabilities p_w^+ and q_w^+ that $e_i = 1$ and $e_i = 0$, respectively, are

$$p_w^+ = \frac{\sum_{j \text{ odd}}^{\leq t} \binom{n-t}{w-j}\binom{t-1}{j-1}}{\sum_{j \text{ odd}}^{\leq t} \binom{n-t}{w-j}\binom{t}{j}}, \quad q_w^+ = \frac{\sum_{j \text{ odd}}^{\leq t} \binom{n-t-1}{w-j-1}\binom{t}{j}}{\sum_{j \text{ odd}}^{\leq t} \binom{n-t}{w-j}\binom{t}{j}}.$$

Let $v_{y,w}^+ = |\{h \in H_w : hc^T \neq 0\}|$. For every bit i, the random variable

$$\frac{1}{v_{y,w}^+} \sum_{h \in H_w} (hc^T \mod 2) h_i$$

is the relative frequency estimate for p_w^+ or q_w^+, depending on whether i is an error position of e. The variance of this random variable is $(\sigma_w^+)^2 = p_w^+(1 - p_w^+)/v_{y,w}^+$. Thus, for H_w large enough, Algorithm 1 allows to recover m.

Algorithm 1. Al Jabri's algorithm for binary statistical decoding.

INPUT: Generator matrix G for an (n, k, t) code, $H_w \subseteq \mathcal{C}^\perp$ and $c \in \{0, 1\}^n$
OUTPUT: $m \in \{0, 1\}^k$ such that $\mathrm{wt}(c - mG) \leq t$

$v \leftarrow \sum_{h \in H_w} (hc^T \mod 2) h \in \mathbb{Z}^n$

Choose $I = \{$positions of the k smallest entries of $v\}$ s.t. $G_{\cdot I}$ is invertible

Return $m \leftarrow c_I G_{\cdot I}^{-1}$

Overbeck improved this algorithm in two ways. First, even error detections are used as well, allowing to extract significantly more information from a given set

H_w. Second, the algorithm is no longer restricted to a fixed value of w. Instead it allows a range for w, and the information extracted from the different sets H_w is combined in the end.

In case of an even error detection, the corresponding probabilities p_w^- and q_w^- are given by

$$p_w^- = \frac{\sum_{2\leq j \text{ even}}^{\leq t} \binom{n-t}{w-j}\binom{t-1}{j-1}}{\sum_{j \text{ even}}^{\leq t} \binom{n-t}{w-j}\binom{t}{j}}, \quad q_w^- = \frac{\sum_{j \text{ even}}^{\leq t} \binom{n-t-1}{w-j-1}\binom{t}{j}}{\sum_{j \text{ even}}^{\leq t} \binom{n-t}{w-j}\binom{t}{j}}.$$

Consequently, $v_{y,w}^- = |\{h \in H_w : hc^T = 0\}|$, and the relative frequency estimates are given by

$$\frac{1}{v_{y,w}^-} \sum_{h \in H_w} (1 - hc^T \mod 2) h_i.$$

Algorithm 2 summarizes the improved algorithm. Note that v is defined as $v = \sum_{w-b}^{B} a_w v_w + \sum_{w=b}^{B} a_{w+B} v_{w+B}$, where each $a_i \in \{0,1\}$, i.e. not all partial results need to be combined in the end.

Algorithm 2. Overbecks's improved algorithm for binary statistical decoding

---

INPUT: Generator matrix G for an (n,k,t) code $\mathcal{C}$, $H = \bigcup_{w=b}^{B} H_w \subseteq \mathcal{C}^\perp$ and $c \in \{0,1\}^n$
OUTPUT: $m \in \{0,1\}^k$ such that $\mathrm{wt}(c - mG) \leq t$

Let $\mathbf{1} = (1, \ldots, 1) \in \{0,1\}^n$
for $w = b \to B$ **do**
  $(\sigma_w^+)^2 = p_w^+(1-p_w^+)v_{y,w}^+$
  $(\sigma_w^-)^2 = p_w^-(1-p_w^-)v_{y,w}^-$

  $v_w \leftarrow \sum_{h \in H_w} (hc^T \mod 2)(h - p_w^+ \mathbf{1})/\sigma_w^+ \in \mathbb{R}^n$
  $v_{w+B} \leftarrow -\sum_{h \in H_w} (1 - hc^T \mod 2)(h - p_w^- \mathbf{1})/\sigma_w^- \in \mathbb{R}^n$
end for

for all binary combinations v of the different v_i **do**
  Choose $I = \{$positions of the k smallest entries of $v\}$ s.t. $G_{\cdot I}$ is invertible
  $m \leftarrow c_I G_{\cdot I}^{-1}$
  if $\mathrm{wt}(c - mG) \leq t$ **then**
    Return m
  end if
end for

3.2 Statistical Decoding over $\mathbb{F}_q$ (for $q > 2$)

The first thing to note when considering codes over non-binary fields $\mathbb{F}_q$ is that the error positions of the secret vector e now take values in $\mathbb{F}_q \backslash \{0\}$. However we are only interested to find k error-free positions such that the corresponding generator matrix $G_{\cdot I}$ is invertible, so we do not have to find those error values.

Secondly, we note that the definition of odd error detection needs to be changed, since $hc^T \in \mathbb{F}_q\backslash\{0\}$ as well: We define odd error detection at bit i as the case when $hc^T \neq 0$ and $h_i \neq 0$. The reason is that since

$$\forall x \in \mathbb{F}_q\backslash\{0\} : h(xc)^T = x(hc^T) \quad \text{and} \quad hc^T \neq 0 \Leftrightarrow x(hc^T) \neq 0, \tag{1}$$

all values of hc^T have the same probability, and they are independent of the value of h_i.

Finally, the original algorithm adds up the vectors h in order to compute the relative frequencies of p_w^+ and q_w^+. Doing the same over $\mathbb{F}_q$ would disturb these frequencies because entries of h greater than 1 bias the computation. Instead, we add $\Theta(h) = (\theta(h_1), \ldots, \theta(h_{n-k}))$, where

$$\theta : \mathbb{F}_q \rightarrow \mathbb{F}_q, x \mapsto \begin{cases} 0 & x = 0 \\ 1 & \text{else} \end{cases}.$$

In order to proceed, we introduce some notation. Consider the case where e and a vector h are both non-zero in exactly i bits. In contrast to the binary case, we don't have the equivalence i even $\Leftrightarrow hc^T = 0$. Since every non-zero value of hc^T has the same probability (and occurs the same number of times when $H_w = \mathcal{C}^\perp$), the quantities

$$\mathfrak{C}(q,i) = \left[\frac{(q-1)^i}{q}\right] \tag{2}$$

$$\mathfrak{C}'(q,i) = (q-1)^i - \left[\frac{(q-1)^i}{q}\right] \cdot (q-1) \tag{3}$$

reflect the relative frequencies of non-zero and zero values of hc^T, respectively, where $[\,]$ denotes rounding to the nearest integer. They are well-defined since $(q-1)^i/q \in \mathbb{N} + \{0.5\}$ can only occur for $q = 2$.

Using equation (1), we can calculate the respective probabilities.

$$p_w^{++} = \frac{\sum_{j=1}^{t} \mathfrak{C}(q,j)\binom{t-1}{j-1}\binom{n-t}{w-j}(q-1)^{w-j}}{\sum_{j=1}^{t} \mathfrak{C}(q,j)\binom{t}{j}\binom{n-t}{w-j}(q-1)^{w-j}} \tag{4}$$

$$q_w^{++} = \frac{\sum_{j=1}^{t} \mathfrak{C}(q,j)\binom{t-1}{j}\binom{n-t}{w-j}(q-1)^{w-j}}{\sum_{j=1}^{t} \mathfrak{C}(q,j)\binom{t}{j}\binom{n-t}{w-j}(q-1)^{w-j}} \tag{5}$$

$$p_w^{--} = \frac{\sum_{j=0}^{t} \mathfrak{C}'(q,j)\binom{t-1}{j-1}\binom{n-t}{w-j}(q-1)^{w-j}}{\sum_{j=0}^{t} \mathfrak{C}(q,j)\binom{t}{j}\binom{n-t}{w-j}(q-1)^{w-j}} \tag{6}$$

$$q_w^{--} = \frac{\sum_{j=0}^{t} \mathfrak{C}'(q,j)\binom{t-1}{j}\binom{n-t}{w-j}(q-1)^{w-j}}{\sum_{j=0}^{t} \mathfrak{C}(q,j)\binom{t}{j}\binom{n-t}{w-j}(q-1)^{w-j}} \tag{7}$$

Consequently, we redefine

$$v_{y,w}^{++} = |\{h \in H_w : hc^T \neq 0\}| \tag{8}$$

$$v_{y,w}^{--} = |\{h \in H_w : hc^T = 0\}| \tag{9}$$

Since we sum up $\Theta(h)$ (instead of h), the variance of v remains unchanged, i.e

$$\sigma_w^{++} = p_w^{++}(1 - p_w^{++})v_{y,w}^{++}.$$

Algorithm 3 is the generalized version for statistical decoding over $\mathbb{F}_q$.

Algorithm 3. Generalized algorithm for statistical decoding over a non-binary field $\mathbb{F}_q$.

INPUT: Generator matrix G for an (n, k, t) code $\mathcal{C}$, $H = \bigcup_{w=b}^{B} H_w \subseteq \mathcal{C}^\perp$ and $c \in \mathbb{F}_q^n$
OUTPUT: $m \in \mathbb{F}_q^k$ such that $\mathrm{wt}(c - mG) \leq t$

Let $\mathbf{1} = (1, \ldots, 1) \in \{0,1\}^n$
for $w = b \to B$ **do**
  $(\sigma_w^{++})^2 = p_w^{++}(1 - p_w^{++})v_{y,w}^{+}$
  $(\sigma_w^{--})^2 = p_w^{--}(1 - p_w^{--})v_{y,w}^{-}$

  Let $H_w^+ = \{h \in H_w : hc^T \neq 0\}$ and $H_w^- = H_w \backslash H_w^+$
  $v_w^+ \leftarrow \sum_{h \in H_w^+} (\Theta(h) - p_w^{++}\mathbf{1})/\sigma_w^{++} \in \mathbb{R}^n$
  $v_{w+B}^+ \leftarrow -\sum_{h \in H_w^-} (\Theta(h) - p_w^{--}\mathbf{1})/\sigma_w^{--} \in \mathbb{R}^n$
end for

for all binary combinations v^+ of the different v_i^+ **do**
  Choose $I = \{$positions of the k smallest entries of $v\}$ s.t. $G_{\cdot I}$ is invertible
  $m \leftarrow c_I G_{\cdot I}^{-1}$
  if $\mathrm{wt}(c - mG) \leq t$ **then**
    Return m
  end if
end for

3.3 Exploiting Additional Structure

Many types of additional structure have been proposed in code-based cryptography in order to reduce the public key size or to increase efficiency. Algorithm 3 allows to exploit various types of such structures. We will give two examples and briefly describe the corresponding techniques:

(Quasi-)cyclic matrices. In the last years, quasi-cyclic (QC) matrices have been used in many proposals in code-based cryptography (e.g. [2,5]). A QC matrix is a block matrix, where each block is cyclic. A cyclic matrix is a matrix where every row is a cyclic shift of the previous. Since the first row of every block of a QC matrix generates the full block, this allows a compact representation, decreasing the size of the public key. We will describe the technique using cyclic matrices, but it applies to QC matrices as well, and also to other types of structured matrices like quasi-dyadic matrices.

Let $\gamma(v)$ denote the cyclic shift of vector v. A cyclic code $\mathcal{C}$ allows to choose a cyclic parity check matrix H. Therefore, for every $h \in \mathcal{C}^\perp$, $\gamma(h) \in \mathcal{C}^\perp$.

This means that we can restrict the precomputed matrix H_w to vectors that are not cyclic shifts of one another, and in the course of running Algorithm 3, test $h \in H_w$ as well as all cyclic shifts of h against the vector c. As a result, while the run time of the actual algorithm is unchanged, the size of the precomputed set H_w can be decreased by a factor of up to n.

Regular words. Regular words of length n and weight t are defined as words consisting of t blocks of length n/t, where each block has a weight of 1. They have been used to increase the efficiency of some code-based schemes, e.g. the FSB hash function [1]; Bernstein et al. showed how to improve information set decoding using the regular words structure [4].

If it is known that the solution is a regular word, the decoding algorithm can be modified as follows. When choosing the set I, we add the additional condition that I must not contain those indices corresponding to a whole block; in other words, for all i with $1 \le i \le t$,

$$\left\{ \frac{(i-1)n}{t} + 1, \ldots, \frac{in}{t} \right\} \subsetneq I.$$

If the values in v are such that a this would happen, the largest value of v in this block is ignored and the index of the next smallest value of v is added to I instead. This modification slightly increaes the chance of decoding successfully, or to achieve the same success probability with a slightly smaller size of H_w.

4 Experimental Results

In this section, we present experimental results of statistical decoding over $\mathbb{F}_q$. In order to estimate the success probability of Algorithm 3, we need to analyze the required size of the set H_w.

In [6], the authors generalize the weight distribution results from [8] to random codes over $\mathbb{F}_q$. Therefore, we have an upper bound for the size of H_w:

$$|H_w| \le \binom{n}{w} (q-1)^w q^{-k}.$$

In order to compute the success probability $\mathcal{P}$, there needs to be a value δ such that the following conditions hold (these conditions were introduced in [13]):

1. For every error position i:

$$v_i > (p_w^{++} - \delta) v_{y,w}^{++}.$$

2. There are at least k non-error positions j such that:

$$v_j < (p_w^{++} - \delta) v_{y,w}^{++}.$$

In the case of codes over $\mathbb{F}_q$, we can now assume that $v_{y,w}^{++} \approx \frac{q-1}{q}|H_w|$. Thus the probability $\mathcal{P}$ that a certain δ satisfies the first condition is

$$\mathcal{P} = \Phi(\delta/\sigma_w^{++})^t = \Phi\left(\delta\sqrt{\frac{(q-1)|H_w|}{qp_w^{++}(1-p_w^{++})}}\right)^t,$$

where Φ refers to the standard normal distribution. Therefore, we get the following condition on $|H_w|$:

$$(\Phi^{-1}(\mathcal{P}^{1/t}))^2\delta^{-2}\frac{q}{q-1}p_w^{++}(1-p_w^{++}) \le |H_w| \le \binom{n}{w}(q-1)^w q^{-k}. \qquad (10)$$

We can assume that half the values of v_j, for j the non-error positions, are below the mean of $p_w^{++}v_{y,w}^{++}$. Any δ satisfying both conditions above will probably be smaller than $|p_w^{++} - q_w^{++}|$. Thus, we expect a success probability of 0.95^t when a set of size

$$|H_w| \approx 2.72\frac{q}{q-1} \cdot \frac{p_w^{++}(1-p_w^{++})}{(p_w^{++} - q_w^{++})^2}$$

is used (since $\Phi^{-1}(0.95)^2 \approx 2.72$). Note that this size is a factor of $\frac{q}{2(q-1)}$ greater compared with the binary case.

In Table 1 we present experimental results obtained using our implementation in Maple.

Table 1. Experimental results of using Algorithm 3 to decode t errors in an (n,k) code over $\mathbb{F}_q$. We ran several thousand decoding attempts, each using a sample of size $|H| = |\bigcup_{w=b}^{B} H_w| = 100$.

(n,k,t)	q	b	B	Successful decodings
(64, 40, 4)	3	44	46	30.6%
	5	51	53	29.8%
	7	56	58	36.1%
	11	58	60	35.8%
	13	59	61	42.7%
	53	61	63	29.4%
(128, 72, 8)	3	84	88	18.1%
	5	100	104	22.9%
	7	108	112	23.0%
	11	115	119	32.1%
	13	117	121	27.8%
	53	123	127	37.8%

The results show that in many cases our algorithm decodes successfully, even though the number of sample vectors $|H|$ was not very large. Also, the success probability can be increased by using a larger weight spectrum $B-b$. However, this increases the complexity of testing all binary combinations v^+. Note that the success probability seems to be independent of the field size q; this is to be expected, since we are only searching for the (non-)error *positions*, not their values. This is an advantage compared with other algorithms like information set decoding, where the algorithm complexity grows significantly with q (more than the impact of q-ary arithmetic, which applies in our case as well).

Also, note that larger field sizes require larger values w when computing the sets H_w. This is due to the fact that the weight distribution of codes over different fields is not identical. For the above fields $\mathbb{F}_q$ with $q \in \{3, 5, 7, 11, 53\}$, the weight distributions are shown in Figure 1. Those distributions are derived from Cheung's result that in an (n, k) code over $\mathbb{F}_q$, the ratio of codewords of weight u to words of weight u is very close to

$$q^{-(n-k)}. \tag{11}$$

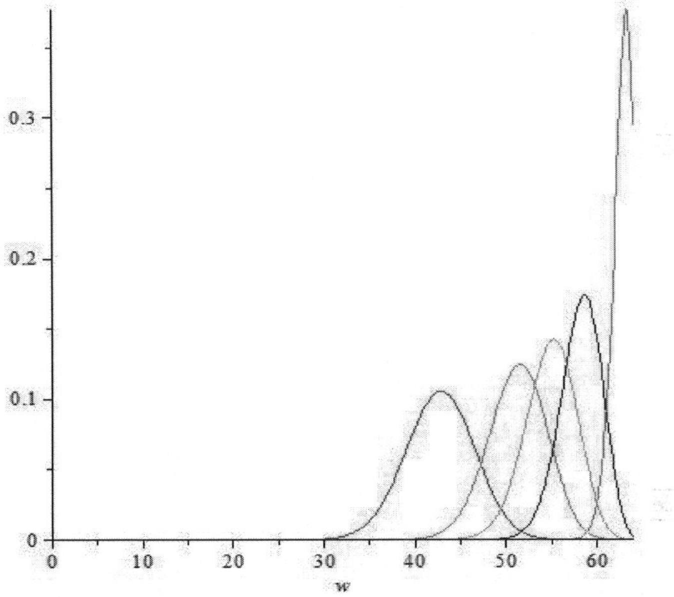

Fig. 1. Weight distribution of (64, 40) codes over different fields $\mathbb{F}_q$

The optimal choice of b and B is difficult to compute: since vectors h of smaller weight can provide more information about the error positions of e, a smaller set H_w is sufficient to achieve a given success probability, but it is more difficult to precompute this set if there exist fewer vectors of this weight in the code. A good value (or range of values) can be estimated using Equations (10) and (11).

4.1 Comparison with ISD

Information set decoding (ISD) is based on a decoding algorithm by Prange [15]. Improved versions of this attack, e.g. [14] achieve complexities close to theoretical lower bounds [11,12].

For those parameters typically used today in code-based cryptography, ISD i much faster than statistical decoding. However, the complexity of ISD increase significantly with the field size q. To estimate the value of q for which statistica decoding becomes faster than ISD, we will compare our algorithm with the on in [14].

In the case of statistical decoding, the largest part of the complexity is due t the generation of the sample sets H_w, so we will restrict our analysis to this. Ou algorithm is not fully optimized; for example, the sets H_w are sampled essentiall randomly, instead of using a generalized version of ISD to sample the vectors We will therefore estimate the total work factor of statistical decoding by

$$\mathrm{WF}_{\mathrm{SD}} \approx \frac{2n(n-k)|H|}{F \cdot P},$$

where $H = \cup_w H_w$, F is the fraction of codewords c with $b \leq \mathrm{wt}(c) \leq B$, and I is the success probability of decoding. The factor of $2n(n-k)$ reflects the fac that our sampling algorithm requires $n(n-k)$ multiplications and additions.

Note that both algorithms estimate the number of q-ary operations (insteac of binary operations), so the results are comparable.

For the $(64, 40)$ code over $\mathbb{F}_3$, ISD requires $2^{13.9}$ operations, compared witl $2^{20.2}$ for our algorithm. Increasing q, we find that ISD is slower than statistica decoding for $q \geq 1201$.

In the case of the $(128, 72)$ code and $q = 3$, the number of operations is $2^{18.}$ for ISD and $2^{22.0}$ for statistical decoding. Here, $q \geq 233$ is sufficient to make ou algorithm the more efficient one.

5 Conclusion

In this paper we have generalized Overbeck's (binary) statistical decoding algo rithm [13] to codes over non-binary fields $\mathbb{F}_q$. Our algorithm was able to decod a large part of the instances successfully. This probability can be increased b using a larger number of sample vectors $|H|$ or a larger weight spectrum $B - b$ but this increases the overall complexity of the algorithm. The success probabil ity showed to be independent of the field size q, making it especially interestin for short codes over large fields $\mathbb{F}_q$.

In addition to that, we showed how knowledge about the structure of th underlying code or about the solution can be used to increase the efficiency o the algorithm.

As further work, we propose to analyze if other types of structure can b exploited as well. Expecially the code structure seems to be very promising, fo example if the underlying code is a Goppa code.

References

1. Augot, D., Finiasz, M., Sendrier, N.: A Family of Fast Syndrome Based Crypto graphic Hash Functions. In: Dawson, E., Vaudenay, S. (eds.) Mycrypt 2005. LNCS vol. 3715, pp. 64–83. Springer, Heidelberg (2005); Cited on page 223

2. Berger, T.P., Cayrel, P.-L., Gaborit, P., Otmani, A.: Reducing Key Length of the McEliece Cryptosystem. In: Preneel, B. (ed.) AFRICACRYPT 2009. LNCS, vol. 5580, pp. 77–97. Springer, Heidelberg (2009); Cited on page 222
3. Bernstein, D.J., Lange, T., Peters, C.: Ball-collision decoding. Cryptology ePrint Archive, Report 2010/585 (2010), `http://eprint.iacr.org/2010/585.pdf`; Cited on page 217
4. Bernstein, D.J., Lange, T., Peters, C., Schwabe, P.: Faster 2-regular information-set decoding. Cryptology ePrint Archive, Report 2011/120 (2011), `http://eprint.iacr.org/`; Cited on page 223
5. Cayrel, P.-L., Véron, P., El Yousfi Alaoui, S.M.: Improved code-based identification scheme. In: SAC 2010 (2010), `http://arxiv.org/abs/1001.3017v1`; Cited on page 222
6. Cheung, K.-M.: The weight distribution and randomness of linear codes. TDA Progress Report 42–97, Communications Systems Research Section (1989); Cited on page 223
7. Al Jabri, A.: A Statistical Decoding Algorithm for General Linear Block Codes. In: Honary, B. (ed.) Cryptography and Coding 2001. LNCS, vol. 2260, pp. 1–8. Springer, Heidelberg (2001); Cited on page 217
8. MacWilliams, F.J., Sloane, N.J.A.: The theory of error-correcting codes. North-Holland Mathematical Library, vol. 16 (1977); Cited on page 223
9. McEliece, R.J.: A public-key cryptosystem based on algebraic coding theory. DNS Progress Report, pp. 114–116 (1978); Cited on page 217
10. Minder, L., Sinclair, A.: The extended k-tree algorithm. In: SODA, pp. 586–595 (2009); Cited on page 217
11. Niebuhr, R., Cayrel, P.-L., Bulygin, S., Buchmann, J.: On lower bounds for Information Set Decoding over $\mathbb{F}_q$. In: SCC 2010, RHUL, London, UK, pp. 143–158 (2010); Cited on pages 217 and 225
12. Niebuhr, R., Cayrel, P.-L., Bulygin, S., Buchmann, J.: On lower bounds for Information Set Decoding over $\mathbb{F}_q$ and on the effect of Partial Knowledge. Submitted to Math in CS, SCC 2010 (2011); Cited on page 225
13. Overbeck, R.: Statistical Decoding Revisited. In: Batten, L.M., Safavi-Naini, R. (eds.) ACISP 2006. LNCS, vol. 4058, pp. 283–294. Springer, Heidelberg (2006); Cited on pages 217, 223, and 226
14. Peters, C.: Information-set decoding for linear codes over $\mathbb{F}_q$. In: Sendrier, N. (ed.) PQCrypto 2010. LNCS, vol. 6061, pp. 81–94. Springer, Heidelberg (2010); Cited on pages 225 and 226
15. Prange, E.: The use of information sets in decoding cyclic codes. IRE Transactions on Information Theory, 5–9 (1962); Cited on page 225
16. Shor, P.W.: Algorithms for quantum computation: discrete logarithms and factoring, pp. 124–134. IEEE Press (1994); Cited on page 217

High-Speed Hardware Implementation of Rainbow Signature on FPGAs

Shaohua Tang[1], Haibo Yi[1], Jintai Ding[2,3], Huan Chen[1], and Guomin Chen[1]

[1] School of Computer Science & Engineering,
South China University of Technology, Guangzhou, China
shtang@IEEE.org, {haibo.yi87,sarlmolapple}@gmail.com, huangege@qq.com
[2] Department of Applied Mathematics,
South China University of Technology, Guangzhou, China
[3] Department of Mathematical Sciences,
University of Cincinnati, OH, USA
jintai.ding@mail.uc.edu

Abstract. We propose a new efficient hardware implementation of Rainbow signature scheme. We enhance the implementation in three directions. First, we develop a new parallel hardware design for the Gauss-Jordan elimination, and solve a 12×12 system of linear equations with only 12 clock cycles. Second, a novel multiplier is designed to speed up multiplication of three elements over a finite field. Third, we design a novel partial multiplicative inverter to speed up the multiplicative inversion of finite field elements. Through further other minor optimizations of the parallelization process and by integrating the major optimizations above, we build a new hardware implementation, which takes only 198 clock cycles to generate a Rainbow signature, a new record in generating digital signatures and four times faster than the 804-clock-cycle Balasubramanian-Bogdanov-Carter-Ding-Rupp design with similar parameters.

Keywords: Multivariate Public Key Cryptosystems (MPKCs), digital signature, Rainbow, finite field, Field-Programmable Gate Array (FPGA), Gauss-Jordan elimination, multiplication of three elements.

1 Introduction

Due to the fast growth of broad application of cryptography, the use of secure and efficient hardware architectures for implementations of cryptosystems receives considerable attention. In terms of asymmetric cryptosystems, most schemes currently used are based on the hardness of factoring large numbers or discrete logarithm problems. However, a potential powerful quantum computer could put much of currently used public key cryptosystems in jeopardy due to the algorithm by Peter Shor [1].

Multivariate Public Key Cryptosystems (MPKCs) [2] is one of main families of public key cryptosytsems that have the potential to resist the attacks by

B.-Y. Yang (Ed.): PQCrypto 2011, LNCS 7071, pp. 228–243, 2011.

quantum computation. They are based on the difficulty of the problem of solving multivariate quadratic equations over finite fields, which is in general NP-hard.

The focus of this paper is to further speed up hardware implementation of Rainbow signature generation (without consideration of the area cost). The Oil-Vinegar family of Multivariate Public Key Cryptosystems consists of three families: balanced Oil-Vinegar, unbalanced Oil-Vinegar and Rainbow [3], a multi-layer construction using unbalanced Oil-Vinegar at each layer. There have been some previous works to efficiently implement multivariate signature schemes, e.g. TTS on a low-cost smart card [4], minimized multivariate PKC on low-resource embedded systems [5], some instances of MPKCs [6], SSE implementation of multivariate PKCs on modern x86 CPUs [7]. Currently the best hardware implementations of Rainbow signature are:

1. A parallel hardware implementation of Rainbow signature scheme [8], the fastest work (not best in area utilization), which takes 804 clock cycles to generate a Rainbow signature;
2. A hardware implementation of multivariate signatures using systolic arrays [9], which optimizes in terms of certain trade-off between speed and area.

In generation of Rainbow signature, the major computation components are: 1. Multiplication of elements in finite fields; 2. Multiplicative inversion of elements in finite fields; 3. Solving system of linear equations over finite fields. Therefore, we focus on further improvement in these three directions.

Our contributions. In terms of multiplication over finite fields, we improve the multiplication according to the design in [10]. In terms of solving system of linear equations, our improvements are based on a parallel Gaussian elimination over $GF(2)$ [11], a systolic Gaussian elimination for computing multiplicative inversion [12], and a systolic Gauss-Jordan elimination over $GF(2^n)$ [13], and develop a new parallel hardware design for the Gauss-Jordan elimination to solve a 12×12 system of linear equations with only 12 clock cycles. In terms of multiplicative inversion, we design a novel partial multiplicative inverter based on Fermat's theorem.

Through further other minor optimizations of the parallelization process and by integrating the major optimizations above, we build a new hardware implementation, which takes only 198 clock cycles to generate a Rainbow signature, a new record in generating digital signatures and four times faster than the 804-clock-cycle Balasubramanian-Bogdanov-Carter-Ding-Rupp design [8] with similar parameters.

We test and verify our design on a Field-Programmable Gate Array (FPGA), the experimental results confirm our estimates.

The rest of this paper is organized as follows: in Section 2, we present the background information used in this paper; in Section 3, the proposed hardware design for Rainbow signature scheme is presented; in Section 4, we implement our design in a low-cost FPGA and experimental results are presented; in Section 5, the implementation is evaluated and compared with other hardware implementations; in Section 6, conclusions are summarized.

2 Background

2.1 Definitions

A finite field, $GF(2^8)$, including its additive and multiplicative structure, is denoted by k; The number of variables used in the signature construction, which is also equal to the signature size, is denoted by n.

For a Rainbow scheme, the number of Vinegar variables used in the ith layer of signature construction is denoted by v_i; the number of Oil variables used in the ith layer of signature construction is denoted by o_i, and $o_i = v_{i+1} - v_i$; the number of layers is denoted by u, a message (or the hash value of a message) is denoted by Y; the signature of Rainbow is denoted by X'; O_i is a set of Oil variables in the the ith layer; S_i is a set of Vinegar variables in the the ith layer.

Rainbow scheme belongs to the class of Oil-Vinegar signature constructions. The scheme consists of a quadratic system of equations involving Oil and Vinegar variables that are solved iteratively. The Oil-Vinegar polynomial can be represented by the form

$$\sum_{i \in O_l, j \in S_l} \alpha_{ij} x_i x_j + \sum_{i,j \in S_l} \beta_{ij} x_i x_j + \sum_{i \in S_{l+1}} \gamma_i x_i + \eta. \quad (1)$$

2.2 Overview of Rainbow Scheme

Rainbow scheme consists of four components: private key, public key, signature generation and signature verification.

Private Key. The private key consists of two affine transformations ${L_1}^{-1}$, ${L_2}^{-1}$ and the center mapping F, which is held by the signer. L_1: $k^{n-v_1} \to k^{n-v_1}$ and L_2: $k^n \to k^n$ are two randomly chosen invertible affine linear transformations. F is a map consists of $n - v_1$ Oil-Vinegar polynomials. F has $u - 1$ layers of Oil-Vinegar construction. The first layer consists of o_1 polynomials where $\{x_i | i \in O_1\}$ are the Oil variables, and $\{x_j | j \in S_1\}$ are the Vinegar variables. The lth layer consists of o_l polynomials where $\{x_i | i \in O_l\}$ are the Oil variables, and $\{x_j | j \in S_l\}$ are the Vinegar variables.

Public Key. The public key consists of the field k and the $n - v_1$ polynomial components of $\overline{F}$, where $\overline{F} = L_1 \circ F \circ L_2$.

Signature Generation. The message is defined by $Y = (y_1, ..., y_{n-v_1}) \in k^{n-v_1}$, and the signature is derived by computing ${L_2}^{-1} \circ F^{-1} \circ {L_1}^{-1}(Y)$.

Therefore, first we should compute $\overline{Y'} = {L_1}^{-1}(Y)$, which is a computation of an affine transformation (i.e. vector addition and matrix-vector multiplication).

Next, to solve the equation $\overline{Y'} = F$, at each layer, the v_i Vinegar variables in the Oil-Vinegar polynomials are randomly chosen and the variables at upper layer are chosen as part of the Vinegar variables. After that, the Vinegar variables are substituted into the multivariate polynomials to derive a set of linear equations with only Oil variables of that layer. If these equations have a solution, we move to next layer. Otherwise, a new set of Vinegar variables should be chosen. This procedure for each successive layer is repeated until the last layer. In this step, we obtain a vector $\overline{X} = (\overline{x_1}, ..., \overline{x_n})$. The computation of this part consists of multivariate polynomial evaluation and solving system of linear equations.

Finally, we compute $X' = {L_2}^{-1}(\overline{X}) = ({x_1}', ..., {x_n}')$. Then X' is the signature for messages Y.

It can be observed that in Rainbow signature generation, two affine transformations are computed by invoking vector addition and matrix-vector multiplication, multivariate polynomials are required to be evaluated, and system of linear equations are required to be solved.

Signature Verification. To verify the authenticity of a signature X', $\overline{F}(X') = Y'$ is computed. If $Y' = Y$ holds, the signature is accepted, otherwise rejected. In this paper, we only work on the signature generation not signature verification.

Parameters of Rainbow Signature. We adopt the parameters of Rainbow signature suggested in [14] for practical applications to design our hardware, which is also implemented in [9]. This is a two-layer scheme which has a security level above 2^{80}. There are 17 random-chosen Vinegar variables and 12 Oil variables in the first layer, and 1 random-chosen Vinegar variables and 12 Oil variables in the second layer. The parameters are shown in Table 1.

Table 1. Parameters of Rainbow in Proposed Hardware Design

Parameter	Rainbow
Ground field size	$GF(2^8)$
Message size	24 bytes
Signature size	42 bytes
Number of layers	2
Set of variables in each layer	(17, 12), (1, 12)

3 Proposed Hardware Design for Rainbow Signature

3.1 Overview of the Hardware Design

The flowchart to generate Rainbow signature is illustrated in Fig. 1. It can be observed that Rainbow signature generation consists of computing affine transformations, polynomial evaluations and solutions for system of linear equations.

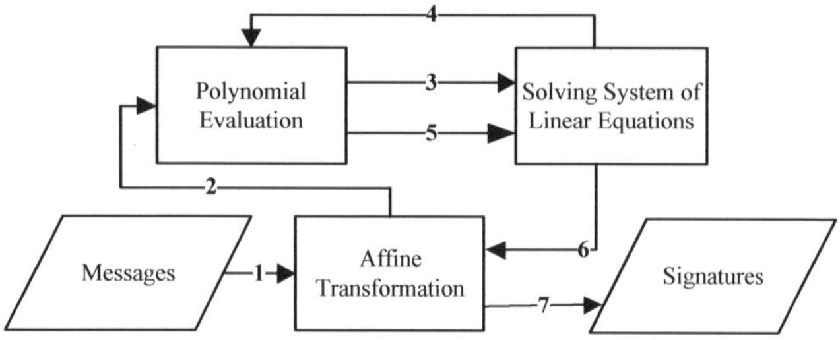

Fig. 1. The Flowchart to Generate Rainbow Signature

3.2 Choice of Irreducible Polynomial for the Finite Field

The choice of the irreducible polynomial for the finite field k is a critical part of our hardware design, since it affects the efficiency of the operations over the finite field. The irreducible polynomials for $GF(2^8)$ over $GF(2)$ can be expressed as 9-bit binary digits with the form $x^8+x^k+...+1$, where $0 < k < 8$ and the first bit and the last bit are valued one. There are totally 16 candidates. We evaluate the performance of the multiplications based on these irreducible polynomials respectively.

By comparing the efficiency of signature generations basing on different irreducible polynomials, $x^8 + x^6 + x^3 + x^2 + 1$ is finally chosen as the irreducible polynomial in our hardware design.

3.3 Efficient Design of Multiplication of Three Elements

In Rainbow signature generation, we notice that there exist not only multiplication of two elements but also multiplication of three elements. An optimized design of the multiplier can dramatically improve the overall hardware execution efficiency.

Therefore, we design new implementation to speed up multiplication of three elements based on the multiplication of two elements [10]. The new design is based on a new observation that, in multiplication of three elements over $GF(2^8)$, it is much faster to multiply everything first than perform modular operation than the other way around. This is quite anti-intuitive and it works only over small fields. This idea, in general, is not applicable for large fields.

Suppose $a(x) = \sum_{i=0}^{7} a_i x^i$, $b(x) = \sum_{i=0}^{7} b_i x^i$ and $c(x) = \sum_{i=0}^{7} c_i x^i$ are three elements in $GF(2^8) = GF(2)[x]/f(x)$, and

$$d(x) = a(x) \times b(x) \times c(x)(\mathrm{mod}(f(x))) = \sum_{i=0}^{7} d_i x^i \tag{2}$$

is the expected multiplication result, where $f(x)$ is the irreducible polynomial.

First, we compute v_{ij} for $i = 0, 1, ..., 21$ and $j = 0, 1, ..., 7$ according to x^i mod $f(x) = \sum_{j=0}^{7} v_{ij}x^j$. Next, we compute S_i for $i = 0, 1, ..., 21$ via $S_i = \sum_{j+k+l=i} a_j b_k c_l$. After that, we compute d_i for $i = 0, 1, ..., 7$ via $d_i = \sum_{j=0}^{21} v_{ji}S_j$. Finally, the multiplication result of $a(x) \times b(x) \times c(x) \; mod \; f(x)$ is $\sum_{i=0}^{7} d_i x^i$.

3.4 Efficient Design of Partial Multiplicative Inversion

The multiplicative inverse over finite fields is a crucial but time-consuming operation in multivariate signature. An optimized design of the inverter can really help to improve the overall performance. Since multiplicative inversion is only used in solving system of linear equations, we do not implement a fully multiplicative inverter but adopt a partial inverter based on Fermat's theorem in our design.

Suppose $f(x)$ is the irreducible polynomial and β is an element over $GF(2^8)$, where $\beta = \beta_7x^7 + \beta_6x^6 + \beta_5x^5 + \beta_4x^4 + \beta_3x^3 + \beta_2x^2 + \beta_1x + \beta_0$. According to the Fermat's theorem, we have $\beta^{2^8} = \beta$, and $\beta^{-1} = \beta^{2^8-2} = \beta^{254}$. Since $2^8 - 2 = 2 + 2^2 + 2^3 + 2^4 + 2^5 + 2^6 + 2^7$, then $\beta^{-1} = \beta^2\beta^4\beta^8\beta^{16}\beta^{32}\beta^{64}\beta^{128}$.

We can then construct the logic expressions of these items.

$$\begin{aligned}\beta^{2^i} = \beta_7x^{2^i\times7} + \beta_6x^{2^i\times6} + \beta_5x^{2^i\times5} + \beta_4x^{2^i\times4} + \\ \beta_3x^{2^i\times3} + \beta_2x^{2^i\times2} + \beta_1x^{2^i} + \beta_0,\end{aligned} \tag{3}$$

The computation of $x^{2^i\times j}$ should be reduction modulo the irreducible polynomial, where $i = 1, 2, ..., 7$ and $j = 0, 1, ..., 7$, then β^{2^i} is transformed into the equivalent form. For instance, $\beta^{2^i} = \beta'_7x^7 + \beta'_6x^6 + \beta'_5x^5 + \beta'_4x^4 + \beta'_3x^3 + \beta'_2x^2 + \beta'_1x + \beta'_0$.

We adopt the three-input multiplier described in Section 3.3 to design the partial inverter, where $ThreeMult(v1, v2, v3)$ stands for multiplication of three elements and $v1$, $v2$, $v3$ are operands and S_1, S_2 are the multiplication results.

$$\begin{aligned}S_1 &= ThreeMult(\beta^2, \beta^4, \beta^8),\\ S_2 &= ThreeMult(\beta^{16}, \beta^{32}, \beta^{64}).\end{aligned} \tag{4}$$

We call the triple (S_1, S_2, β^{128}) the partial multiplicative inversion of β. Below we will present how we adopt partial inversion in solving system of linear equations.

3.5 Optimized Gauss-Jordan Elimination

We propose a parallel variant of Gauss-Jordan elimination for solving a system of linear equations with the matrix size 12×12. The optimization and parallelization of Gauss-Jordan elimination can enhance the overall performance of solving system of linear equations.

Algorithm and Architecture. We give a straightforward description of the proposed algorithm of the parallel variant of Gauss-Jordan elimination in Algorithm 1, where *operation*(i) stands for operation performed in the i-th iteration, and $i = 0, 1, ..., 11$. The optimized Gauss-Jordan elimination with 12 iterations consists of pivoting, partial multiplicative inversion, normalization and elimination in each iteration.

We enhance the algorithm in four directions. First, multiplication of three elements is computed by invoking three-input multipliers designed in Section 3.3. Second, we adopt a partial multiplicative inverter described in Section 3.4 in our design. Third, the partial multiplicative inversion, normalization and elimination are designed to perform simultaneously. Fourth, during the elimination in the i-th iteration, we simultaneously choose the right pivot for the next iteration, namely if element $a_{i+1,i+1}$ of the next iteration is zero, we swap the $(i+1)$-th row with another j-th row with the nonzero element a_{ji}, where $i, j = 0, 1, ..., 11$. The difference from usual Gauss-Jordan elimination is that the usual Gauss-Jordan elimination choose the pivot after the elimination, while we perform the pivoting during the elimination. In other words, at the end of each iteration, by judging the computational results in this iteration, we can decide the right pivoting for the next iteration. By integrating these optimizations, it takes only one clock cycle to perform one iteration.

Algorithm 1. Solving a system of linear equations $Ax = b$ with 12 iterations, where A is a 12×12 matrix

```
var
    i: Integer;
begin
    i := 0;
    Pivoting(i = 0);
    repeat
        Partial_inversion(i), Normalization(i), Elimination(i);
        Pivoting(i+1);
        i:= i+1;
    until i = 12
end.
```

The proposed architecture is depicted in Fig. 2 with matrix size 12×12, where a_{ij} is the element located at the i-th row and j-th column of the matrix.

There exist three kinds of cells in the architecture, namely I, N_l, and E_{kl}, where $k = 1, 2, ..., 11$ and $l = 1, 2, ..., 12$. The I cell is for partial multiplicative inversion. As described in 3.4, two three-input multipliers are included in the I cell for computed partial multiplicative inversion. The N_l cells are for normalization. And the E_{kl} cells are for elimination. The architecture consists of one I cell, 12 N_l cells and 132 E_{lk} cells.

The matrixes depicted in Fig. 2 are used only to illustrate how the matrix changes. The left-most matrix is the one in the first clock cycle while the i-th matrix is the one in the i-th clock cycle. In the first clock cycle, the left-most matrix is sent to the architecture. a_{00} is sent to I cell for partial multiplicative inversion. The first row is sent to N_l for normalization. And the other rows except the first row are sent to E_{lk} for elimination. In this clock cycle, one iteration of Gauss-Jordan elimination is performed and the matrix has been updated. In the following clock cycles, the pivot element is sent to I cell for partial multiplicative inversion. The pivot row is sent to N_l for normalization. And the other rows except the pivot row are sent to E_{lk} for elimination. It can be observed that the system of linear equations with matrix size 12×12 can be solved with 12 clock cycles.

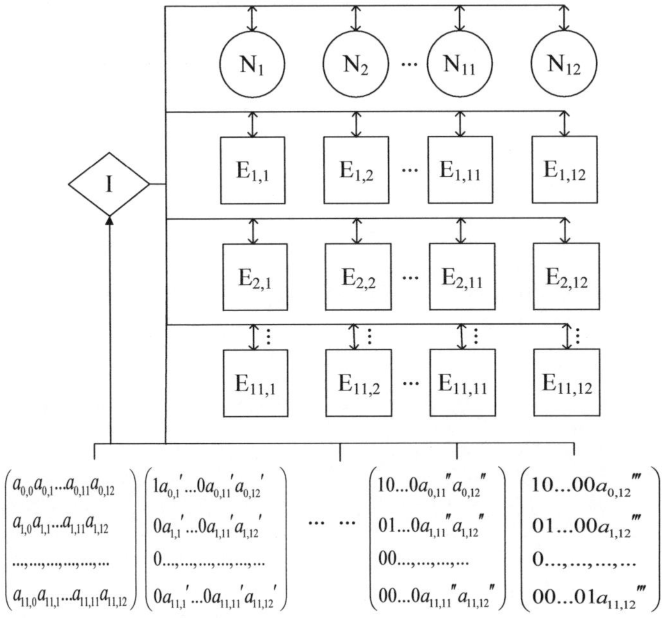

Fig. 2. Proposed Architecture for Parallel Solving System of Linear Equations with Matrix Size 12×12

Pivoting Operation. If the pivot a_{ii} of the i-th iteration is zero, we should find a nonzero element a_{ji} in the pivot column, i.e, the i-th column, as the new pivot element, where $i, j = 0, 1, ..., 11$. Then the computational results of the j-th row is sent to the N_l cells for normalization as the new pivot row. At the same time, the computational results of the i-th row is sent to the E_{jl} cells for elimination. In this way, we can ensure that the pivot element is nonzero in a new iteration. Therefore, the I cell, the N_l cells and the E_{kl} cells can execute simultaneously.

An example of pivoting is shown in Fig. 3. Before the second iteration, the second row is the pivot row but the pivot element is zero. The fourth row can be chosen as the new pivot row since a_{31} is nonzero. Then a_{31} is sent to I

cell for partial multiplicative inversion. The fourth row is sent to N_l cells for normalization, and then the other rows including the second row are sent to E_{1l} cells for elimination. Therefore, the computation of one iteration can be performed with one clock cycle.

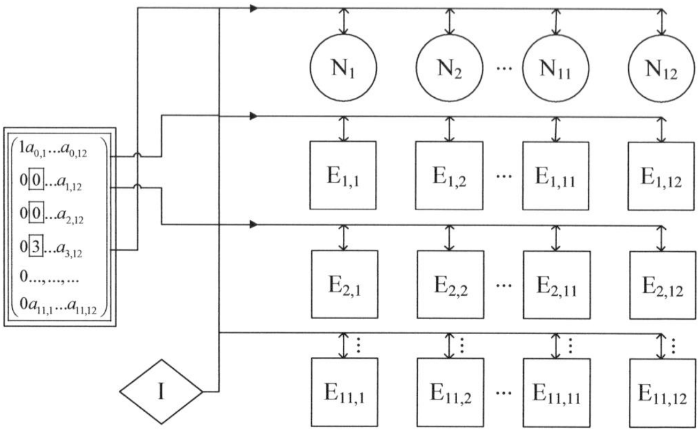

Fig. 3. Pivoting in Solving System of Linear Equations

Normalizing Operation. The normalizing operation invokes multiplicative inversions and multiplications, then we can enhance the implementation in two aspects.

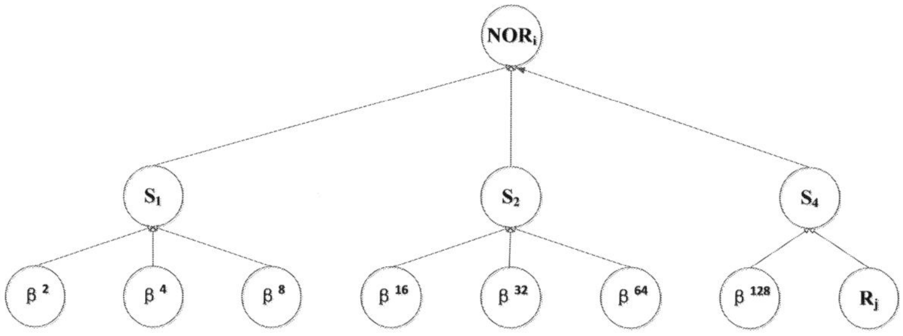

Fig. 4. Optimized Normalization in Solving System of Linear Equations

First, the multiplicative inverse β^{-1} over $GF(2^8)$ is optimized to the multiplication of 7 elements due to $\beta^{-1} = \beta^2\beta^4\beta^8\beta^{16}\beta^{32}\beta^{64}\beta^{128}$, as mentioned in Section 3.4.

Second, a new multiplier is designed to speed up the multiplication of three elements that denoted by $ThreeMult(v1, v2, v3)$, where $v1$, $v2$ and $v3$ are operands, while the multiplication of two elements is defined by $TwoMult(v1, v2)$.

The schematic diagram of normalization is shown in Fig. 4, where R_i for the i-th element in the pivot row, and NOR_i for the normalizing result, respectively. Then, we have the expressions

$$\begin{aligned} S_1 &= ThreeMult(\beta^2, \beta^4, \beta^8), \\ S_2 &= ThreeMult(\beta^{16}, \beta^{32}, \beta^{64}), \\ S_4 &= TwoMult(\beta^{128}, R_i), \\ NOR_i &= ThreeMult(S_1, S_2, S_4). \end{aligned} \tag{5}$$

S_1 and S_2 are executed in I cell for partial multiplicative inversion while S_4 and NOR_i are executed in N_i cells for normalization. Thus one two-input multiplier as well as another three-input multiplier are included in N_i cells. Since S_1, S_2 and S_4 can be implemented in parallel in each iteration, the critical path of normalizing consists of only two multiplications of three elements.

Eliminating Operation. The schematic diagram of normalization is shown in Fig. 5, where R_j stands for the j-th element in the pivot row, C_i for the i-th element in the pivot column, and ELI_{ij} is the eliminated result of a_{ij}.

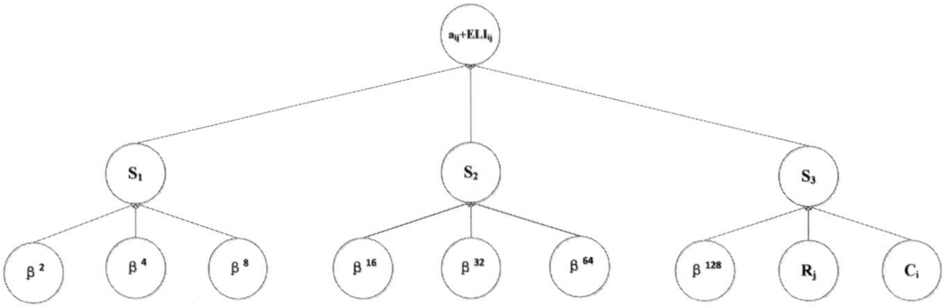

Fig. 5. Optimized Elimination in Solving System of Linear Equations

Then, we have the expressions

$$\begin{aligned} S_1 &= ThreeMult(\beta^2, \beta^4, \beta^8), \\ S_2 &= ThreeMult(\beta^{16}, \beta^{32}, \beta^{64}), \\ S_3 &= ThreeMult(\beta^{128}, R_j, C_i), \\ ELI_{ij} &= a_{ij} + ThreeMult(S_1, S_2, S_3). \end{aligned} \tag{6}$$

S_1 and S_2 are executed in I cell for partial multiplicative inversion while S_3 and ELI_{ij} are executed in E_{ij} cells for elimination. Thus two three-input multipliers and one adder are included in E_{ij} cells. Since S_1, S_2 and S_3 can be implemented in parallel in each iteration, the critical path of elimination consists of only two multiplications of three elements and one addition.

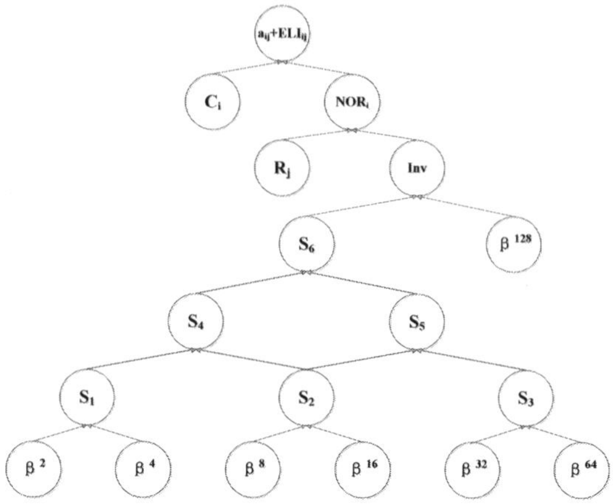

Fig. 6. Original Design of Gauss-Jordan Elimination

Overall Optimization. By integrating the optimizations above, Fig. 7 shows that the critical path of our design is reduced from five multiplications and one addition to two multiplications and one addition, compared with the original principle of Gauss-Jordan elimination illustrated in Fig. 6.

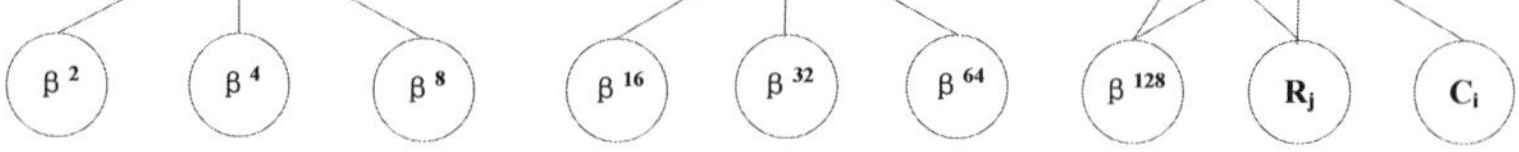

Fig. 7. Optimized Design of Gauss-Jordan Elimination

Therefore, our design takes one clock cycle to perform the operations in each iteration of solving system of linear equations. In the end, it takes only 12 clock cycles to solve a system of linear equations where the matrix size is 12×12.

3.6 Designs of Affine Transformations and Polynomial Evaluations

${L_1}^{-1}$: $k^{24} \to k^{24}$ and ${L_2}^{-1}$: $k^{42} \to k^{42}$ affine transformations are computed by invoking vector addition and vector-multiplication over a finite field. Two-layer Oil-Vinegar constructions including 24 multivariate polynomials are evaluated by invoking multiplication over a finite field. Thus multiplication over a finite field is

Table 2. Number of Multiplications in ${L_1}^{-1}$, ${L_2}^{-1}$ Affine Transformations and Polynomial Evaluations

Components	Number of multiplications
${L_1}^{-1}$ transformation	576
The first 12 polynomial evaluations	6324
The second 12 polynomial evaluations	15840
${L_2}^{-1}$ transformation	1764
Total	24504

the most time-consuming operation in these computations. Table 2 summarizes the numbers of multiplications in two affine transformations and polynomial evaluations. The number of multiplications of the components of polynomial evaluations is summarized in Table 3.

Table 3. Number of Multiplications in Components of Polynomial Evaluations

	The first layer	The second layer
V_iO_j	2448	4320
V_iV_j	3672	11160
V_i	204	360
Total	6324	15840

4 Implementations and Experimental Results

4.1 Overview of Our Implementation

Our design is programmed in VHDL and implemented on a EP2S130F1020I4 FPGA device, which is a member of ALTERA Stratix II family. Table 4 summarizes the performance of our implementation of Rainbow signature measured in clock cycles, which shows that our design takes only 198 clock cycles to generate a Rainbow signature. In other words, our implementation takes 3960 ns to generate a Rainbow signature with the frequency of 50 MHz. All the experimental results mentioned in this section are extracted after place and route.

Table 4. Running Time of Our Implementation in Clock Cycles

Steps	Components	Clock cycles
1	${L_1}^{-1}$ transformation	5
2	The first 12 polynomial evaluations	45
3	The first round of solving system of linear equations	12
4	The second 12 polynomial evaluations	111
5	The second round of solving system of linear equations	12
6	${L_2}^{-1}$ transformation	13
	Total	198

4.2 Implementation of Multiplier, Partial Inverter and LSEs Solver

Our multipliers and partial inverter can execute a multiplication and partial multiplicative inversion over $GF(2^8)$ within one clock cycle respectively. As mentioned in Section 3.5, the critical path of each iteration of optimized Gauss-Jordan elimination includes two multiplications and one addition. Since there exist some overlaps in two serial multiplications, one iteration of optimized Gauss-Jordan elimination can be computed in 20 ns with one clock cycle. Therefore, it takes 12 clock cycles to solve a system of linear equations of matrix size 12×12, which is 240 ns with a frequency of 50 MHz.

Table 5. FPGA Implementations of the Multiplier, Partial Inverter and Optimized Gauss-Jordan Elimination over $GF(2^8)$

Components	Multiplier	Partial inverter	Gauss-Jordan elimination
Combinational $ALUTs$	37	22	21718
Dedicated logic registers	0	0	1644
Clock cycles	1	1	12
Running time (ns)	10.768	9.701	240

Table 5 is extracted after place and route of multiplication, partial multiplicative inversion and optimized Gauss-Jordan elimination over $GF(2^8)$. Three different kinds of cells included in our proposed architecture have been described and their resource consumptions are given in Table 6.

Table 6. The Resource Consumptions for Each Cell in the Proposed Architecture for Solving System of Linear Equations

Cell	Use	Two-input multiplier	Three-input multiplier	Adder
I cell	Partial inversion	0	2	0
N cell	Normalization	1	1	0
E cell	Elimination	0	2	1

4.3 Implementation of Transformations and Polynomial Evaluations

The affine transformations ${L_1}^{-1}$ and ${L_2}^{-1}$ invoke vector addition and matrix-vector multiplication over $GF(2^8)$. Table 7 shows that two affine transformations take 18 clock cycles, which is 360 ns with a frequency of 50 MHz, where the second and fourth columns are the performance of vector additions using L_1 offset and L_2 offset respectively and the third and fifth columns are the performance of matrix-vector multiplications using the matrixes of ${L_1}^{-1}$ and ${L_2}^{-1}$ respectively.

Table 8 illustrates that polynomial evaluations takes 156 clock cycles, which is 3120 ns with a frequency of 50 MHz, where the second, third and fourth columns are the performances of components of multivariate polynomials, respectively.

Table 7. Clock Cycles and Running Time of Two Affine Transformations

Components	L_1 offset	${L_1}^{-1}$	L_2 offset	${L_2}^{-1}$	Total
Clock cycles	1	4	1	12	18
Running time (ns)	20	80	20	240	360

Table 8. Clock Cycles and Running Time of Polynomial Evaluations

Components	V_iO_j	V_iV_j	V_i	Total cycles	Total time
The first layer	17	26	2	45	900 ns
The second layer	30	78	3	111	2220 ns

Note here that our implementation focuses solely on speeding up the signing process, and, in terms of area, we compute the size in gate equivalents (GEs), about 150,000 GEs, which is 2-3 times the area of [8].

5 Comparison with Related Works

We compare the implementations of solving system of linear equations and Rainbow signature generation with related works by the following tables, which clearly demonstrate the improvements of our new implementation.

Table 9. Comparison of Solving System of Linear Equations with Matrix Size 12×12

Scheme	Clock cycles
Original Gauss-Jordan elimination	1116
Original Gaussian elimination	830
Wang-Lin's Gauss-Jordan elimination [12]	48
B. Hochet's Gaussian elimination [13]	47
A Bogdanov's Gaussian elimination [11]	24
Implementation in this paper	12

Table 10. Performance Comparison of Signature Schemes

Scheme	Clock cycles
en-TTS [5]	16000
Rainbow (42,24) [9]	3150
Long-message UOV [9]	2260
Rainbow [8]	804
Short-message UOV [9]	630
This paper	198

6 Conclusions

We propose a new optimized hardware implementation of Rainbow signature scheme, which can generate a Rainbow signature with only 198 clock cycles, a new record in generating digital signatures.

Our main contributions include three parts. First, we develop a new parallel hardware design for the Gauss-Jordan elimination, and solve a 12×12 system of linear equations with only 12 clock cycles. Second, a novel multiplier is designed to speed up multiplication of three elements over finite fields. Third, we design a novel partial multiplicative inverter to speed up the multiplicative inversion of finite field elements. Through further other minor optimizations of the parallelization process and by integrating the major optimizations above, we build a new hardware implementation, which takes only 198 clock cycles to generate a Rainbow signature, four times faster than the 804-clock-cycle Balasubramanian-Bogdanov-Carter-Ding-Rupp design [8] with similar parameters. Our implementation focuses solely on speeding up the signing process not area utilization.

The optimization method of three-operand multiplier, partial multiplicative inverter, and LSEs solver proposed can be further applied to various applications like matrix factorization, matrix inversion, and other multivariate PKCs.

Acknowledgement. This work is supported by National Natural Science Foundation of China under Grant No. 61170080 and 60973131, and supported by Guangdong Province Universities and Colleges Pearl River Scholar Funded Scheme (2011). This paper is also supported by the Fundamental Research Funds for the Central Universities of China under Grant No.2009ZZ0035 and No.2011ZG0015.

References

1. Shor, P.W.: Polynomial-time algorithms for prime factorization and discrete logarithms on a quantum computer. SIAM Review 41(2), 303–332 (1999)
2. Ding, J., Schmidt, D.: Multivariate public key cryptosystems. In: Advances in Information Security, vol. 25, Springer, Heidelberg (2006)
3. Ding, J., Schmidt, D.: Rainbow, a New Multivariable Polynomial Signature Scheme. In: Ioannidis, J., Keromytis, A.D., Yung, M. (eds.) ACNS 2005. LNCS, vol. 3531, pp. 164–175. Springer, Heidelberg (2005)
4. Yang, B.-Y., Chen, J.-M., Chen, Y.-H.: TTS: High-Speed Signatures on a Low-Cost Smart Card. In: Joye, M., Quisquater, J.-J. (eds.) CHES 2004. LNCS, vol. 3156, pp. 371–385. Springer, Heidelberg (2004)
5. Yang, B.-Y., Cheng, C.-M., Chen, B.-R., Chen, J.-M.: Implementing Minimized Multivariate PKC on Low-Resource Embedded Systems. In: Clark, J.A., Paige, R.F., Polack, F.A.C., Brooke, P.J. (eds.) SPC 2006. LNCS, vol. 3934, pp. 73–88. Springer, Heidelberg (2006)
6. Chen, A.I.-T., Chen, C.-H.O., Chen, M.-S., Cheng, C.-M., Yang, B.-Y.: Practical-Sized Instances of Multivariate PKCs: Rainbow, TTS, and ℓIC-derivatives. In: Buchmann, J., Ding, J. (eds.) PQCrypto 2008. LNCS, vol. 5299, pp. 95–108. Springer, Heidelberg (2008)

7. Chen, A.I.-T., Chen, M.-S., Chen, T.-R., Cheng, C.-M., Ding, J., Kuo, E.L.-H., Lee, F.Y.-S., Yang, B.-Y.: SSE Implementation of Multivariate PKCs on Modern X86 CPUs. In: Clavier, C., Gaj, K. (eds.) CHES 2009. LNCS, vol. 5747, pp. 33–48. Springer, Heidelberg (2009)
8. Balasubramanian, S., Bogdanov, A., Rupp, A., Ding, J., Carter, H.W.: Fast multivariate signature generation in hardware: The case of Rainbow. In: 16th International Symposium on Field-Programmable Custom Computing Machines, pp. 281–282 (April 2008)
9. Bogdanov, A., Eisenbarth, T., Rupp, A., Wolf, C.: Time-Area Optimized Public-Key Engines: $\mathcal{MQ}$-Cryptosystems as Replacement for Elliptic Curves? In: Oswald, E., Rohatgi, P. (eds.) CHES 2008. LNCS, vol. 5154, pp. 45–61. Springer, Heidelberg (2008)
10. Mastrovito, E.: VLSI Designs for Multiplication over Finite Fields $GF(2^m)$. In: Huguet, L., Poli, A. (eds.) AAECC 1987. LNCS, vol. 356, Springer, Heidelberg (1989)
11. Bogdanov, A., Mertens, M.C., Paar, C., Pelzl, J., Rupp, A.: A parallel hardware architecture for fast Gaussian elimination over $GF(2)$. In: 14th Annual IEEE Symposium on Field-Programmable Custom Computing Machines, pp. 237–248. IEEE (2006)
12. Wang, C.-L., Lin, J.-L.: A systolic architecture for computing inverses and divisions in finite fields $GF(2^m)$. IEEE Transactions on Computers 42(9), 1141–1146 (1993)
13. Hochet, B., Quinton, P., Robert, Y.: Systolic Gaussian elimination over $GF(p)$ with partial pivoting. IEEE Transactions on Computers 38(9), 1321–1324 (1989)
14. Ding, J., Yang, B.-Y., Chen, C.-H.O., Chen, M.-S., Cheng, C.-M.: New Differential-Algebraic Attacks and Reparametrization of Rainbow. In: Bellovin, S.M., Gennaro, R., Keromytis, A.D., Yung, M. (eds.) ACNS 2008. LNCS, vol. 5037, pp. 242–257. Springer, Heidelberg (2008)

Wild McEliece Incognito

Daniel J. Bernstein[1], Tanja Lange[2], and Christiane Peters[3]

[1] Department of Computer Science
University of Illinois at Chicago, Chicago, IL 60607–7045, USA
djb@cr.yp.to

[2] Department of Mathematics and Computer Science
Technische Universiteit Eindhoven, P.O. Box 513, 5600 MB Eindhoven, Netherlands
tanja@hyperelliptic.org

[3] Department of Mathematics
Technical University of Denmark, 2800 Kgs. Lyngby, Denmark
c.p.peters@mat.dtu.dk

Abstract. The wild McEliece cryptosystem uses wild Goppa codes over finite fields to achieve smaller public key sizes compared to the original McEliece cryptosystem at the same level of security against all attacks known. However, the cryptosystem drops one of the confidence-inspiring shields built into the original McEliece cryptosystem, namely a large pool of Goppa polynomials to choose from.

This paper shows how to achieve almost all of the same reduction in key size while preserving this shield. Even if support splitting could be (1) generalized to handle an unknown support set and (2) sped up by a square-root factor, polynomial-searching attacks in the new system will still be at least as hard as information-set decoding.

Furthermore, this paper presents a set of concrete cryptanalytic challenges to encourage the cryptographic community to study the security of code-based cryptography. The challenges range through codes over $\mathbf{F}_2, \mathbf{F}_3, \ldots, \mathbf{F}_{32}$, and cover two different levels of how much the wildness is hidden.

Keywords: McEliece cryptosystem, Niederreiter cryptosystem, Goppa codes, wild Goppa codes, list decoding.

1 Introduction

The McEliece cryptosystem [15] is based on classical Goppa codes (corresponding to genus-0 AG codes) over $\mathbf{F}_2$. A code is built using a Goppa polynomial

* This work was supported by the Cisco University Research Program, by the National Institute of Standards and Technology under grant 60NANB10D263, by the Danish Council for Independent Research under the Technology and Production Sciences (FTP) grant 11-105325, and by the European Commission under Contract ICT-2007-216676 ECRYPT II. This work was started while the third author was with Technische Universiteit Eindhoven and continued during her employment at the University of Illinois at Chicago. Permanent ID of this document: cd39ef08c48d12b29da6b9db66559c41. Date: 2011.09.10.

B.-Y. Yang (Ed.): PQCrypto 2011, LNCS 7071, pp. 244–254, 2011.

$g \in \mathbf{F}_{2^m}[x]$ for some integer m. For $\deg(g) = t$ the code can correct t errors. Generalizations of the McEliece cryptosystem (or equivalently the Niederreiter cryptosystem [16]) using Goppa codes over larger fields $\mathbf{F}_q$ were investigated but not found to offer advantages for small q, since it was believed that for a code built from a polynomial g of degree t one could correct only $\lfloor t/2 \rfloor$ errors.

Peters showed in [18] that, despite this reduced error-correction capacity, codes over $\mathbf{F}_{31}$ offer advantages in key size compared to codes over $\mathbf{F}_2$ while maintaining the same security level against all attacks known. However, codes over smaller fields such as $\mathbf{F}_3$ were still not competitive in key size with codes over $\mathbf{F}_2$.

In [6] we introduced the "wild McEliece" cryptosystem, using Goppa codes over $\mathbf{F}_q$ built on polynomials of the form g^{q-1}. These codes have a better error-correction capacity: they can correct up to $\lfloor qt/2 \rfloor$ errors for $\deg(g) = t$. The extra factor $q/(q-1)$ makes "larger tiny fields" attractive and bridges the gap between $\mathbf{F}_2$ and $\mathbf{F}_{31}$. That paper contains cryptosystem parameters that minimize key size for different finite fields, subject to the requirement of achieving 128-bit security against information-set-decoding attacks.

This key-size optimization for 128-bit security reduces the number of irreducible polynomials g below 2^{128} for $q \geq 11$, and below 2^{30} for $q \geq 31$. Enumerating all possibilities for g thus becomes more efficient than performing information-set decoding. The parameters were intentionally chosen this way in [6]; otherwise the key-size benefit of wild McEliece would disappear as q grows.

In McEliece's original proposal, a large space of possibilities for g is the primary shield against structural attacks. There are secrets other than g, specifically a random support permutation P and a random invertible matrix S, but Sendrier's support-splitting algorithm [21] quickly computes both P and S given g and the public key. The cost of breaking McEliece's system is thus at most a small multiple of the number of choices of g: the attacker checks each possibility for g with the support-splitting algorithm.

This attack fails against [6], because there is another shield in [6]: a secret support set. In McEliece's original proposal, the support set was all of $\mathbf{F}_{2^m}$; however, one can define Goppa codes using smaller support sets. We chose parameters in [6] so that there are more than 2^{256} possible support sets. There is no known attack against the McEliece system with secret support sets, even if the Goppa polynomial is *published*; in particular, the support-splitting algorithm uses the support set as input.

However, a secret support set has far less history than a secret choice of g, and therefore cannot inspire as much confidence. One can reasonably worry that there is a generalization of support-splitting that handles many support sets more efficiently than separately trying each possible support set. Parameters relying on secret support sets were marked with biohazard symbols in [6].

In this paper we hide the wild codes in two ways, achieving almost all of the key-size benefit of wild McEliece without sacrificing the confidence provided by a large space of polynomials. First, we consider codes built on polynomials $f \cdot g^{q-1}$; for $\deg(f) = s$ and $\deg(g) = t$ these codes can correct up to $\lfloor (s + qt)/2 \rfloor$ errors.

A small extra factor f makes the space of polynomials too large to search. Second, we use subcodes as suggested by Berger and Loidreau in [2]. The combination of these defenses leads to slightly larger key sizes but requires the attacker to see simultaneously through subcodes, secret support sets, and a huge set of polynomials.

This paper also announces a web page of code-based crypto challenges and presents some sample challenges. The challenges cover finite fields as large as $\mathbf{F}_{32}$ and start with training challenges that should be easy to break but still can show which attacks are faster than others. We originally considered issuing challenges with several different wildness percentages, ranging from 100% wild codes (g^{q-1}) to 50% wild codes (fg^{q-1} with $\deg(f) \approx (q-1)\deg(g)$) and beyond, but we decided to focus on percentages close to 100%, since those are adequate to prevent polynomial enumeration. For each set of parameters, a public key and a ciphertext are presented.

Acknowledgement. The authors are grateful to Peter Beelen for interesting discussions and in particular for allowing us to use his suggestion of the extra factor f as a way to hide the wildness of g^{q-1}.

2 An Extra Shield for Wild Goppa Codes

Fix a prime power q; a positive integer m; a positive integer $n \le q^m$; an integer $t < n/m$; distinct elements $a_1, \ldots, a_n$ in $\mathbf{F}_{q^m}$; and a polynomial $g(x)$ in $\mathbf{F}_{q^m}[x]$ of degree t such that $g(a_i) \neq 0$ for all i.

We denote the linear code consisting of all words $c = (c_1, \ldots, c_n)$ in $\mathbf{F}_{q^m}^n$ satisfying

$$\sum_{i=1}^{n} \frac{c_i}{x - a_i} \equiv 0 \pmod{g(x)} \tag{2.1}$$

by $\Gamma_{q^m}(a_1, \ldots, a_n, g)$; this is a special case of a generalized Reed–Solomon code over $\mathbf{F}_{q^m}$ having dimension $n - t$.

The *Goppa code* $\Gamma_q(a_1, \ldots, a_n, g)$ with *Goppa polynomial* $g(x)$ and *support* $a_1, \ldots, a_n$ is the restriction of $\Gamma_{q^m}(a_1, \ldots, a_n, g)$ to the field $\mathbf{F}_q$, i.e., the set of elements $(c_1, \ldots, c_n)$ in $\mathbf{F}_q^n$ that satisfy (2.1); this code $\Gamma_q(a_1, \ldots, a_n, g)$ has *dimension* at least $n - mt$ and *minimum distance* at least $t + 1$. These codes were introduced in [11] and [12].

Goppa codes can be decoded by any decoder for generalized Reed–Solomon codes. For example, Berlekamp's algorithm corrects $\lfloor t/2 \rfloor$ errors; see, e.g., [3]. Note that $t+1$ is a lower bound for the minimum distance. There are Goppa codes whose minimum distance is much larger. Binary Goppa codes have minimum distance at least $2t + 1$ as shown in [11], and allow fast decoding of t errors. The standard t-error decoding algorithm for binary Goppa codes, in the typical case that g is monic and irreducible, is Patterson's algorithm from [17]. There are polynomial-time list-decoding algorithms that decode more errors; for more information and references see, e.g., [4], [1], and [5].

In this paper we use the McEliece cryptosystem, the Niederreiter cryptosystem, etc. with codes of the form $\Gamma_q(a_1,\ldots,a_n,fg^{q-1})$, where f and g are coprime squarefree monic polynomials.

If $g=1$ then $\Gamma_q(a_1,\ldots,a_n,fg^{q-1})$ is the squarefree Goppa code $\Gamma_q(a_1,\ldots,a_n,f)$; these are, for $q=2$, the traditional codes used in the McEliece cryptosystem. If $f=1$ then $\Gamma_q(a_1,\ldots,a_n,fg^{q-1})$ is the *wild Goppa code* $\Gamma_q(a_1,\ldots,a_n,g^{q-1})$, which we proposed in [6] for the *wild McEliece cryptosystem*; what makes these codes interesting is that they can correct $\lfloor qt/2\rfloor$ errors, or even slightly more using list decoding.

The Goppa code with polynomial fg^{q-1} has dimension at least $n-m(s+(q-1)t)$, where s is the degree of f and t is the degree of g. Theorem 2.1 below says that fg^q gives the same Goppa code. It follows that $\Gamma_q(a_1,a_2,\ldots,a_n,fg^{q-1})$ has minimum distance at least $s+qt+1$. One can plug fg^q into (e.g.) the alternant decoder described in [6, Section 5] to efficiently decode $\lfloor(s+qt)/2\rfloor$ errors, or into the list-decoder described in [5] to efficiently decode more errors.

Theorem 2.1 is a special case of a theorem of Sugiyama, Kasahara, Hirasawa, and Namekawa [24]. To keep this paper self-contained we give a streamlined proof here, generalizing the streamlined proof for $f=1$ from [6].

Theorem 2.1. *Let q be a prime power. Let m be a positive integer. Let n be an integer with $1\le n\le q^m$. Let $a_1,a_2,\ldots,a_n$ be distinct elements of $\mathbf{F}_{q^m}$. Let f and g be coprime monic polynomials in $\mathbf{F}_{q^m}[x]$ that both do not vanish at any of $a_1,\ldots,a_n$. Assume that g is squarefree. Then $\Gamma_q(a_1,a_2,\ldots,a_n,fg^{q-1})=\Gamma_q(a_1,a_2,\ldots,a_n,fg^q)$.*

Proof. If $\sum_i c_i/(x-a_i)=0$ in $\mathbf{F}_{q^m}[x]/(fg^q)$ then certainly $\sum_i c_i/(x-a_i)=0$ in $\mathbf{F}_{q^m}[x]/(fg^{q-1})$.

Conversely, consider any $(c_1,c_2,\ldots,c_n)\in\mathbf{F}_q^n$ such that $\sum_i c_i/(x-a_i)=0$ in $\mathbf{F}_{q^m}[x]/(fg^{q-1})$; i.e., fg^{q-1} divides $\sum_i c_i/(x-a_i)$ in $\mathbf{F}_{q^m}[x]$. We need to show that fg^q divides $\sum_i c_i/(x-a_i)$ in $\mathbf{F}_{q^m}[x]$, in particular that g^q divides $\sum_i c_i/(x-a_i)$ in $\mathbf{F}_{q^m}[x]$. Find an extension k of $\mathbf{F}_{q^m}$ so that g splits into linear factors in $k[x]$. Then $\sum_i c_i/(x-a_i)=0$ in $k[x]/g^{q-1}$, so $\sum_i c_i/(x-a_i)=0$ in $k[x]/(x-r)^{q-1}$ for each factor $x-r$ of g. The elementary series expansion

$$\frac{1}{x-a_i}=-\frac{1}{a_i-r}-\frac{x-r}{(a_i-r)^2}-\frac{(x-r)^2}{(a_i-r)^3}-\cdots$$

then implies

$$\sum_i\frac{c_i}{a_i-r}+(x-r)\sum_i\frac{c_i}{(a_i-r)^2}+(x-r)^2\sum_i\frac{c_i}{(a_i-r)^3}+\cdots=0$$

in $k[x]/(x-r)^{q-1}$; i.e., $\sum_i c_i/(a_i-r)=0$, $\sum_i c_i/(a_i-r)^2=0$, $\ldots$, $\sum_i c_i/(a_i-r)^{q-1}=0$. Now take the qth power of the equation $\sum_i c_i/(a_i-r)=0$, and use the fact that $c_i\in\mathbf{F}_q$, to obtain $\sum_i c_i/(a_i-r)^q=0$. Work backwards to see that $\sum_i c_i/(x-a_i)=0$ in $k[x]/(x-r)^q$.

By hypothesis g is the product of its distinct linear factors $x - r$. Therefore g^q is the product of the coprime polynomials $(x - r)^q$, and $\sum_i c_i/(x - a_i) = 0$ in $k[x]/g^q$; i.e., $\sum_i c_i/(x - a_i) = 0$ in $\mathbf{F}_{q^m}[x]/g^q$. Finally, f is coprime to g^q, so $\sum_i c_i/(x - a_i) = 0$ in $\mathbf{F}_{q^m}[x]/(fg^q)$. □

3 Attacks and Defenses

Generic attacks against code-based cryptosystems are those whose hardness depends only on the code parameters q, n, k and the number w of errors. For $q > 2$ the most efficient generic attack stated in the literature is the generalized information-set-decoding attack described in [18]. As far as we know, generic attacks are the largest threat against the wild McEliece system and the wild McEliece incognito system, when parameters are chosen sensibly.

The extra factor f described in the previous section allows us to increase the number of Goppa polynomials so that an attacker cannot enumerate all polynomials f and g of the given degrees in less time than performing information-set decoding. We actually suggest increasing the number of polynomials to the square of this, in case there is some square-root attack against the space of polynomials. We also retain the defense used in [6], namely choosing the support as a secret proper subset of $\mathbf{F}_{q^m}$, again with the number of possibilities being the square of the cost of information-set decoding.

One might think that the factorizability of fg^{q-1} is somehow analogous to the concatenated structure attacked in [20]. However, one cannot even begin the attack of [20] without finding low-weight words in the dual code. We have checked in examples of various sizes that the dual code of $\Gamma_q(a_1, \ldots, a_n, fg^{q-1})$ does not have words of low weight, so attacks of this type do not apply. Note that any severe problem with factorizability would also break the original McEliece system, since every polynomial can be factored over a suitable extension field.

To make structural attacks even harder we suggest using an idea of Berger and Loidreau [2] which they introduced in an attempt to protect Generalized Reed-Solomon (GRS) codes, namely to add ℓ additional rows to the parity-check matrix. There are $\binom{k}{\ell}_q = \frac{(1-q^k)(1-q^{k-1})\cdots(1-q^{k-\ell+1})}{(1-q)(1-q^2)\cdots(1-q^\ell)}$ subspaces of dimension ℓ in a k-dimensional code over $\mathbf{F}_q$; this is a very large number even for $\ell = 1$. Wieschebrink showed in [25] that the structure of GRS can still be detected despite the extra defense, but the attack relies strongly on properties of GRS and does not seem to carry over to wild Goppa codes and their close relatives.

We emphasize that these defenses have very low cost, only slightly increasing the size of the public key compared to pure wild McEliece. The effect of [2] is that in systematic form the public key has $(n - k + \ell)(k - \ell)$ entries instead of $(n - k)k$; this is a negligible effect for small ℓ. The small effect of f on the key size is illustrated with optimized numerical examples in Section 5. There are even cases (e.g., 2^{100} security for $q = 31$) where the improved granularity of fg^{q-1} allowed our computations to find *smaller* keys for fg^{q-1} than for g^{q-1} at the same security level.

4 Challenges

We have created a spectrum of cryptanalytic challenges as a way to measure and focus progress in attacking our proposals. Each challenge consists of a public key and a ciphertext; we challenge the readers to find a matching plaintext or even to find the secret keys. Our challenges are online at `http://pqcrypto.org/wild-challenges.html`. We intend to keep this web page up to date to show

- any solutions (plaintexts) sent to us — with credit to the first solver of each challenge, and with as much detail as the solver is willing to provide regarding how the challenge was cryptanalyzed;
- any secret keys sent to us — again with credit to the first solver of each challenge;
- cryptanalytic benchmarks — measurements of the speed of publicly available cryptanalytic software for the smaller challenges, as a way for the community to verify and demonstrate improvements in attack algorithms;
- predictions — estimates of how difficult the larger challenges will be to break.

Our challenges, like the RSA Factoring Challenge (see [19] and [26]) and the Certicom ECC Challenges (see [8]), cover a wide range of levels of difficulty. The smallest challenges require only a small amount of computer time; the larger challenges increase rapidly in difficulty. However, we did not imitate (e.g.) the 1000× increase in difficulty between the ECC2K-108 and ECC2K-130 challenges in [8]. That increase has kept the list of solved ECC2K challenges static since 2000, not reflecting the impact of more than a decade of advances in computer technology; we prefer smaller gaps between challenges.

Each of our challenges is labelled by (1) "wild McEliece" for [6], or "wild Mceliece incognito" for this paper; (2) a field size q; (3) a key size expressed in kilobytes. Each challenge also has public parameters m, n, s, t chosen as discussed below. After choosing these parameters we built the challenge as follows:

- Choose a secret sequence of n distinct elements $a_1, \ldots, a_n$ of $\mathbf{F}_{q^m}$.
- Choose a secret irreducible polynomial g of degree t in $\mathbf{F}_{q^m}[x]$. If g has any of $a_1, \ldots, a_n$ as roots, repeat this step. (This can occur only for $t = 1$.)
- Choose a secret irreducible polynomial f of degree s in $\mathbf{F}_{q^m}[x]$. If f has any of the $a_1, \ldots, a_n$ as roots, repeat this step. (In principle we should, but we did not, also check for the rare possibility that $s = t$ and $f = g$.)
- Write down an $(n - k) \times n$ parity-check matrix H for the Goppa code $\Gamma_q(a_1, \cdots, a_n, fg^{q-1})$, where $k = n - m(s + (q - 1)t)$.
- Row-reduce H so that it begins with an $(n - k) \times (n - k)$ identity matrix and continues with an $(n - k) \times k$ public key. If this fails (i.e., the first $n - k$ columns of H are not invertible), go back to the first step.
- Choose a secret plaintext. Here we use the Niederreiter variant [16]: a plaintext is a random element of $\mathbf{F}_q^n$ of Hamming weight w, where $w = \lfloor (s + (q - 1)t)/2 \rfloor$. (This can be made CCA2-secure with negligible loss of efficiency, by techniques analogous to the techniques of [14].) For simplicity we do not use list decoding here. We also do not use the Berger–Loidreau defense.

– Multiply the secret plaintext by the row-reduced H, obtaining a public ciphertext in $\mathbf{F}_q^{n-k}$.
– As a verification step, use the secret key to legitimately decrypt the ciphertext, and then check that the result matches the original plaintext.
– Throw away all the secret information, leaving only the ciphertext and the public key.

We wrote a script in the Sage computer-algebra system [23] to do all this, relying on Sage's random-number generator to produce all secrets; the Sage documentation indicates that the random-number generator is cryptographic. This script appears on our web page. The script was designed mainly as a reference implementation, easy to understand and easy to verify; it was not designed for speed. However, we did incorporate a few standard speedups (such as a balanced product tree inside interpolation in generalized Reed–Solomon decryption) to reduce the time spent generating challenges.

We formatted each challenge as a text file containing cryptosystem parameters, a ciphertext, and a public key. For example, here is our 20kB "wild McEliece incognito" challenge for $q = 13$, except that in the actual file there are various additional numbers in place of the dots:

```
kilobytes = 19.9869819590563
q = 13
m = 3
n = 472
s = 7
t = 3
u = 43
k = 343
w = 23
ciphertext = [7, 4, 12, 7, 7, ..., 2, 8, 10, 5, 0]
recovered_plaintext_using_secret_key = True
pubkeycol129 = [9, 11, 0, 4, 9, ..., 4, 12, 8, 1, 3]
pubkeycol130 = [5, 4, 12, 7, 2, ..., 6, 12, 5, 11, 12]
...
pubkeycol471 = [0, 1, 11, 3, 6, ..., 11, 12, 4, 11, 3]
```

In this example there are approximately 2^{1644} possible sets $\{a_1, \ldots, a_{472}\}$, and approximately 2^{107} possible pairs (f, g). This challenge has wildness percentage 84% because $\deg(g^{q-1}) = 36$ accounts for 84% of $u = \deg(fg^{q-1}) = 43$. The ciphertext is a column vector containing $n - k = 129$ elements of $\mathbf{F}_{13}$. This column vector is a sum of nonzero coefficients times $w = 23$ columns chosen secretly from the 472 columns of the row-reduced H; these 472 columns consist of $k = 343$ public-key columns shown in the challenge file, and 129 extra columns containing an identity matrix.

The public key in this challenge has $343 \cdot 129 \cdot \log(13)/\log 2 \approx 163733$ bits of information, slightly below the advertised "20kB" (163840 bits). A standard radix-13 integer encoding of the matrix would fit into 163734 bits but would take some work to uncompress. Packing each 129-entry column separately into 478 bits would consume 163954 bits. A standard 4-bit encoding of $\mathbf{F}_{13}$ would consume only slightly more space, 21.6kB.

The generalized information-set-decoding attack introduced by Peters in [18] will break this challenge in roughly 2^{53} bit operations. This is obviously feasible.

As another example, our 40kB "wild McEliece" challenge for $q = 31$ has $m = 2$, $n = 666$, $s = 0$, $t = 2$, $k = 546$, and $w = 31$. In this case security relies critically on the defense suggested in [6]: there are only about 2^{19} possible polynomials g, but there are almost 2^{850} possible support sets. Information-set decoding will break this challenge in roughly 2^{89} bit operations.

As a final example, our 20kB "wild McEliece" challenge for $q = 3$ has $m = 6$, $n = 729$, $s = 0$, $t = 16$, $k = 537$, and $w = 24$. In this case there is only 1 possible set $\{a_1, \ldots, a_{729}\}$, namely all of $\mathbf{F}_{3^6}$, but there are approximately 2^{148} possible polynomials g. Information-set decoding will break this challenge in roughly 2^{54} bit operations. Does knowing the support help any attack?

We considered a huge number of possible parameters m, n, s, t for each challenge, and many parameters for the attack in [18], subject to the key-size constraint $k(n-k)\log_2 q \le 8192K$, where K is the specified number of kilobytes in the key. We assigned a security level 2^b to (m, n, s, t) according to an approximation to the cost of the attack in [18]. For the "wild McEliece incognito" challenges we rejected (m, n, s, t) whenever the number of polynomials was below 2^{2b}, and we also rejected (m, n, s, t) whenever the number of support sets was below 2^{2b}. For the "wild McEliece" challenges we did not require these defenses separately: we allowed the product of the number of polynomials and the number of support sets to be as small as 2^{2b}. Subject to these constraints we selected (m, n, s, t) for each challenge to maximize b. This procedure matches how we would expect parameters to be chosen in practice.

5 Parameters

In this section we propose parameters (n, k, s, t) for the McEliece cryptosystem using codes $\Gamma = \Gamma_q(a_1, \ldots, a_n, fg^{q-1})$ that provide 2^{128} security against information-set decoding and that have more than 2^{256} choices of fg^{q-1}. Our parameter search uses the analysis of information-set decoding in [18]. We chose the code length n, the degree s of f, the degree t of g and the dimension $k = n - \lceil \log_q n \rceil (s + (q-1)t)$ of Γ to minimize the key size $\lceil (n-k)k\log_2 q \rceil$ for 128-bit security when w errors are added. Table 5.1 gives an overview. The last column of the table shows the "wildness percentage" p, i.e., the contribution of g^{q-1} to the Goppa polynomial, measured in terms of how its degree relates to the overall degree.

Figure 5.1 illustrates for $q = 13$ that, given a particular key size, higher wildness percentages generally add extra security against information-set decoding. The figure compares Goppa codes with no correction factor (100% wild) to Goppa codes where the degrees of f and g^{q-1} are balanced (50% wild), and to Goppa codes without the wild trick (0% wild). We emphasize that adding our

Table 5.1. Optimized parameters (n, k, s, t) for wild Goppa codes over $\mathbf{F}_q$ achieving 128-bit security when introducing $w = \lfloor (s + qt)/2 \rfloor$ errors.

q	key size	n	k	s	t	w	p
3	186 kB	2136	1492	0	46	69	100%
4	210 kB	2252	1766	0	27	54	100%
5	191 kB	1878	1398	0	24	60	100%
7	170 kB	1602	1186	8	16	60	92%
8	187 kB	1628	1204	8	14	60	92%
9	205 kB	1668	1244	10	12	59	91%
11	129 kB	1272	951	17	9	58	84%
13	142 kB	1336	1033	17	7	54	83%
16	157 kB	1328	1010	16	6	56	85%
17	162 kB	1404	1113	17	5	51	82%
19	169 kB	1336	1015	17	5	56	84%
23	183 kB	1370	1058	16	4	54	85%
25	189 kB	1314	972	18	4	59	84%
27	200 kB	1500	1218	42	2	48	55%
29	199 kB	1390	1081	19	3	53	82%
31	88 kB	856	626	25	3	59	78%
32	89 kB	852	618	24	3	60	79%

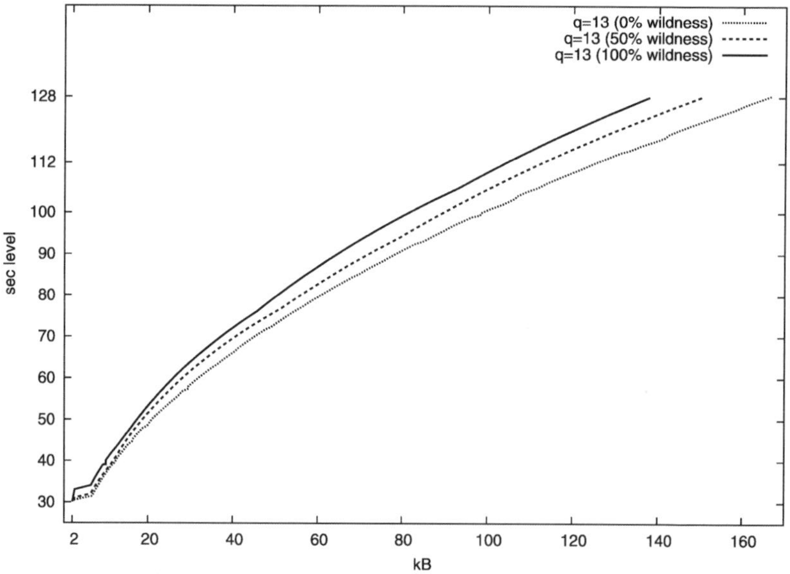

Fig. 5.1. Security levels attained for wild McEliece keys with different wildness percentages for $q = 13$.

shield against polynomial-searching attacks does not require dropping the wildness percentage from 100% all the way down to 50%; the parameters suggested in Table 5.1 typically have very small extra factors f, profiting from the higher error-correction capability induced by g^{q-1}.

References

[1] Augot, D., Barbier, M., Couvreur, A.: List-decoding of binary Goppa codes up to the binary Johnson bound (2010), http://arxiv.org/abs/1012.3439; Citations in this document: §2

[2] Berger, T.P., Loidreau, P.: How to mask the structure of codes for a cryptographic use. Designs, Codes and Cryptography 35, 63–79 (2005), MR 2006d:94038, http://www.springerlink.com/index/JR001118R1567U13.pdf; Citations in this document: §1, §3, §3

[3] Berlekamp, E.R.: Algebraic coding theory. Aegean Park Press (1984) ISBN 0894120638; Citations in this document: §2

[4] Bernstein, D.J.: List decoding for binary Goppa codes. In: IWCC [10], pp. 62–80 (2011), http://cr.yp.to/papers.html#goppalist; Citations in this document: §2

[5] Bernstein, D.J.: Simplified high-speed high-distance list decoding for alternant codes. In: PQCrypto [27], pp. 200–216 (2011), http://cr.yp.to/papers.html#simplelist; Citations in this document: §2, §2

[6] Bernstein, D.J., Lange, T., Peters, C.: Wild McEliece. In: SAC 2010 [7], pp. 143–158 (2011), http://eprint.iacr.org/2010/410; Citations in this document: §1, §1, §1, §1, §1, §1, §2, §2, §2, §3, §4, §4

[7] Biryukov, A., Gong, G., Stinson, D.R. (eds.): Selected areas in cryptography — 17th international workshop, SAC 2010, Waterloo, Ontario, Canada, August 12-13, 2010, revised selected papers. Lecture Notes in Computer Science, vol. 6544. Springer, Heidelberg (2011); See [6]

[8] Certicom: Certicom ECC Challenge (1997), http://www.certicom.com/images/pdfs/cert_ecc_challenge.pdf; Citations in this document: §4, §4

[9] Charpin, P. (ed.): Livre des résumés — EUROCODE 94, Abbaye de la Bussière sur Ouche, France (October 1994); See [20]

[10] Chee, Y.M., Guo, Z., Ling, S., Shao, F., Tang, Y., Wang, H., Xing, C. (eds.): Coding and cryptology — third international workshop, IWCC 2011, Qingdao, China, May 30-June 3, 2011, proceedings. Lecture Notes in Computer Science, vol. 6639. Springer, Heidelberg (2011); See [4]

[11] Goppa, V.D.: A new class of linear error correcting codes. Problemy Peredachi Informatsii 6, 24–30 (1970); Citations in this document: §2, §2

[12] Goppa, V.D.: Rational representation of codes and (L, g)-codes. Problemy Peredachi Informatsii 7, 41–49 (1971); Citations in this document: §2

[13] Kim, K. (ed.): Public key cryptography: proceedings of the 4th international workshop on practice and theory in public key cryptosystems (PKC 2001) held on Cheju Island, February 13-15, 2001. Lecture Notes in Computer Science, vol. 1992. Springer, Heidelberg (2001); See [14]

[14] Kobara, K., Imai, H.: Semantically secure McEliece public-key cryptosystems — conversions for McEliece PKC. In: PKC 2001 [13], pp. 19–35 (2001), MR 2003c:94027; Citations in this document: §4

[15] McEliece, R.J.: A public-key cryptosystem based on algebraic coding theory. JPL DSN Progress Report, 114–116 (1978), http://ipnpr.jpl.nasa.gov/progress_report2/42-44/44N.PDF; Citations in this document: §1

[16] Niederreiter, H.: Knapsack-type cryptosystems and algebraic coding theory. Problems of Control and Information Theory 15, 159–166 (1986); Citations in this document: §1, §4

[17] Patterson, N.J.: The algebraic decoding of Goppa codes. IEEE Transactions on Information Theory 21, 203–207 (1975); Citations in this document: §2

[18] Peters, C.: Information-set decoding for linear codes over $\mathbf{F}_q$. In: PQCrypto 2010 [22], pp. 81–94 (2010), http://eprint.iacr.org/2009/589; Citations in this document: §1, §3, §4, §4, §4, §5

[19] RSA Laboratories: The RSA Factoring Challenge (1991), http://www.rsa.com/rsalabs/node.asp?id=2092; Citations in this document: §4

[20] Sendrier, N.: On the structure of a randomly permuted concatenated code. In: EUROCODE 94 [9], pp. 169–173 (1994); Citations in this document: §3, §3

[21] Sendrier, N.: Finding the permutation between equivalent linear codes: the support splitting algorithm. IEEE Transactions on Information Theory 46, 1193–1203 (2000), MR 2001e:94017, http://hal.inria.fr/docs/00/07/30/37/PDF/RR-3637.pdf; Citations in this document: §1

[22] Sendrier, N. (ed.): Post-quantum cryptography, third international workshop, PQCrypto, Darmstadt, Germany, May 25-28, 2010. Lecture Notes in Computer Science, vol. 6061. Springer, Heidelberg (2010); See [18], [25]

[23] Stein, W. (ed.): Sage Mathematics Software (Version 4.4.3). The Sage Group (2010), http://www.sagemath.org; Citations in this document: §4

[24] Sugiyama, Y., Kasahara, M., Hirasawa, S., Namekawa, T.: Further results on Goppa codes and their applications to constructing efficient binary codes. IEEE Transactions on Information Theory 22, 518–526 (1976); Citations in this document: §2

[25] Wieschebrink, C.: Cryptanalysis of the Niederreiter public key scheme based on GRS subcodes. In: PQCrypto 2010 [22], pp. 61–72 (2010); Citations in this document: §3

[26] Wikipedia: RSA Factoring Challenge — Wikipedia, The Free Encyclopedia. accessed 01 July 2011 (2011), http://en.wikipedia.org/wiki/RSA_Factoring_Challenge; Citations in this document: §4

[27] Yang, B.-Y. (ed.): Post-quantum cryptography, Fourth international workshop, PQCrypto, Taipei, Taiwan, November 29-December 02, 2011. Lecture Notes in Computer Science, vol. 7071. Springer, Heidelberg (2011); See [5]

A New Spin on Quantum Cryptography: Avoiding Trapdoors and Embracing Public Keys

Lawrence M. Ioannou[1,2] and Michele Mosca[1,2,3]

[1] Institute for Quantum Computing
[2] Department of Combinatorics and Optimization, University of Waterloo, 200 University Avenue, Waterloo, Ontario, N2L 3G1, Canada
[3] Perimeter Institute for Theoretical Physics
31 Caroline Street North, Waterloo, Ontario, N2L 2Y5, Canada

Abstract. We give new arguments in support of *signed quantum key establishment*, where quantum cryptography is used in a public-key infrastructure that provides the required authentication. We also analyze more thoroughly than previous works the benefits that quantum key establishment protocols have over certain classical protocols, motivated in part by the various objections to quantum key establishment that are sometimes raised. Previous knowledge of quantum cryptography on the reader's part is not required for this article, as the definition of "quantum key establishment" that we use is an entirely classical and black-box characterization (one need only trust that protocols satisfying the definition exist).

Quantum cryptography[1] has been promoted as a more secure alternative to public-key cryptography based on computational assumptions (see the abstract of Ref. [1] for a typical example). However, an opposing view is sometimes voiced by classical cryptographers and computer security specialists questioning whether quantum cryptography is really a practical way to achieve security against quantum computers, also known as *quantum resistance*. Several detailed analyses have appeared that consider the benefits and disadvantages of quantum cryptography in comparison to classical alternatives [2,3,4,5]. The present article contributes to the dialogue in a way that we hope is very palatable to the community of quantum-questioning cryptographers: we give new arguments in support of *signed quantum key establishment*, where quantum cryptography is used in a public-key infrastructure that provides the required authentication.

We also analyze more thoroughly than previous works the benefits that quantum key establishment (QKE) protocols have over certain classical protocols, motivated in part by the various objections to QKE that have been put forward (for example, in Ref. [5]). Some of those objections follow.[2]

---

[1] Note that quantum cryptography includes many protocols that this paper does not discuss. We use the term "quantum cryptography" here as a synonym for "quantum key establishment", often called "quantum key distribution" or "QKD".

[2] We have stated these objections in our own words.

B.-Y. Yang (Ed.): PQCrypto 2011, LNCS 7071, pp. 255–274, 2011.

- **Objection 1**: Quantum computers are not known to be able to break all classical public-key cryptosystems, such as the McEliece cryptosystem or those based on lattice problems; so we can just upgrade to these quantum-resistant cryptosystems and forget quantum cryptography—that way, we'd retain all the benefits of a public-key infrastructure.
- **Objection 2**: If all of classical public-key cryptography is found to be easily breakable, then we might as well revert to using our best symmetric-key cryptography, including block ciphers like AES, which we all agree is quantum resistant; quantum cryptography would require symmetric shared initial keys anyway in this case, so it wouldn't gain us anything.
- **Objection 3**: We don't need any means of key distribution, let alone a quantum mechanical one—let's just exchange a lifetime's worth of symmetric keying material at the start. If for whatever reason we do need new keys, see Objection 4.
- **Objection 4**: We don't need any means of generating independent secret key over telecommunication links—let's just use a trusted courier each time we need independent secret key.

We address all of these objections.

Not Quantum Cryptography Again. Like in pro-quantum-cryptography articles that have come before this, we assume here that the universe is quantum mechanical, so that, at a minimum, the secret key generated by a secure key-establishment protocol must be secure against an adversary able to perform probabilistic-polynomial-time computations on a quantum computer. As well, as stated by Stebila et al. [4], we "expect the costs and challenges of using [QKE] to decrease to the point where [such] systems can be deployed affordably and their behaviour can be certified." In fact, most of the advantages of quantum cryptography that we point out here have been noted by Paterson et al. [2] or Stebila et al. [4].

Despite these similarities to previous works, our analysis contains distinct new features: it

- suggests a new way to define the classes of classical and QKE protocols, in order to aid their comparison,
- deals properly with the option of using trusted couriers instead of QKE, by distinguishing between in-band and out-of-band actions,
- uses the weakest possible notion of "security" in a quantum universe (i.e. computational security), and therefore does not focus on information-theoretic security—for its own sake—as an advantage of QKE over computationally-secure classical alternatives,
- provides a finer-grained analysis of the computational assumptions underlying the classical alternatives to QKE,
- highlights a property (we call it "nonattributability") of QKE that has received little attention in the literature, and
- supports a recommendation that is both theoretically and practically sound, which both sides of the "quantum debate" can agree upon.

Generally, we hope the reader finds this article to benefit from a more precise cryptographic analysis, despite its more limited scope in taking an idealized view and thus not discussing the more technological or economical aspects of QKE (including side-channel attacks). In other words, this paper studies the value of the QKE primitive assuming it is available in practice and is as cost-effective as any type of "in-band" classical key establishment (see Definition 1).[3] We adopt the same foundational approach that Goldreich does in Refs. [7,8]. This basically means that, when reviewing which computational assumptions are known to be necessary or sufficient for certain cryptographic primitives, we ignore those assumptions (and the schemes based on them) that are ad hoc: we deal only in fundamental computational assumptions, in particular, one-way functions and trapdoor predicates.

But the foregoing analysis is not as complete as it could be. In particular, we do not treat the distributed authenticated key establishment problem (i.e., in a network setting and where simultaneous, multiple key establishment sessions among many pairs of users are considered) as rigorously as it deserves (e.g. [9,10]). That is, we implicitly assume that *point-to-point*[4] unauthenticated key establishment protocols (whether they be key transport protocols or key agreement protocols[5]) and message-authentication protocols (whether they be digital signature schemes or message authentication codes) may be combined in such a way as to form robust *distributed* authenticated key establishment protocols, without stating the details of how this combining—especially with regard to authentication—actually works.[6] This deficiency is manifest in the definition of

[3] The practical availability of the QKE primitive between a typical real-world Alice and Bob is a very non-trivial assumption. For a fairly recent status report on practical QKE systems, one can see Ref. [6], where it is evident that key-rate, distance and availability remain serious obstacles for most practical applications today. In the cases that one believes that QKE could in principle add value, one will need to do an in depth analysis of the various costs and practical limitations before deciding whether in some particular practical situation QKE will be the preferred alternative. Weighing the costs against the value depends on many parameters which vary widely from place to place and over time, and analyzing this broad spectrum is beyond the scope of this paper.

[4] By "point-to-point" protocols or key establishment systems we mean those that presume a unique pair of honest participants in the protocol; in other words, Alice and Bob are fixed.

[5] Recall that a *key transport protocol* is a key establishment protocol where the final secret key is generated by one party and sent to the other party (using some kind of encryption mechanism). By contrast, a *key agreement protocol* is a key establishment protocol where both parties contribute to the generation of the final secret key. See Ref. [11] for more details.

[6] We follow Ref. [11] in our use of the terms "authenticated (key establishment)" and "unauthenticated (key establishment)". In this convention, the word "(un)authenticated" describes the *guaranteed condition* of the final shared key resulting from the protocol. We note that this convention is the opposite of that in Ref. [8], where "(un)authenticated" describes the *a priori assumption* on the (classical) communication channel used in the protocol.

"security" that we use (Definition 3): it only refers to privacy of the secret key and not its integrity; we take authentication *in a network-setting* for granted (for both classical and quantum networks). Thus, analyzing point-to-point key establishment systems is sufficient for our scope and, for such systems, integrity of the established secret key is obtained either by assumption (in the case of unauthenticated key establishment) or by the message-authentication protocols used to authenticate the classical communication channel (in the case of authenticated key establishment). Our omission of the analysis of distributed QKE in no way is meant to imply that the problem is trivial—we believe it is an important open problem, which to our knowledge has not been addressed in any previous works.

As a final note to the reader, we stress that previous knowledge of quantum cryptography is not required for this article. The definition of "QKE" that we use is an entirely classical and black-box characterization (one need only trust that protocols satisfying the definition exist).

Key establishment. We are ultimately interested in authenticated key establishment (or AKE), since, in practice, it is usually not a reasonable assumption that the classical channel connecting Alice and Bob is authenticated a priori. But we shall also consider unauthenticated key establishment (or UKE), because, as well as being useful as a building block for AKE systems, it is an often-considered cryptographic primitive in more foundational works, e.g., Ref. [27] (see Remark 4). We now make some precise definitions.

A (point-to-point) AKE *system* consists of two probabilistic-polynomial-time (quantum) computers, called "Alice" and "Bob", that

- are preloaded with classical *initial keys*, k_A (stored on Alice) and k_B (stored on Bob), which are pre-distributed out of band (see Definition 1) in an authenticated and, where necessary (for example, when the keys are symmetric), private fashion, and
- are connected by two insecure channels, one quantum and one classical, variously monitored or controlled by an adversarial probabilistic-polynomial-time (quantum) computer, called "Eve", and
- together execute a particular (point-to-point) AKE *protocol*, the specification π of which is preloaded authentically but is not secret, and
- which results in Alice and Bob computing outputs s_A and s_B, respectively, such that either $s_A = s_B = \perp$, which corresponds to Alice and Bob aborting the protocol, or s_A and s_B are bit-strings, in which case, if $s_A = s_B$, then the *secret key* $s := s_A$ is defined.

When the initial keys are symmetric ($k_A = k_B$), we may use k to denote each one, i.e., $k = k_A = k_B$; if the initial keys are asymmetric ($k_A \neq k_B$), then

$$k_A = (x_A, y_B) \tag{1}$$

$$k_B = (x_B, y_A), \tag{2}$$

where (x_A, y_A) is Alice's private-public key-pair and (x_B, y_B) is Bob's private-public key-pair. We will say more about asymmetric (public-key) cryptography later on.

Definition 1 (In band/out of band). *The term "in band" describes actions carried out in the normal course of telecommunications strictly via remote signalling across communication channels. The term "out of band" is used to mean "not in band" and describes communication via non-digital/manual means as opposed to via standard telecommunication devices.*

Remark 2 (Classical channel). *Strictly speaking, there is no need for a dedicated classical channel between Alice and Bob, since classical information can be sent along the quantum channel. However, the well-known* QKE *protocols (i.e., those based on the ones in Refs [12,13]) clearly distinguish the classical from the quantum communication; in particular, it suffices that only the classical communication is authenticated in order for the secret key to be authenticated at the end of the protocol (whereas, one could imagine a quantum protocol where the quantum communication also needs to be authenticated). In line with this distinction, we assume separate quantum and classical channels.*

A (point-to-point) UKE *system* is defined similarly to an AKE system, with only the following differences:

- Alice and Bob possess no initial keys and
- the classical channel is assumed to be authenticated, i.e., Eve is assumed only to passively monitor the classical channel (but she can still totally control the quantum channel), and
- π is a (point-to-point) UKE *protocol.*

We also need to define conditions under which a key establishment protocol is secure or, more specifically, quantum-resistant. We would like a definition that applies equally well to both quantum and fully classical protocols, i.e., all protocols allowed in the above frameworks. Since we take authentication for granted (as explained above), the following security definition is sufficient for both AKE and UKE systems. Call a key establishment protocol *perfectly secure* if, for any algorithm for Eve, we have that (1) $s_A = s_B$, (2) if $s_A \neq \perp$ then s_A is uniformly distributed and independent of Eve's state, and (3) if Eve does not interfere with the protocol (where we assume otherwise perfect channels), then $s_A \neq \perp$. Let $\mathcal{I}$ be an *ideal key establishment system* that implements a perfectly secure protocol. Let $\mathcal{R}(\pi)$ be a *real key establishment system* that uses protocol π. Let n be the minimum length of the secret key s if Alice and Bob do not abort. Consider a probabilistic-polynomial-time (quantum) *distinguisher* running in time polynomial in n, that interacts with either $\mathcal{I}$ or $\mathcal{R}(\pi)$ and then outputs a guess bit B; the distinguisher has access to Eve's system and the outputs s_A and s_B.

Definition 3 (Quantum-resistant key-establishment protocol (with respect to privacy)). *Assuming the above definitions, a point-to-point key-establishment protocol* π *is* quantum-resistant (with respect to privacy) *if, for any such distinguisher, the quantity*

$$|\Pr[B=1|\mathcal{I}] - \Pr[B=1|\mathcal{R}(\pi)]| \tag{3}$$

is negligible for all sufficiently large n*, where* $\Pr[B=1|\mathcal{I}]$ *and* $\Pr[B=1|\mathcal{R}(\pi)]$ *are the probabilities that* $B=1$ *when the distinguisher interacts with* $\mathcal{I}$ *and* $\mathcal{R}(\pi)$*, respectively.*

We give this (semi-formal) definition for completeness; we refer the reader to Refs [14,15,7,16] for how to rigorize such a definition.

As a final specification of our basic setup, it will be helpful to define the *classical communication* c in a key establishment protocol. For classical protocols, the classical communication is all the communication between Alice and Bob. For arbitrary (quantum) protocols, defining the classical communication is a bit more subtle; we refrain from giving a formal definition here (for the sake of the reader who may be unfamiliar with quantum measurement). Rather, for the quantum protocols we care about, it suffices to define the classical communication tautologically as the classical communication specified in the protocol, since these protocols clearly and naturally distinguish the classical and quantum information sent between Alice and Bob.

The contenders. Below are listed and defined two main classes of point-to-point UKE protocols as well as the five main classes of point-to-point AKE protocols that are considered in the literature when evaluating the usefulness of quantum cryptography in comparison to classical techniques for key establishment. These classes, as defined, do not cover all conceivable protocols, but do cover all the ones that are usually considered (which suffices here). In defining these classes, we restrict to quantum-resistant protocols (because the universe is quantum). It will help to view the quantities k_A, k_B, k, s, and c introduced above as random variables. For example, in the case of symmetric initial keys, the quantity k may be viewed as a uniformly distributed random variable in $\{0,1\}^\ell$, for some fixed $\ell \in \mathbb{Z}^{>0}$ that determines the length of the initial keys.

Unauthenticated key establishment protocols:

- *Classical* UKE (c-UKE)—This class includes any quantum-resistant and totally classical UKE protocol. It includes unauthenticated key transport protocols based on public-key encryption (but not those based on symmetric-key encryption).
- *Quantum* UKE (q-UKE)—This class includes any quantum-resistant UKE protocol such that, whenever Eve has not interfered with the protocol, the secret key s is independent of the classical communication c, i.e., for all values c' of the classical communication and all values s' of the secret key,

$$\Pr[s=s'|c=c'] = \Pr[s=s']. \tag{4}$$

It includes (some versions of) the well-known QKE protocols and can easily be shown not to include any classical protocols.[7]

Remark 4 (Secret key agreement). *The cryptographic primitive realized by protocols in* c-UKE *is usually referred to as* secret key agreement *(or sometimes just* secret agreement*) in the literature. Note that this primitive is also realized by protocols in* q-UKE.

Authenticated key establishment protocols:

- *Out-of-band key establishment* (OOB)—This class includes any AKE protocol where Alice and Bob are preloaded with the secret key out of band, i.e.,

$$s = k_A = k_B. \tag{5}$$

  It includes protocols that employ a trusted courier. The initial keys in such protocols are typically much larger than in protocols belonging to the classes below.
- *Pseudorandom generator expansion* (PGE)—This class includes any quantum-resistant and totally classical AKE protocol not in OOB that uses symmetric initial keys where Alice and Bob establish a secret key that is efficiently computable from the initial keys, i.e., there exists a deterministic-polynomial-time classical algorithm A such that

$$s = A(\pi, k). \tag{6}$$

  It includes protocols that use a pseudorandom generator to expand the initial keys into a secret key.
- *Weak classical* AKE (wc-AKE)—This class includes any quantum-resistant and totally classical AKE protocol in neither PGE nor OOB that uses symmetric initial keys. Note such protocols have the property that the secret key is efficiently computable from the initial keys and the communication, i.e., there exists a deterministic-polynomial-time classical algorithm A such that

$$s = A(\pi, k, c). \tag{7}$$

  The class includes authenticated key transport protocols based on symmetric-key encryption.

[7] We note that not all versions of the well-known QKE protocols satisfy this definition. We sketch a proof of the latter fact that no purely classical protocol can be quantum resistant and satisfy (4). Let r_A and r_B be binary strings encoding the private local randomness that Alice and Bob respectively use in the protocol. Consider the sequence $c_1, c_2, \ldots$ of messages passed between Alice and Bob. Each c_i places constraints on the values of r_A and r_B. Since, at the end of the protocol, the secret key s is uniquely determined, it must be that r_A and r_B are determined by the classical communication c up to implying a unique s, i.e., $H(s|c) = 0$, where H is the Shannon entropy. For any two random variables X and Y, $H(X|Y) = H(X)$ if and only if X and Y are independent [17]. Therefore, if (4) holds, then $H(s) = H(s|c) = 0$, so that s is a constant and thus the protocol is not quantum resistant.

- *Strong*[8] *classical* AKE (sc-AKE)—This class includes any quantum-resistant and totally classical AKE protocol, where Alice and Bob establish an authenticated secret key s that is not functionally dependent on the initial keys k_A and k_B, i.e., there exists a deterministic-polynomial-time classical algorithm A such that

$$s = A(\pi, r_A, r_B), \tag{8}$$

where r_A and r_B are (random variables representing) the private local random choices of Alice and Bob respectively (made independently of the initial keys). It includes authenticated key transport protocols based on public-key encryption (but not those based on symmetric-key encryption); more generally, it includes the "authenticated version" of any quantum-resistant UKE protocol, where the initial keys are used (only) to authenticate all the communication of the protocol (see Remark 9).
- *Quantum* AKE (q-AKE)—This class includes any quantum-resistant AKE protocol such that, whenever Eve has not interfered with the protocol, the secret key s is independent of the initial keys and the classical communication c, i.e., for all values k'_A and k'_B of the initial keys and all values c' of the classical communication and all values s' of the secret key,

$$\Pr[s = s' | k_A = k'_A, k_B = k'_B, c = c'] = \Pr[s = s']. \tag{9}$$

It includes the authenticated versions of the q-UKE-protocols and can easily be shown not to include any classical protocols (similarly to the class q-UKE).

Remark 5 (Possible emptiness of classical classes). *Of the classes of in-band key establishment protocols, only* q-UKE *and* q-AKE *are known to be nonempty.*

Remark 6 (Key pre-distribution v. dynamic key establishment). *The union of the classes* OOB *and* PGE *contains protocols referred to collectively as* key pre-distribution schemes *[11], which is why we label these two classes differently. Note that there is no need to authenticate the in-band communication in these protocols because there is none. Protocols that are not key pre-distribution schemes are said to accomplish* dynamic key establishment.

Remark 7 (Definition of sc-AKE). *The class* sc-AKE *may contain protocols that use the "quantum public-key cryptosystems" in Ref. [19], since the model does not stipulate how initial keys are derived (i.e., they could be derived using a quantum computer).*

[8] Our use of the word "strong" differs from that in Ref. [18], where a key establishment protocol is secure only if it remains secure under the reveal of any subset of the initial (also called "long-term") and ephemeral keys that does not contain both the initial and ephemeral keys of one of the parties. The protocols of the class we define here need only remain secure under the reveal of the initial keys. Indeed, the "strong" of Ref. [18] is stronger than ours.

Remark 8 (Definition of q-AKE). *The class* q-AKE *may contain protocols obeying physical theories other than quantum theory.*

Remark 9 (UKE implies AKE). *Note that if π is in* c-UKE*, then π naturally gives rise to a protocol in* sc-AKE *when combined with a secure classical message-authentication protocol. A similar statement holds for* q-UKE *and* q-AKE*.*

We subdivide the classes sc-AKE and q-AKE by the type of initial keys—either symmetric or public—used in the particular key establishment protocol, i.e., we have the following disjoint unions

$$\mathsf{sc\text{-}AKE} = \mathsf{sc\text{-}AKE_{sym}} \cup \mathsf{sc\text{-}AKE_{pub}} \tag{10}$$

$$\mathsf{q\text{-}AKE} = \mathsf{q\text{-}AKE_{sym}} \cup \mathsf{q\text{-}AKE_{pub}}. \tag{11}$$

Table 1 summarizes the different classes by the various categories.

Table 1. The different classes of key establishment protocols

	UKE	AKE	
key pre-distribution	-	OOB	out-of-band
	-	PGE	in-band
dynamic key establishment	-	wc-AKE	
	c-UKE	sc-AKE	
	q-UKE	q-AKE	

Apples and Oranges. The class OOB is included in the above list (and in the following analysis) largely for completeness; it is not technically considered a key establishment protocol. Out-of-band protocols for key establishment need not employ any fundamental cryptographic primitives and cannot provide the same essential functionality that in-band protocols do, i.e., generating new secret key in band. The generally accepted view is that out-of-band key establishment is the most secure way to establish potentially very long secret keys, but that well-implemented in-band protocols typically provide either a more feasible solution in particular applications or a more cost-effective solution in the long term. Because we are making the (reasonable) assumption that QKE will be cost-effective in the future, it reasonably follows that, in at least some cases, it will also be more cost-effective than out-of-band key establishment in the future. We mean to challenge here previous comments made by Bernstein [5], that trusted couriers perform equally as well as QKE systems insofar as their ability to generate entropy in the cryptographic system (from Eve's point of view). The distinction between in-band and out-of-band entropy generation is an important one (cost-wise), and it is impossible to generate entropy in band using classical cryptography alone.

Computational assumptions. We would like to closely examine the fundamental computational assumptions that underlie the various kinds of key establishment protocols. To do this, we start by recalling the following well-known theorems.[9]

Theorem 10 **([7]).** *Pseudorandom generators exist if and only if one-way functions exist.*

Theorem 11 **([8]).** *Symmetric-key encryption schemes exist if and only if one-way functions exist.*

Theorem 12 **([20]).** *Public-key encryption schemes exist if and only if trapdoor predicates exist.*

Theorem 13 **([21]).** *Information-theoretically-secure symmetric-key message authentication codes exist.*

Theorem 14 **([22,23]).** *Public-key signature schemes exist if and only if one-way functions exist.*

Theorem 15 **([24]).** *Information-theoretically-secure* q-UKE*-protocols exist.*

Because we are assuming a quantum universe, one-way functions and trapdoor predicates[10] in this article (if they exist) are secure against an adversary with a quantum computer, but are still assumed to be efficiently computable on a classical computer; also, trapdoors are still considered to be classical objects.[11] We also note that Theorems 10, 11, 12, and 14 hold with respect to *black-box reductions*: if the theorem states that X implies Y, then Y can be constructed from X, only using X as a black box, i.e., the reduction does not rely on the specifics of how X works; furthermore, the security reduction is also a black-box one, i.e., an algorithm for breaking X can be constructed from a black box for

[9] The following theorems and other similar statements should be interpreted as follows. A statement of the form "Cryptographic objects of type Y exist if cryptographic objects of type X exist" means "If there exists an object of type X, then there exists an object of type Y such that breaking the object of type Y implies breaking the object of type X." Such a statement may also be phrased, "X implies Y".

[10] Informally, the predicate $B(x) \in \{0,1\}$ is a(n) *(unapproximable) trapdoor predicate* if anyone can find an x such that $B(x) = 0$ or a y such that $B(y) = 1$ efficiently on a classical computer, but only one who knows the trapdoor can, given z, compute $B(z)$ efficiently on a quantum computer (this notion was introduced in Ref. [20]). Note that one can use a trapdoor predicate for public-key encryption: the bit b is encrypted as any x such that $B(x) = b$.

[11] One could consider "one-way/trapdoor quantum functions", where the input and output of the functions are classical or quantum, and the functions only need to be computable efficiently on a quantum computer. We stick to classical one-way functions and trapdoor predicates that are quantum resistant, candidates of which are, e.g., the trapdoor predicates underlying some lattice-based cryptosystems (see Ref. [25] for more examples).

breaking Y. Non-black-box theorems of this sort are also possible (for example, see Ref. [26]), but are rarely required for these kinds of results, and indeed are not required for the theorems we quote. This is lucky, since it guarantees us that the theorems still hold with respect to a quantum universe.

Table 2. Minimal known fundamental computational assumptions sufficient for the existence of key establishment protocols in each class

Protocol class	Computational assumptions
OOB	none
PGE	one-way functions
wc-AKE	one-way functions
c-UKE/sc-AKE	trapdoor predicates
q-UKE/q-AKE$_{sym}$	none
q-AKE$_{pub}$	one-way functions

The theorems establish the minimal fundamental computational assumptions known to be sufficient for the existence of protocols by class, summarized in Table 2. Public-key encryption implies one-way functions [8]. Thus, the classes c-UKE and sc-AKE require the strongest assumption in the table—the existence of trapdoor predicates—which reflects the fact that it is not known how to construct any protocol in these classes without relying on (or implying) public-key encryption.[12] To facilitate our discussion, we summarize this point as the following conjecture:

Conjecture 16 (Classical secret key agreement implies public-key encryption). *Every protocol in* c-UKE *implies a trapdoor predicate (with respect to a possibly-non-black-box reduction).*

Safest Fair Comparison. Most articles on quantum cryptography that appeared in the 1990s and early 2000s stressed the fact that q-AKE$_{sym}$ (respectively, q-UKE) is the only known class of in-band AKE (respectively, UKE) protocols that requires no computational assumptions. But implicitly discarding all computational assumptions in this way makes it impossible to have a serious discussion about the relative merits of classical and quantum protocols for key establishment (since any classical key-establishment protocol requires some computational assumption). So, suppose we give classical cryptography a fighting chance: suppose we allow only *the weakest computational assumption necessary for in-band classical key establishment*—one-way functions.

[12] One might declare Table 2 misleading, since, for example, Theorem 14 is usually regarded merely as a plausibility result: the construction of a signature scheme from an arbitrary one-way function is relatively very inefficient. To address this issue, we note that reasonably practical constructions are known for pseudorandom generators, symmetric-key encryption schemes, and signature schemes from one-way *permutations* [7,8]. Thus, even restricting to reasonably practical schemes, the class sc-AKE still requires the assumption of a primitive possessing a trapdoor property, as far as we know.

There is good reason to do this. Trapdoor predicates seem to be inherently less secure than one-way functions in general. Firstly, trapdoor predicates easily imply one-way functions [8], whereas the converse is believed not to be true. As some evidence for this, we note that it has been shown in Ref. [27] that, with respect to black box reductions (and with respect to a classical universe), one-way functions are not sufficient (even) to imply secret key agreement (see Remark 4; but we have not checked that this theorem holds with respect to a quantum universe—in general, such classical black-box no-go theorems need not). Secondly, using the equivalences stated in Theorem 11 and Theorem 12, it seems far more likely that an efficient algorithm would be found for breaking a public-key cryptosystem (i.e. computing a trapdoor predicate) than breaking a symmetric-key cryptosystem (i.e. inverting a one-way function without the trapdoor property), because the public-key cryptosystem possesses more structure in order to embed a trapdoor into the encryption "function". Quantum computers are firmly believed not to be able to invert all one-way functions efficiently; we state this as a conjecture:

Conjecture 17 (One-way functions exist). *Quantum-resistant one-way functions (computable in polynomial-time on a classical computer) exist.*

We do not mean to suggest that quantum-resistant trapdoor predicates do not exist (we don't know). We do suggest, though, that the added structure of trapdoor predicates makes it much more likely that algorithms for the underlying problems will improve at a more unpredictable rate: plain one-way functions are less risky.

Even allowing one-way functions, we see that QKE has advantages over classical systems, beyond unconditional security.

Advantages of QKE assuming (only) one-way functions. Most of the advantages below have appeared elsewhere in the literature in one form or another, but our presentation is motivated differently. The following four advantages are not intended to be totally independent; indeed, each is just a qualitatively different consequence of the fact that the secret key is independent of both the initial keys and classical communication in QKE (and that we have taken sc-AKE-protocols out of the picture).

- Advantage 1: Improved security against reveal of initial keys

In classical cryptography, the physical nature of a cryptosystem and protocol leads to the consideration of different types of attacks, some more serious or more technologically difficult to mount than others. Similarly, adversaries are often categorized by their power, for example, *passive* adversaries are considered only to be able to read certain data that is sent along a channel, whereas *active* adversaries are assumed to have complete control over the channel. It is also relevant to consider precisely *when* Eve may become active; a *delayed* adversary is one that remains passive until the key establishment protocol completes, but is active immediately afterwards.

The physical nature of a QKE system leads to the consideration of new kinds of attacks and adversaries. Because of the two different channels used, Eve can now operate differently on these two channels.[13] Thus an adversary can be defined by whether it is passive, delayed, or active on the classical and quantum channels respectively; e.g., (p,p) means "passive on both channels" and (a,d) means "active on the classical channel and delayed on the quantum channel".

With these terms in place, Table 3 shows how q-AKE-protocols have advantages over the other classical protocols that also assume (at most) one-way functions, for certain types of adversary; the table indicates whether secure key can be established when the initial keys have been revealed. For any situation where an immediate active attack is not deployed for whatever reason (e.g. not technologically feasible, or not a high priority at the time), a passive adversary who knows the initial keys loses the ability to compromise the secret key later should she become an active attacker later. Note that if "sc-AKE" appeared in the left-most column of the table, the corresponding row of "yes"/"no" values would look the same as the row corresponding to the class q-AKE.

Table 3. Security against reveal of initial keys. The entries (yes/no) of the chart indicate whether the secret key generated from the key establishment protocol is secure under the reveal of either Alice's or Bob's initial key for the given adversary (see the main text for an explanation of the notation used to define the adversaries). The class sc-AKE does not appear, since we are not assuming trapdoor predicates (and there is no known sc-AKE-scheme that does not imply trapdoor predicates).

	(p,p)	(d,d)	(a,p)	(a,d)	(a,a)
OOB	no	no	no	no	no
PGE	no	no	no	no	no
wc-AKE	no	no	no	no	no
q-AKE	yes	yes	yes	yes	no

Note that, in order to break a q-AKE-protocol—or, more precisely, break the cryptosystem that comprises the q-AKE-protocol—Eve, knowing all the initial keys, can mount an active and sustained "man-in-the-middle" attack; furthermore, for a $\mathsf{q\text{-}AKE}_{\mathsf{sym}}$-system, the active attack must occur during the first instance of the protocol (as any subsequent instance will use different and independent initial keys). In large networks, this may pose a considerable challenge for Eve, depending on when she learns the initial keys and whether the connections among users are fixed or ad-hoc.

Remark 18 (Perfect forward secrecy). *Note that Advantage 1 is different from perfect forward secrecy,* a much *weaker notion referring to whether secret*

[13] We define "passive" on the quantum channel to mean having no access, since it is difficult to formulate a definition of "read only" for a quantum channel. Measurement, which seems necessary for reading, is an active process.

keys established in past sessions *(with old initial keys no longer stored on Alice and Bob) are secure once* current *initial keys are revealed. While* q-AKE-*protocols certainly have perfect forward secrecy, Bernstein [5] has noted that well-implemented* PGE-*protocols do, too.*

- Advantage 2: Reduced dependence on out-of-band actions

Because a q-AKE$_{\mathsf{sym}}$-protocol generates secret key that is independent of the initial keys and the classical communication, initial keys can be smaller in the q-AKE$_{\mathsf{sym}}$-protocol than in an OOB-protocol, i.e., less initial entropy is needed to prime the system. Also, a q-AKE$_{\mathsf{sym}}$-system may require fewer subsequent out-of-band actions for refreshing initial keys, compared to PGE- and wc-AKE-systems (at the very least because the latter are more vulnerable to initial-key-reveal attacks—see above).

- Advantage 3: Reduced dependence on trusted third parties

In a network, key establishment can be done in a mediated fashion, via a trusted *key distribution centre*, whose job is to give *session keys* to Alice and Bob so that they may communicate securely. As part of the setup, every user in the network, including Alice and Bob, shares an initial key (established out of band) with the key distribution centre; in principle, these initial keys may be asymmetric or symmetric. An example of such a system is Kerberos, where the initial keys are symmetric, and, upon request by either Alice or Bob, the key distribution centre generates a symmetric key and sends it (encrypted using the initial keys) to Alice and Bob, who then use it to encrypt and decrypt messages between each other.

Quantum key establishment may also be done in a mediated fashion, so that the channels connecting Alice to Bob go through a key distribution centre, which gives Alice and Bob a session key to be used as a symmetric initial key in a q-AKE$_{\mathsf{sym}}$-protocol.

If trapdoor predicates are not assumed to exist, then any classical mediated key establishment system must use symmetric initial keys; this is because the key distribution centre must send keys to Alice and Bob, and these keys must be, at least partially, encrypted (assuming the key distribution centre is not to play an active part in the communication between Alice and Bob). Similarly, the session keys must be symmetric keys, too.

Comparing any classical mediated key establishment system to one where Alice and Bob use their symmetric session keys as initial keys in a q-AKE$_{\mathsf{sym}}$-protocol, we see that, in the quantum case, Alice and Bob do not need to trust the key distribution centre after their key establishment protocol is complete. By contrast, in the classical case, the key distribution centre must always be trusted, since it knows the keys that Alice and Bob use to communicate securely. As well, Alice and Bob may be able to decouple themselves completely from the key distribution centre after their first q-AKE$_{sym}$-session. Thus, any compromise of the key distribution centre after the first q-AKE$_{sym}$-session does not necessarily affect Alice and Bob.

- Advantage 4: Long-term security from short-term security

The secret key generated by any q-AKE-protocol will be information-theoretically secure even if the authentication algorithm is broken in the short term—as long as the break occurs after the key establishment protocol is completed. We may refer to this as "conditional information-theoretic security". This allows for the use of authentication algorithms that are perhaps less secure in the long term but are easier to manage with regard to initial keys, i.e., public-key algorithms. Note that any q-AKE$_{pub}$-system has the extra advantage over a q-AKE$_{sym}$-system that it is less susceptible to running out of authentication key due to noise or eavesdropping, because there is no practical limit on how many classical messages may be authenticated. In other words, using public-key authentication guards against at least one type of denial-of-service attack.

Also, Alice and Bob may not need to rely on the same type of authentication used for the first q-AKE-session for subsequent q-AKE-sessions, i.e., for the first session, Alice and Bob may execute a q-AKE$_{pub}$-protocol, but, for all subsequent sessions (in principle, i.e., in the absence of sufficiently heavy adversarial action or noise), they may execute a q-AKE$_{sym}$-protocol. Two potential advantages of such a two-phase system are that (1) subsequent key establishment sessions may run faster (since the symmetric-key algorithms may be more efficient than public-key algorithms for the required level of security) and (2) subsequent key establishment sessions may not need to rely on any computational assumptions.

If quantum computers can be assumed not to exist in the short term, i.e., for the service-lifetime of the public keys, then one can even use public-key signature schemes whose security relies on the assumption of hardness of factoring and the discrete logarithm problem for classical computers.

We believe that its ability to derive long-term from short-term security, also known as *everlasting security*,[14] may be the most attractive aspect of QKE systems from a security perspective.

The baby... The advent of public-key cryptography revolutionized secure telecommunications, by vastly simplifying the problems of key distribution and key management: Alice and Bob no longer needed to pre-share a symmetric key. Instead, Alice could publish her own public key, and that would be sufficient for her to receive encrypted messages from anyone who got a hold of it.

Of course, "publishing" a public key is easier said than done, but public-key cryptography helps solve this problem, too. A signature scheme can be used

[14] The term "everlasting security" has been used in the context of the bounded storage model (see, e.g., Ref. [28]), where, e.g., it describes the case where encryption is secure even if the adversary, at some later time, learns the pre-shared symmetric key, as long as, at the time of transmission of the ciphertext, the adversary has bounded storage capability (see Ref. [29]). The term seems equally well suited to QKE.

in conjunction with a network of trusted third parties to help Bob be certain that he has Alice's legitimate public key.[15] This is probably the reason Rivest [31] wrote, "The notion of a digital signature may prove to be one of the most fundamental and useful inventions of modern cryptography."

...the bathwater. There is a price to pay for the advantages of a public-key infrastructure. Security necessarily depends on assumptions about the hardness of certain mathematical problems; proofs that such problems are actually hard seem to be beyond the reach of theoretical computer scientists.

After Peter Shor discovered an efficient quantum algorithm for factoring and computing discrete logarithms in 1994, QKE protocols, the earliest of which dates back to 1984, received renewed interest. Most literature on QKE that appeared in the 1990s and early 2000s focussed on protocols in the class $\mathsf{q\text{-}AKE_{sym}}$. And rightfully so: it is remarkable that symmetric initial keys can be expanded into much larger, independent, and information-theoretically secure secret keys in band by exploiting quantum mechanics. As such, these articles, through their reference to Shor's discovery, may have been seen as suggesting that all computational assumptions should be jettisoned at the earliest opportunity—for who knew what problems might next succumb to the power of a quantum computer?

A new spin on quantum cryptography. It was known (though perhaps not widely) that insisting on unconditional security was not the only way forward in order to ensure reasonable security against quantum attacks. It was evident that public-key signature schemes could be used to authenticate the classical channel in a QKE protocol, and that such a system would have some attractive features; this idea first appeared in the literature in Ref. [2]. Indeed, in light of Theorem 14 and Table 2, and assuming Conjecture 17 is true, this idea becomes rather more striking:

- *Quantum cryptography is the only known way to achieve (quantum-resistant) private communication in a public-key infrastructure with the minimal computational assumptions.*

(If in addition Conjecture 16 is true, then the word "known" can be dropped.) In other words, with some abuse of the metaphor, quantum cryptography potentially allows us to throw out some of the bathwater—i.e., primitives with a trapdoor property—while keeping most of the baby—i.e., authenticated encryption without symmetric initial keys—and no classical scheme is known to accomplish this. At the very least, quantum cryptography certainly allows us

[15] On the Internet, this works as follows. Bob's web-browser comes from the manufacturer pre-loaded with the public key of a trusted third party Charlie. When Bob wants to communicate with Alice, she shows Bob a certificate which contains her purported public key and Charlie's signature of the certificate, which also contains Alice's name (and other uniquely identifying and publicly-agreed-upon details about Alice). Bob checks that Alice's public key is valid by verifying Charlie's signature using the pre-loaded public key. In this context, signature schemes are said to offer "manageable persistence" (via digital signature) of the binding of a name and a key [30].

to sidestep the question of the necessity of trapdoor predicates for secret key agreement (or trapdoor functions for trapdoor predicates [32]). We view this as strengthening the case for signed QKE.

If public-key encryption exists... If trapdoor predicates do exist and are secure in the long term, we note that Advantages 1 through 4 can variously be achieved by sc-AKE-protocols to at least some degree. However, in this case, QKE protocols may have other advantages over classical ones. Because the secret key s generated in a q-AKE-protocol is independent of the classical communication c, there is no mathematical way to connect these two quantities or—attribute—the secret key to Alice's and Bob's publicly readable discussion; we say that the secret key is *nonattributable.*[16]

There are two ways in which a secret key may be considered attributable: it is attributable to Alice's and Bob's public discussion (through its dependence on the classical communication) and it is attributable to Alice and/or Bob (because they participated in the classical communication). For the former way, we just use the term *attributable* to describe the secret key; for the latter way, we say the secret key is *party-attributable.* If the classical communication is authenticated via a signature scheme, then the secret key may be party-attributable in a provable way, or *provably party-attributable.* If the secret key is subsequently used in an encryption scheme to encrypt a plaintext, then we say that the plaintext is (party- or provably party-) attributable whenever the secret key is.

Because q-AKE-protocols do not produce an attributable secret key, a q-$\mathsf{AKE}_{\mathsf{pub}}$-protocol may be used in composition with a one-time pad encryption scheme, and then the secret key (and hence the plaintext) would never be attributable. No totally classical scheme can achieve the same thing, i.e., non-party-attributable, public-key, secure communication.

For symmetric-key ciphers where the bit-length of the secret key is much smaller than the bit-length the message (e.g., AES), the cipher itself provides a subroutine for recognizing the secret key (i.e., if a candidate secret key s' decrypts the ciphertext to something sensible, then with high probability s' equals the actual secret key). If the secret key was produced by a sc-$\mathsf{AKE}_{\mathsf{pub}}$-protocol, then the secret key (and hence the plaintext) are provably party-attributable given the secret

16 In Ref. [33], Beaver discusses "deniability" (see Refs [34,35]) of QKE, which is similar to nonattributability. However, in that paper, it is assumed that Alice and Bob keep a record of their qubit-measurement outcomes (often called "raw key bits") made during the protocol and that, if Alice and Bob are to deny that a particular secret key was established, this record must be consistent with any measurements made by an eavesdropper, i.e., someone who is forcing Alice or Bob to reveal the secret key (or the plaintext encrypted by it). We assume that Alice and Bob do not keep such records and that it is sufficient that the forcer cannot provide evidence that attributes a particular secret key to the classical communication; any measurement on the quantum channel that the forcer made is not publicly verifiable, so we do not view its outcome as part of the public record. In other words, in our model, Alice and Bob need not provide evidence to support their (tacit) denial. Incidentally, Beaver concludes that the standard QKE protocols do not provide deniability in his model.

key; however, if the secret key was produced by a $\mathsf{q\text{-}AKE}_{\mathsf{pub}}$-protocol, it is not attributable at all. This is a potential advantage of using QKE to generate AES keys.

Closing Remarks. Recall the objections to QKE that we listed earlier (see Page 255). We have addressed Objection 4 early on, by highlighting the fundamental distinction between in-band and out-of-band key establishment protocols. We believe there exist (or will exist) applications where in-band generation of entropy is desirable.

Objections 2 and 3 both propose using (potentially very long) symmetric initial keys in OOB or PGE protocols. We have presented a considerable list of advantages that QKE has over these protocols.

Objection 1 is the strongest one, but it relies on the computational assumption of a trapdoor predicate, which (until any lower bounds are proven) incurs risk when public-key encryption is used for long-term secrets. The field of quantum algorithms is still relatively young, so it is probably unwise to assume any particular candidate trapdoor predicate with a particular set of parameters is secure (the recent discovery of a subexponential-time quantum algorithm for elliptic curve isogenies supports this perspective [36]). However, in addition to these standard counter-arguments for Objection 1, we have shown that QKE may offer the benefit of nonattributability in scenarios where no purely classical scheme can. We also note that it is conceivable that, in the future, a q-AKE-system may be more efficient (i.e. have a higher secret key rate) than a sc-AKE-system, as public-key encryption is known to be rather slow. As well, q-AKE-systems may be more cost-effectively resistant to side-channel attacks, which are notoriously difficult to defend against in the classical world.

The debate on the merits of QKE may have suffered from a focus on unconditional security, which may have given the impression that it is of no value to practical cryptography. The message from classical cryptographers has been loud and clear: the pre-sharing of symmetric keys is costly and thus to be avoided in the majority of key-establishment applications: e.g., Paterson et al. [2] wrote, "[Quantum key establishment], when unconditionally secure, does not solve the problem of key distribution. Rather, it exacerbates it, by making the pre-establishment of symmetric keys a requirement." They also wrote, "It is likely that using [QKE] with public key authentication [...] has security benefits [...]. However, [QKE] loses much of its appeal in [this setting], as the overall system security is no longer guaranteed by the laws of quantum physics alone." Our article is completely in accordance with the former comment and, with regard to the latter comment, expands on the "benefits" of signed QKE in order to bolster its "appeal". As such, we hope to have firmed up the middle ground between unconditionally-secure QKE and computationally-secure classical key establishment in the "quantum debate".

Acknowledgements. We are indebted to Alfred Menezes for contributing many ideas to this paper. We thank Douglas Stebila for alerting us to the open problem of distributed quantum key distribution. We thank Phillip R. Kaye and Renato Renner for fruitful discussions. L. M. Ioannou was supported by QuantumWorks, MITACS, and IQC. M. Mosca was supported by NSERC, CFI, CIFAR, Ontario-MRI, CRC, OCE, QuantumWorks and MITACS.

References

1. Bennett, C.H., Shor, P.W.: Privacy in a quantum world. Science 284(5415), 747–748 (1999)
2. Paterson, K.G., Piper, F., Schack, R.: Quantum cryptography: a practical information security perspective. In: Zukowski, M., Kilin, S., Kowalik, J. (eds.) Quantum Communication and Security (2007)
3. Alleaume, R., Bouda, J., Branciard, C., Debuisschert, T., Dianati, M., Gisin, N., Godfrey, M., Grangier, P., Länger, T., Leverrier, A., Lütkenhaus, N., Painchault, P., Peev, M., Poppe, A., Pornin, T., Rarity, J., Renner, R., Ribordy, G., Riguidel, M., Salvail, L., Shields, A., Weinfurter, H., Zeilinger, A.: Secoqc white paper on quantum key distribution and cryptography, arXiv:quant-ph/0701168 (2007)
4. Stebila, D., Mosca, M., Lutkenhaus, N.: The case for quantum key distribution. In: Zukowski, M., Kilin, S., Kowalik, J. (eds.) Proceedings of QuantumComm 2009 Workshop on Quantum and Classical Information Security, vol. 36 (2009)
5. Bernstein, D.: Cost-benefit analysis of quantum cryptography. In: Workshop on Classical and Quantum Information Assurance Foundations and Practice, Schloss Dagstuhl (July 2009), `http://www.dagstuhl.de/Materials/index.en.phtml?09311`
6. Lütkenhaus, N., Shields, A.J.: Focus on quantum cryptography: Theory and practice. New Journal of Physics 11(4), 045005
7. Goldreich, O.: Foundations of cryptography (Volume I): Basic tools. Cambridge University Press, Cambridge (2001)
8. Goldreich, O.: Foundations of cryptography (Volume II): Basic applications. Cambridge University Press, Cambridge (2004)
9. Bellare, M., Rogaway, P.: Entity Authentication and Key Distribution. In: Stinson, D.R. (ed.) CRYPTO 1993. LNCS, vol. 773, pp. 232–249. Springer, Heidelberg (1994)
10. Canetti, R., Krawczyk, H.: Analysis of key-exchange protocols and their use for building secure channels. Cryptology ePrint Archive, Report 2001/040 (2001), `http://eprint.iacr.org/2001/040`
11. Menezes, A.J., van Oorschot, P., Vanstone, S.: Handbook of Applied Cryptography. CRC Press LLC, Boca Raton (1996)
12. Bennett, C.H., Brassard, G.: Quantum cryptography: Public key distribution and coin tossing. In: Proceedings of IEEE International Conference on Computers, Systems and Signal Processing, pp. 175–179. IEEE Press, New York (1984)
13. Ekert, A.K.: Quantum cryptography based on Bell's theorem. Phys. Rev. Lett. 67(6), 661–663 (1991)
14. Müller-Quade, J., Renner, R.: Composability in quantum cryptography. New Journal of Physics 11(8), 085006
15. Canetti, R.: Universally composable security: A new paradigm for cryptographic protocols. Cryptology ePrint Archive, Report 2000/067 (2000), `http://eprint.iacr.org/`
16. Nielsen, M., Chuang, I.: Quantum Computation and Quantum Information. Cambridge University Press, Cambridge (2000)
17. Stinson, D.R.: Cryptography: Theory and Practice. CRC Press LLC, Boca Raton (1995)
18. LaMacchia, B.A., Lauter, K., Mityagin, A.: Stronger security of authenticated key exchange. In: Susilo, W., Liu, J.K., Mu, Y. (eds.) ProvSec 2007. LNCS, vol. 4784, pp. 1–16. Springer, Heidelberg (2007)

19. Okamoto, T., Tanaka, K., Uchiyama, S.: Quantum Public-Key Cryptosystems. In: Bellare, M. (ed.) CRYPTO 2000. LNCS, vol. 1880, p. 147. Springer, Heidelberg (2000)
20. Goldwasser, S., Micali, S.: Probabilistic encryption. Journal of Computer and Systems Sciences 28(2), 270–299 (1984)
21. Wegman, M.N., Lawrence Carter, J.: New hash functions and their use in authentication and set equality, pp. 265–279 (1981)
22. Naor, M., Yung, M.: Universal one-way hash functions and their cryptographic applications. In: Proceedings of the 21st Annual ACM Symposium on Theory of Computing (1989)
23. Rompel, J.: One-way functions are necessary and sufficient for secure signatures. In: STOC 1990: Proceedings of the Twenty-Second Annual ACM Symposium on Theory of Computing (1990)
24. Renner, R.: Security of quantum key distribution. PhD thesis, Swiss Federal Institute of Technology (2005)
25. Bernstein, D.J., Buchmann, J., Dahmen, E.: Post Quantum Cryptography (2008)
26. Goldreich, O., Micali, S., Wigderson, A.: Proofs that yield nothing but their validity or all languages in np have zeroknowledge proofs. Journal of the ACM (1991)
27. Impagliazzo, R., Rudich, S.: Limits on the Provable Consequences of One-Way Permutations. In: Goldwasser, S. (ed.) CRYPTO 1988. LNCS, vol. 403, pp. 8–26. Springer, Heidelberg (1990)
28. Cachin, C., Maurer, U.M.: Unconditional Security Against Memory-Bounded Adversaries. In: Kaliski Jr., B.S. (ed.) CRYPTO 1997. LNCS, vol. 1294, pp. 292–306. Springer, Heidelberg (1997)
29. Ding, Y.Z., Rabin, M.O.: Hyper-Encryption and Everlasting Security. In: Alt, H., Ferreira, A. (eds.) STACS 2002. LNCS, vol. 2285, pp. 1–26. Springer, Heidelberg (2002)
30. Adams, C., Lloyd, S.:
31. Rivest, R.L.: Cryptography. In: Handbook of Theoretical Computer Science, pp. 717–755. Elsevier (1990)
32. Gertner, Y., Malkin, T., Reingold, O.: On the impossibility of basing trapdoor functions on trapdoor predicates. In: IEEE Press (ed.) Proc. 42nd Annual IEEE Symposium on the Foundations of Computer Science (FOCS 2001), pp. 126–135 (2001)
33. Beaver, D.: On Deniability in Quantum Key Exchange. In: Knudsen, L.R. (ed.) EUROCRYPT 2002. LNCS, vol. 2332, pp. 352–367. Springer, Heidelberg (2002)
34. Canetti, R., Dwork, C., Naor, M., Ostrovsky, R.: Deniable Encryption. In: Kaliski Jr., B.S. (ed.) CRYPTO 1997. LNCS, vol. 1294, pp. 90–104. Springer, Heidelberg (1997)
35. Klonowski, M., Kubiak, P., Kutyłowski, M.: Practical Deniable Encryption. In: Geffert, V., Karhumäki, J., Bertoni, A., Preneel, B., Návrat, P., Bieliková, M. (eds.) SOFSEM 2008. LNCS, vol. 4910, pp. 599–609. Springer, Heidelberg (2008)
36. Childs, A., Jao, D., Soukharev, V.: Constructing elliptic curve isogenies in quantum subexponential time (2010) (in preparation)

A Security Analysis of Uniformly-Layered Rainbow

Revisiting Sato-Araki's Non-commutative Approach to Ong-Schnorr-Shamir Signature towards PostQuantum Paradigm

Takanori Yasuda[1] and Kouichi Sakurai[1,2]

[1] Institute of Systems, Information Technologies and Nanotechnologies
[2] Department of Informatics, Kyushu University

Abstract. In 1984, Ong, Schnorr and Shamir proposed an efficient signature scheme (OSS signature scheme) using a bivariate quadratic equation. Its security was believed to be based on the difficulty of integer factorization. However, an efficient attack without integer factorization was subsequently found. In 2008, Hashimoto and Sakurai proposed an extended scheme (HS scheme), based on OSS signature scheme that used multivariate and non-commutative ring. HS scheme uses a composite number as a modulus in the same manner as OSS signature scheme.

In this paper, we redefine HS scheme in such a way that it deals with not only integers modulo a composite number, but also elements of a finite field. In the case of a finite field, it becomes a scheme in the multivariate public key cryptosystem. In fact, its public key is constructed by a version of Rainbow in which all the components in the parameter are equal. (We call such a Rainbow a uniformly-layered Rainbow.) In particular, our scheme is a candidate for post-quantum cryptography. If a non-commutative ring used in the proposed scheme is chosen by the group ring associated to dihedral group, the speed of the signature generation can be accelerated by about 50% in comparison with the corresponding Rainbow. We analyze the security of the extended HS scheme against some attacks and conclude that if its base field is $GF(256)$, then the dimension of a non-commutative ring must be more than 10 in order to be secure.

Keywords: Multivariate Public Key Cryptosystem, Post-quantum cryptography, Digital signature, Rainbow, Non-commutative ring.

1 Introduction

In 1984, Ong, Schnorr and Shamir [14] proposed an efficient signature scheme (OSS signature scheme) using the following bivariate quadratic equation,

$$x^2 + hy^2 \equiv m \bmod N \tag{1}$$

where N is a composite number that cannot easily be factorized. The security of this scheme was supposed to be based on the difficulty of integer factorization.

B.-Y. Yang (Ed.): PQCrypto 2011, LNCS 7071, pp. 275–294, 2011.

However, Pollard and Schnorr proposed an algorithm to solve the equation (1) efficiently without factorizing N. OSS signature scheme was then extended in two ways.

In 1994, Shamir [21] proposed a multivariate variant of OSS signature scheme, called Birational Permutation scheme. However, Coppersmith, Stern and Vaudenary [5] presented an efficient attack by observing a linear combination of components of the public key.

In 1997, Sato and Araki [20] extended from OSS signature scheme using quaternion algebra. Here, they replaced $\mathbb{Z}/N\mathbb{Z}$ in OSS signature scheme using quaternion algebra over $\mathbb{Z}/N\mathbb{Z}$. However, Coppersmith then found two efficient attacks using a special property of quaternion algebra.

In 2008, Hashimoto and Sakurai [10] proposed a new scheme (HS scheme), including the properties of both Birational Permutation scheme and Sato-Araki scheme. In 2010, Uchiyama and Ogura [22] showed that this scheme can be reduced to Rainbow, which is a signature scheme in the multivariate public key cryptosystem (MPKC), and discussed the possibility of forgery in the case of HS scheme with a small size.

Non-commutative rings often appear in cryptography, for example, in Sato-Araki scheme and HS scheme. Quaternion algebras and group rings are known as typical examples of non-commutative rings. They have a complex algebraic structure, which provides for cryptographic applications. There are two areas of cryptographic interest in the study of both non-commutative rings and OSS signature scheme. One is the relationship between security, efficiency and a choice of a non-commutative ring in HS scheme. Another is a reconstruction of the HS scheme as a part of MPKC which is a candidate for a post-quantum cryptosystem. This paper is a report on the latter.

Works related to non-commutative rings, other than Sato-Araki scheme and HS scheme include the non-commutative version of Polly Cracker [19] and the cryptography using the Braid group ([1],[11]).

Birational Permutation scheme can be rewritten as a scheme in MPKC by naturally changing a base ring $\mathbb{Z}/N\mathbb{Z}$ for a Galois field. We apply this method to HS scheme in order to construct a new MPKC scheme. In addition, we consider the security against known MPKC attacks.

By carefully observing how Uchiyama and Ogura reduced HS scheme to Rainbow, it becomes clear that our object, as a cryptosystem, is included in the class of Rainbow in which all components in the parameter are equal. In this paper, we refer to such a Rainbow as a uniformly-layered Rainbow. When Rainbow has 3 layers, in order to be secure, parameters whose components are nearly equal are selected [16]. Therefore, it is worth observing the uniformly-layered Rainbow. The setting of uniformly-layered Rainbow has not yet been studied, and its security is discussed for the first time in this paper. In addition, in the appendix, we try to extend HS scheme, which loosens the condition of non-commutative rings.

In our propose scheme, a finite field with a small order is used as a base ring. In the original HS scheme, the base ring is $\mathbb{Z}/N\mathbb{Z}$. Therefore, the arithmetic operation in our proposed scheme is more efficient than that of the original. Moreover, if we use a non-commutative ring with efficient arithmetic operation, we can achieve a high processing speed.

This paper is organized as follows. §2 briefly reviews Birational Permutation scheme and the attack described by Coppersmith, Stern and Vaudenary. §3 reviews Sato-Araki scheme and the two attacks described by Coppersmith against the scheme. §4 redefines HS scheme as a scheme that deals with finite fields as base rings. §5 analyzes security measures against the attack of Coppersmith, Stern and Vaudenary and the two attacks of Coppersmith. §6 describes how Uchiyama and Ogura reduced (proposed) HS scheme to Rainbow, and gives an example of the reduction. §7 analyzes security against UOV, MinRank and High-Rank attacks, which are attacks against Rainbow. §8 observes the efficiency of signature generation in HS scheme and compare the efficiency of signature generation in HS scheme and the corresponding Rainbow. §9 concludes the paper. In the appendix, we extend HS scheme using rings with involution.

2 Birational Permutation Scheme

In this section, we summarize the attack described by Coppersmith, Stern and Vaudenary against Birational Permutation scheme. We will analyze this attack in the extended HS scheme later. First, we describe Birational Permutation scheme [21].

Let p and q be prime numbers and $N = pq$. Let us assume that the factorization of N is difficult. Let n be a natural number. For $k = 2, 3, \dots, n$, we define $g_k : (\mathbb{Z}/N\mathbb{Z})^n \to \mathbb{Z}/N\mathbb{Z}$ by a homogeneous quadratic polynomial over $\mathbb{Z}/N\mathbb{Z}$ as follows:

$$g_k(x_1, x_2, \dots, x_n) = \sum_{i=1}^{k-1} a_{ik} x_i x_k + \sum_{1 \le i \le j \le k-1} a_{ij} x_i x_j,$$

where $a_{ij} \in \mathbb{Z}/N\mathbb{Z}$. The central map of Birational Permutation scheme is constructed by

$$G = (g_2, g_3, \dots, g_n) : (\mathbb{Z}/N\mathbb{Z})^n \to (\mathbb{Z}/N\mathbb{Z})^{n-1}.$$

The key generation, the signature generation, and the verification of Birational Permutation scheme are described as follows.

Key Generation
The secret key consists of prime numbers p and q, the central map G and two affine (linear) transformations $A_1 : (\mathbb{Z}/N\mathbb{Z})^{n-1} \to (\mathbb{Z}/N\mathbb{Z})^{n-1}$, $A_2 : (\mathbb{Z}/N\mathbb{Z})^n \to (\mathbb{Z}/N\mathbb{Z})^n$. The public key consists of N and the composite map $F = A_1 \circ G \circ A_2 = (f_2, f_3, \dots, f_n) : (\mathbb{Z}/N\mathbb{Z})^n \to (\mathbb{Z}/N\mathbb{Z})^{n-1}$.

Signature Generation
Let $\mathbf{M} \in (\mathbb{Z}/N\mathbb{Z})^{n-1}$ be a message. We compute $\mathbf{A} = A_1^{-1}(\mathbf{M}), \mathbf{B} = G^{-1}(\mathbf{A})$ and $\mathbf{C} = A_2^{-1}(\mathbf{B})$ in the same order. The signature of the message is $\mathbf{C} \in (\mathbb{Z}/N\mathbb{Z})^n$. Here, $G^{-1}(\mathbf{A})$ represents an element of the preimage of $\mathbf{A}$.

Verification
If $F(\mathbf{C}) = \mathbf{M}$ then the signature is accepted, otherwise it is rejected.

2.1 Attack against Birational Permutation Scheme

It is believed that solving general equations over $\mathbb{Z}/N\mathbb{Z}$ is more difficult than doing so over a finite field. The security of Birational Permutation scheme was based on the difficulty of solving the problem over $\mathbb{Z}/N\mathbb{Z}$. However, Coppersmith, Stern and Vaudenary gave an efficient algorithm [5] to compute A_2, a part of the secret key, without solving equations over $\mathbb{Z}/N\mathbb{Z}$.

For simplicity, assume that A_2 are linear transformations. We write A, B for the matrix expression of the linear parts of A_1, A_2, respectively, and g_k, f_k ($k = 2, 3, \ldots, n$) are denoted by

$$g_k(\mathbf{x}) = \mathbf{x}^T G_k \mathbf{x}, \quad f_k = \mathbf{x}^T F_k \mathbf{x} \quad (\mathbf{x} = (x_1, \ldots, x_n)^T), \tag{2}$$

for some $F_k, G_k \in \mathbb{M}(n, \mathbb{Z}/N\mathbb{Z})$. ($T$ means the transpose operator.) Since

$$f_k(\mathbf{x}) = \sum_{l=2}^{n} a_{kl} \mathbf{x}^T B^T G_j B \mathbf{x} = \mathbf{x}^T B^T \left(\sum_{l=2}^{n} a_{kl} G_l \right) B \mathbf{x}$$

where $A = (a_{kl})$, we have

$$F_k = B^T \left(\sum_{l=2}^{n} a_{kl} G_l \right) B. \tag{3}$$

For a variable λ and $1 \le k_1, k_2 \le n$, the determinant of $\sum_{l=2}^{n} a_{k_1 l} G_l - \lambda \sum_{l=2}^{n} a_{k_2 l} G_j$ is factored by $(a_{k_1 n} - \lambda a_{k_2 n})^2$. From (3), the determinant of $F_{k_1} - \lambda F_{k_2}$ is also factored by $(a_{k_1 n} - \lambda a_{k_2 n})^2$. Therefore $a_{k_1 n}/a_{k_2 n}$, which is denoted by λ_0, is computed by the public key. By calculating the kernel and the image of $F_{k_1} - \lambda_0 F_{k_2}$, $(\mathbb{Z}/N\mathbb{Z})^n$ is decomposed as

$$(\mathbb{Z}/N\mathbb{Z})^n = B^{-1}((\mathbb{Z}/N\mathbb{Z})^{n-1} \times \{0\}) \oplus B^{-1}(\{0\}^{n-1} \times (\mathbb{Z}/N\mathbb{Z})) \tag{4}$$

Continuing this operation, we finally arrive at a decomposition,

$$(\mathbb{Z}/N\mathbb{Z})^n = B^{-1}((\mathbb{Z}/N\mathbb{Z}) \times \{0\}^{n-1}) \oplus \cdots \oplus B^{-1}(\{0\}^{n-1} \times (\mathbb{Z}/N\mathbb{Z}))$$

by subspaces with rank 1. By rewriting the public key on a basis of the above decomposition, we obtain a system of equations with the same form as the central map, and in this way, we can forge a signature.

3 Sato-Araki Scheme

In this section, we summarize the two attacks described by Coppersmith against Sato-Araki scheme. We will analyze these attacks in the extended HS scheme later.

Sato-Araki scheme [20] uses quaternion algebra over $\mathbb{Z}/N\mathbb{Z}$. Let R be a $\mathbb{Z}/N\mathbb{Z}$-analogue of Hamilton's quaternion algebra. Here, R is defined by

$$R = \mathbb{Z}/N\mathbb{Z} \cdot 1 \oplus \mathbb{Z}/N\mathbb{Z} \cdot i \oplus \mathbb{Z}/N\mathbb{Z} \cdot j \oplus \mathbb{Z}/N\mathbb{Z} \cdot ij,$$

and $i^2 = j^2 = -1$, $ij = -ji$. R is identified with a subring of a matrix ring by the embedding homomorphism,

$$\begin{aligned} R \ni\ & a_0 \cdot 1 + a_1 \cdot i + a_2 \cdot j + a_3 \cdot ij \\ \longmapsto\ & \begin{pmatrix} a_0 + a_1\sqrt{-1} & a_3 + a_2\sqrt{-1} \\ -a_3 + a_2\sqrt{-1} & a_0 - a_1\sqrt{-1} \end{pmatrix} \in \mathbb{M}(2, \mathbb{Z}/N\mathbb{Z}[\sqrt{-1}]). \end{aligned} \tag{5}$$

Note that R is closed by the transpose operation. Sato-Araki scheme is described as follows.

Key Generation
The secret key consists of the primes p, q and $u \in R^{\times}$. The public key consists of N and $h := -(u^T)^{-1}u^{-1} \in R$.

Signature Generation
Let $\mathbf{M} = \mathbf{M}^T \in R$ be a message. Choose $\rho \in R^{\times}$ randomly. We now compute $\mathbf{C}_1 := \rho^{-1}\mathbf{M} + \rho^T$, $\mathbf{C}_2 := u(\rho^{-1}\mathbf{M} - \rho^T) \in R$. $(\mathbf{C}_1, \mathbf{C}_2)$ is a signature.

Verification
If $\mathbf{C}_1^T\mathbf{C}_1 + \mathbf{C}_2^T h\mathbf{C}_2 = 4\mathbf{M}$, then the signature is accepted, otherwise it is rejected.

3.1 Attacks against Sato-Araki Scheme

The security of Sato-Araki scheme was based on the difficulty of solving an equation over R,

$$X^T X + h \equiv 0 \mod N.$$

However, Coppersmith proposed two efficient attacks [4] by using a special property of quaternion algebra, without factorizing N.

Coppersmith's first attack. The first attack proposed by Coppersmith is a chosen message attack. For $i = 1, 2, 3$, let $(\mathbf{C}_1^{(i)}, \mathbf{C}_2^{(i)})$ be signatures for messages $\mathbf{M}_i$. The key of the attack is as follows: For $i = 1, 2, 3$,

$$(\mathbf{C}_1^{(i)})^T u\mathbf{C}_2^{(i)} \text{ are symmetric matrices.} \tag{6}$$

Then these span a subspace $\{\delta = \delta^T \in R\}$ of rank 3 with high probability. One can compute $X \in R$ by satisfying

$$(\mathbf{C}_1^{(i)})^T X \mathbf{C}_2^{(i)} \text{ are symmetric matrices } \ (i = 1, 2, 3),$$

which is determined up to scalars. From the above fact, X is found to be proportional to u. It is not difficult to compute u, a part of the secret key, from X.

Coppersmith's second attack. The second Coppersmith's attack is based on the following algorithm.

Proposition 1 ([2]). *Let N be an odd positive integer and $f(x, y)$ a bivariate quadratic polynomial over $\mathbb{Z}/N\mathbb{Z}$. $\Delta(f)$ denotes the discriminant of f defined as in [2]. If $\gcd(\Delta(f), N) = 1$, then there exists an algorithm which gives a solution to $f(x, y) = 0$ with probability $1 - \epsilon$, and requires $O(\log(\epsilon^{-1} \log N) \log^4 N)$ arithmetic operations on integers of size $O(\log N)$ bits.*

If $x, y \in R$ are written as

$$x = x_0 \cdot 1 + x_1 \cdot i + x_2 \cdot j + x_3 \cdot ij, \quad y = y_0 \cdot 1 + y_1 \cdot i + y_2 \cdot j + y_3 \cdot ij,$$

then the equation over R, which is

$$x^T x + y^T hy = 4\mathbf{M} \tag{7}$$

is rewritten by 4 quadratic equations with respect to 8 variables $x_0, x_1, \ldots, y_3$. By simplifying equation (7) and using the property of quaternion algebra, the problem of solving the system of quadratic equations can be reduced to that of a set of bivariate quadratic equations. Therefore, it is possible to forge a signature based on the above proposition.

4 Redefinition of HS Scheme

In HS scheme, more general non-commutative rings can be utilized than the quaternion algebra used in Sato-Araki scheme. In this section, we describe HS scheme associated to non-commutative rings over a field K or $\mathbb{Z}/N\mathbb{Z}$. In the case of $\mathbb{Z}/N\mathbb{Z}$, the scheme becomes the original HS scheme.

4.1 Non-commutative Rings

Let L be either a field K or $\mathbb{Z}/N\mathbb{Z}$. In this paper, we say that an L-algebra R is a non-commutative ring only if

(1) R is a free module over L with finite rank, and
(2) R is non-commutative.

Example 1 (Quaternion algebra). For $a \in L^{\times}$, a non-commutative ring $Q_L(a)$ is defined as followsF

$$\text{(Set)}\ \ Q_L(a) = L\cdot 1 \oplus L\cdot i \oplus L\cdot j \oplus L\cdot ij,$$
$$\text{(Product)}\ \ i^2 = a,\ j^2 = -1,\ ij = -ji.$$

$Q_L(a)$ is a free module over L with rank 4. $Q_L(a)$ is called a quaternion algebra. When $L = \mathbb{Z}/N\mathbb{Z}$ and $a = -1$, R coincides with the quaternion algebra used in Sato-Araki scheme. If $L = GF(q)$, we write also $Q_q(a) = Q_K(a)$. $Q_L(a)$ is embedded into a matrix ring:

$$\iota : Q_L(a) \ni c_1 + c_2 i + c_3 j + c_4 ij \mapsto \begin{pmatrix} c_1 + c_2 i & c_3 + c_4 i \\ -c_3 + c_4 i & c_1 - c_2 i \end{pmatrix} \in \mathbb{M}(2, L[i]). \quad (8)$$

If $Q_L(a)$ is identified with the image of ι, it is closed by transpose operation.

Let R be a non-commutative ring over L and r be its rank over L. Then there exists an L-linear isomorphism,

$$\phi : L^r \xrightarrow{\sim} R. \quad (9)$$

Using this isomorphism ϕ, an element $\alpha \in R$ can be represented by r elements in L.

4.2 HS Scheme over L

Let R be a non-commutative ring over L of rank r and let us fix ϕ as in (9). For the rest of this paper, let us assume that R is realized as a subring of the matrix ring $\mathbb{M}(s, L)$ for some $s \in \mathbb{N}$, and closed by the transpose operation.

Let $\tilde{n}$ be a positive integer. HS scheme deploys non-commutative multivariate polynomials as a central map:

$$\tilde{g}_k(x_1, \ldots, x_{\tilde{n}}) = \sum_{i=1}^{k-1} x_i^T \alpha_{ij}^{(k)} x_k + \sum_{1\le i,j\le k-1} x_i^T \alpha_{ij}^{(k)} x_j + \sum_{1\le i\le k} \beta_i^{(k)} x_i + \gamma^{(k)} \quad (k = 2, 3, \ldots, \tilde{n}),$$

where $\alpha_{i,j}^{(k)}, \beta_i^{(k)}, \gamma^{(k)} \in R$. Note that $\tilde{g}_k$ is essentially a polynomial of k variables. The central map of HS scheme is constructed by

$$\tilde{G} = (\tilde{g}_2, \ldots, \tilde{g}_{\tilde{n}}) : R^{\tilde{n}} \to R^{\tilde{n}-1}$$

The key generation, the signature generation and the verification are described as follows.

Key Generation The secret key consists of R, the central map $\tilde{G}$, and two affine transformations $A_1 : L^m \to L^m$ $(m = r\tilde{n} - r)$, $A_2 : L^n \to L^n$ $(n = r\tilde{n})$. The public key consists of L and the composed map $\tilde{F} = A_1 \circ \phi^{-\tilde{n}+1} \circ \tilde{G} \circ \phi^{\tilde{n}} \circ A_2 :$

$L^n \to L^m$, which is a system of m quadratic polynomials of n variables over L. In what follows, let us suppose that $\tilde{F}$ is expressed as $\tilde{F} = (\tilde{f}_{r+1}, \ldots, \tilde{f}_n)^T$.

Signature Generation. Let $\mathbf{M} \in L^m$ be a message. We compute $\mathbf{A} = A_1^{-1}(\mathbf{M})$, $\mathbf{B} = G^{-1}(\mathbf{A})$ and $\mathbf{C} = A_2^{-1}(\mathbf{B})$ in the same order. The signature of the message is $\mathbf{C} \in L^n$. Here, $\mathbf{B} = \tilde{G}^{-1}(\mathbf{A})$ is computed by the following procedure.

Step 1. Choose a random element $b_1 \in R$.
Step 2. For $k = 1, \ldots, \tilde{n}$, do the following operation recursively.
Here, $\tilde{g}_k$ is a non-commutative polynomial with respect to $x_1, \ldots, x_k$. By substituting $x_1 = b_1, \ldots, x_{k-1} = b_{k-1}$ into $\tilde{g}_k$, we obtain a non-commutative polynomial, $\bar{g}_k$, of one variable x_k and with at most 1 degree. We compute the solution $b_k \in R$ of

$$\bar{g}_k(x_k) = a_k \tag{10}$$

where $\mathbf{A} = (a_i) \in R^{\tilde{m}}$. (If there is no solution, return to Step 1.)
Step 3. Set $\mathbf{B} = (b_1, \ldots, b_{\tilde{n}})$.

Verification. If $\tilde{F}(\mathbf{C}) = \mathbf{M}$, then the signature is accepted, otherwise it is rejected.

This scheme is denoted by HS(R; $\tilde{n}$).

Remark 1. In general, it is difficult to solve a non-commutative equation (10) directly. However, once a L-basis of R is fixed, we have a new system of (commutative) linear equations with respect to this basis. This system is easy to be solved in general. If R has an efficient arithmetic operation, the equation (10) can be solved more efficiently. For example, in the case of a quaternion algebra $Q_L(a)$, its realization (8) enable us to compute its arithmetic operation efficiently.

5 Security Analysis of HS Scheme

Thus far, we have analyzed [10] the attacks against HS scheme, in the case of $L = \mathbb{Z}/N\mathbb{Z}$, for both Coppersmith, Stern and Vaudenary (CSV) [5] and Coppersmith [4] attacks. In this section, we analyze security countermeasures against these attacks on the HS scheme when $L = K$. Note that some technique to solve equations of multivariate polynomials is available in the security analysis in the case of $L = K$, on the other hand, it is difficult to be applied in the case of $L = \mathbb{Z}/N\mathbb{Z}$. This is the difference between security analysis in the case of $L = K$ and that in the case of $L = \mathbb{Z}/N\mathbb{Z}$.

5.1 Security against CSV Attack

For the public key $F = (f_2, f_3, \ldots, f_n)$, we use notation $F_2, F_3 \ldots, F_n$ as in §2.1. The key step in the CSV attack is to find a linear combination $\Lambda = \sum_i c_i F_i$ of $F_2, \ldots, F_n$ which dose not have full rank. In Birational Permutation scheme, such Λ can be found by solving an equation of polynomial of one variable.

In HS scheme, the attack by finding a matrix which dose not have full rank, like a Λ, can be extended. Moreover, in a more general setting of scheme in MPKC, the CSV attack can be extended, which is called the HighRank attack. We will analyze the security against the HighRank attack in §7.

5.2 Security against Coppersmith's First Attack

Sato-Araki scheme can be translated using HS scheme as $\mathrm{HS}(Q_{\mathbb{Z}/N\mathbb{Z}}(-1); 2)$ with

$$A_1 = \text{Identity map of } \mathbb{Z}/N\mathbb{Z},$$
$$A_2 : (\mathbb{Z}/N\mathbb{Z})^2 \to (\mathbb{Z}/N\mathbb{Z})^2 \text{ given by } A_2(z_1, z_2) = (z_1 + uz_2, z_1 - uz_2)$$

where u is a part of the secret key described as in §3. Because A_1 and A_2 are fixed and expressed by simple transformations, a simple relation (6) holds. Therefore the Coppersmith's first attack is applicable for Sato-Araki scheme. However in HS scheme, we adopt random affine transformations as A_1 and A_2. This makes difficult to find a simple relation like a (6) in HS scheme.

5.3 Security against Coppersmith's Second Attack

In Sato-Araki scheme, the problem to forge a signature is reduced to that to solve the non-commutative equation (7). Moreover this equation is rewritten by 4 quadratic equations with respect to (commutative) 8 variables as in §3.1. Fortunately, the system of these quadratic equations is decomposed into some systems of bivariate quadratic equations, and therefore we can solve the non-commutative equation (7) using Proposition 1. However, in HS scheme, the numbers of non-commutative variables and equations increase. Concretely, we need to solve $r(\tilde{n} - 1)$ quadratic equations with respect to (commutative) $r\tilde{n}$ variables for $\mathrm{HS}(R; \tilde{n})$ where r is the dimension of R over K. Therefore the security against Coppersmith's second attack in HS scheme is reduced to the MQ problem.

6 Reduction of HS Scheme to Rainbow

Uchiyama and Ogura [22] pointed out that the original HS scheme, which is defined over $\mathbb{Z}/N\mathbb{Z}$, can be rewritten using a $\mathbb{Z}/N\mathbb{Z}$-analogue of Rainbow in which the original Rainbow [7] is a multilayer variant of the Unbalanced Oil and Vinegar signature scheme. This implies that attacks against Rainbow also apply to HS scheme.

6.1 Original Rainbow and Its Analogue

To deal with both the original Rainbow and its analogue over a finite field, we prepare Rainbow defined over L which is either K or $\mathbb{Z}/N\mathbb{Z}$.

At first, we define the parameters that determine the layer structure of Rainbow. Let t be the number of layers of Rainbow. Let $v_1, \ldots, v_{t+1}$ be a sequence of positive $t+1$ integers, such that

$$0 < v_1 < v_2 < \cdots < v_t < v_{t+1}.$$

For $h = 1, \ldots, t$, the sets V_h, O_h of the indices of the Vinegar and Oil variables of the h-th layer of Rainbow are defined by

$$V_h = \{1, 2 \ldots, v_h\}, \quad O_h = \{v_h + 1, v_h + 2, \ldots, v_{h+1} - 1, v_{h+1}\}.$$

The number of elements in O_h and V_h are $v_{h+1} - v_i$ and v_i, respectively, and denote $o_h = v_{h+1} - v_h$. Note that the smallest integer in O_1 is $v_1 + 1$. We define $n = v_{t+1}$, which is the maximum number of variables used in Rainbow.

Rainbow consists of t layers of multivariate polynomials of n variables. For $h = 1, 2, \ldots, t$, the h-th layer of Rainbow deploys the following system of o_h multivariate polynomials:

$$g_k(x_1, \ldots, x_n) = \sum_{i \in O_h, j \in V_h} \alpha_{i,j}^{(k)} x_i x_j + \sum_{i,j \in V_h, i \leq j} \beta_{i,j}^{(k)} x_i x_j + \sum_{i \in V_{h+1}} \gamma_i^{(k)} x_i + \eta^{(k)} \qquad (k \in O_h), \tag{11}$$

where $\alpha_{i,j}^{(k)}, \beta_{i,j}^{(k)}, \gamma_i^{(k)}, \eta^{(k)} \in L$. Note that g_k is essentially a polynomial of $v_h + o_h$ variables. We call variables x_i $(i \in O_h)$ and x_j $(i \in V_j)$ the Oil and Vinegar variables, respectively. Then the central map of Rainbow is constructed by

$$G = (g_{v_1+1}, \ldots, g_n) : L^n \to L^{n-v_1}.$$

Note that we can easily compute one of the preimages of G for any element of L^{n-v_1}. For a system of o_h equations for the h-th layer,

$$g_k(b_1, \ldots, b_{v_h}, x_{v_h+1}, \ldots, x_{v_{h+1}}) = a_k \quad (k \in O_h)$$

becomes o_h linear equations of o_h variables for any $(a_{v_h+1}, \ldots, a_{v_{h+1}}) \in L^{o_h}$ and $(b_1, \ldots, b_{v_h}) \in L^{v_h}$. The values of the Oil variables in the h-th layer obtained by solving these linear equations are used for the Vinegar variables in the $(h+1)$-th layer.

Next, we describe the key generation, the signature generation and the verification of Rainbow.

Key Generation. The secret key consists of the central map, G, and two affine transformations $A_1 : L^m \to L^m$ $(m = n - v_1)$, $A_2 : L^n \to L^n$. The public key consists of L, which is either a field, K, or $\mathbb{Z}/N\mathbb{Z}$, and the composed map $F = A_1 \circ G \circ A_2 : L^n \to L^m$, which is a system of m quadratic polynomials of n variables over L. In what follows, let us suppose that F is expressed as $F = (f_{v_1+1}, \ldots, f_n)^{\mathrm{T}}$.

Signature Generation. Let $\mathbf{M} \in L^m$ be a message. We compute $\mathbf{A} = A_1^{-1}(\mathbf{M})$, $\mathbf{B} = G^{-1}(\mathbf{A})$ and $\mathbf{C} = A_2^{-1}(\mathbf{B})$ in the same order. The signature of the message is $\mathbf{C} \in L^n$. Note that $\mathbf{B} = G^{-1}(\mathbf{A})$ can easily be computed by the above property of G.

Verification. If $F(\mathbf{C}) = \mathbf{M}$ then the signature is accepted, otherwise it is rejected.

This scheme is denoted by Rainbow(L; $v_1, o_1, \ldots, o_t$), and we call $v_1, o_1, \ldots, o_t$ a parameter of Rainbow.

6.2 Reduction of HS Scheme to Rainbow

Uchiyama and Ogura wrote down $\phi^{-\tilde{n}+1} \circ \tilde{G} \circ \phi^{\tilde{n}}$ for HS($\mathbb{Z}/N\mathbb{Z}, \tilde{n}$) and showed the following [22].

Proposition 2. *Let R be a non-commutative ring over $\mathbb{Z}/N\mathbb{Z}$ of rank r. Let $\tilde{F}$ be a public key of HS(R; $\tilde{n}$). Then $\tilde{F}$ becomes a public key of Rainbow($\mathbb{Z}/N\mathbb{Z}$; $\overbrace{r, \ldots, r}^{\tilde{n}}$).*

Remark 2. The above proposition defines correspondence between signature schemes,

$$\begin{array}{rccc} & \text{HS}(R;\, \tilde{n}) & \rightsquigarrow & \text{Rainbow}(\mathbb{Z}/N\mathbb{Z};\, \overbrace{r, \ldots, r}^{\tilde{n}}) \\ \text{Secret Key:} & (A_1, \tilde{G}, A_2) & \mapsto & (A_1, \phi^{-\tilde{n}+1} \circ \tilde{G} \circ \phi^{\tilde{n}}, A_2) \\ \text{Public Key:} & \tilde{F} & \mapsto & \tilde{F}. \end{array}$$

Using this notation, we have the following correspondence:

$$\begin{array}{ccc} \text{OSS scheme} & \rightsquigarrow & \text{Rainbow}(\mathbb{Z}/N\mathbb{Z};\, 1, 1), \\ \text{Birational Permutation scheme} & \rightsquigarrow & \text{Rainbow}(\mathbb{Z}/N\mathbb{Z};\, 1, \ldots, 1), \\ \text{Sato-Araki scheme} & \rightsquigarrow & \text{Rainbow}(\mathbb{Z}/N\mathbb{Z};\, 4, 4). \end{array}$$

The argument of Uchiyama and Ogura in [22] is also valid for the case of HS scheme defined over field K. Therefore, we have

Proposition 3. *Let R be a non-commutative ring over K of dimension r. Let $\tilde{F}$ be a public key of HS(R; $\tilde{n}$). Then $\tilde{F}$ becomes a public key of Rainbow(K; $\overbrace{r, \ldots, r}^{\tilde{n}}$).*

Remark 3. The above proposition shows that HS scheme is another way of construction of the uniformly-layered Rainbow. Here, "uniformly-layered" means that all components in the parameter of Rainbow are equal.

6.3 Example of Reduction of HS Scheme

Let $K = GF(13)$ and let us consider a case of $\mathrm{HS}(Q_{13}(2), 3)$ as a toy example of Proposition 3. (See §4.1 for the definition of $Q_{13}(2)$) A linear isomorphism ϕ is defined by

$$\phi : K^4 \ni (c_1, c_2, c_3, c_4) \mapsto c_1 + c_2 i + c_3 j + c_4 ij \in Q_{13}(2),$$

where $i^2 = 2, j^2 = -1$. A central map $\tilde{G} = (\tilde{g}_2, \tilde{g}_3)$ is defined by, for $\mathbf{X} = (X_1, X_2, X_3)^T$,

$$\tilde{g}_2(\mathbf{X}) = \mathbf{X}^T \begin{pmatrix} 1+12i+11j & 1+8j & 0 \\ 0 & 0 & 0 \\ 0 & 0 & 0 \end{pmatrix} \mathbf{X}$$
$$+ (1+3i+9j+8ij, 5+9i+1j+10ij, 0)\mathbf{X} + 9+1i+2j+6ij,$$
$$\tilde{g}_3(\mathbf{X}) = \mathbf{X}^T \begin{pmatrix} 5+1i+2j+11ij & 3+10i+5j+6ij & 11+1i+11j+5ij \\ 9+9i+10j+6ij & 12+8i+4j+11ij & 12+8i+9j+12ij \\ 0 & 0 & 0 \end{pmatrix} \mathbf{X}$$
$$+ (4+11i+3j+5ij, 7+3i+6j+8ij, 12+1i+8ij)\mathbf{X} + 8+11i+9j$$

Two affine transformation $A_1 : K^8 \to K^8$ and $A_2 : K^{12} \to K^{12}$, which are the rest of secret key, are defined by

$$A_1(\mathbf{x}) = L_1\mathbf{y} + v_1, \quad A_2(\mathbf{x}) = L_2\mathbf{x} + v_2,$$

$$L_1 = \begin{pmatrix} 7 & 9 & 3 & 7 & 11 & 11 & 2 & 5 \\ 6 & 2 & 7 & 11 & 5 & 7 & 0 & 5 \\ 0 & 4 & 8 & 5 & 2 & 2 & 5 & 4 \\ 10 & 9 & 5 & 11 & 10 & 3 & 0 & 1 \\ 2 & 5 & 7 & 12 & 3 & 2 & 2 & 12 \\ 10 & 1 & 10 & 4 & 7 & 5 & 7 & 11 \\ 0 & 0 & 11 & 7 & 10 & 2 & 4 & 1 \\ 0 & 12 & 11 & 3 & 7 & 7 & 0 & 11 \end{pmatrix}, \quad L_2 = \begin{pmatrix} 11 & 2 & 7 & 2 & 5 & 12 & 12 & 8 & 9 & 11 & 3 & 4 \\ 9 & 11 & 9 & 9 & 2 & 4 & 12 & 5 & 5 & 2 & 0 & 6 \\ 12 & 1 & 3 & 9 & 7 & 1 & 9 & 9 & 4 & 4 & 6 & 4 \\ 12 & 7 & 7 & 3 & 7 & 3 & 8 & 11 & 7 & 9 & 12 & 5 \\ 7 & 0 & 6 & 3 & 11 & 1 & 3 & 12 & 2 & 7 & 11 & 8 \\ 5 & 4 & 1 & 0 & 6 & 1 & 2 & 7 & 5 & 0 & 11 & 1 \\ 6 & 9 & 8 & 8 & 4 & 0 & 6 & 5 & 1 & 0 & 5 & 4 \\ 10 & 6 & 8 & 10 & 0 & 10 & 10 & 7 & 6 & 11 & 1 & 5 \\ 4 & 8 & 9 & 1 & 6 & 7 & 8 & 11 & 5 & 1 & 12 & 4 \\ 7 & 10 & 0 & 0 & 9 & 11 & 7 & 8 & 2 & 2 & 3 & 6 \\ 7 & 5 & 9 & 3 & 8 & 3 & 12 & 5 & 6 & 7 & 2 & 8 \\ 5 & 0 & 9 & 9 & 4 & 8 & 11 & 2 & 5 & 11 & 8 & 9 \end{pmatrix},$$

$$v_1 = (12, 10, 1, 11, 3, 10, 9, 6)^T, \quad v_2 = (5, 2, 11, 9, 1, 11, 12, 4, 8, 7, 4, 9)^T.$$

Then the public key $\tilde{F} = (\tilde{f}_5, \tilde{f}_6, \ldots, \tilde{f}_{12})$ is described as

$$\tilde{f}_k(\mathbf{x}) = \mathbf{x}^T \overline{A}_k \mathbf{x} + \overline{b}_k \mathbf{x} + \overline{c}_k$$

where $\overline{A}_k, \overline{b}_k, \overline{c}_k$ are defined as follows.

$$\overline{A}_5 = \begin{pmatrix} 7&2&2&7&4&4&10&1&1&12&8&4 \\ 0&2&5&9&2&11&11&8&7&3&2&10 \\ 0&0&8&9&2&8&6&8&5&4&6&0 \\ 0&0&0&4&5&5&2&7&4&7&3&10 \\ 0&0&0&0&9&10&4&3&11&4&9&8 \\ 0&0&0&0&0&1&10&7&7&0&7&10 \\ 0&0&0&0&0&0&8&11&0&10&3&1 \\ 0&0&0&0&0&0&0&11&0&2&3&4 \\ 0&0&0&0&0&0&0&0&3&4&6&0 \\ 0&0&0&0&0&0&0&0&0&3&3&6 \\ 0&0&0&0&0&0&0&0&0&0&7&11 \\ 0&0&0&0&0&0&0&0&0&0&0&5 \end{pmatrix}, \quad \overline{A}_6 = \begin{pmatrix} 12&12&7&11&1&4&6&3&6&1&3&6 \\ 0&10&9&12&11&9&0&4&9&11&8&0 \\ 0&0&7&11&10&12&2&3&6&1&5&7 \\ 0&0&0&8&1&2&8&8&12&1&9&8 \\ 0&0&0&0&4&7&6&3&4&10&12&5 \\ 0&0&0&0&0&10&7&11&8&2&1&11 \\ 0&0&0&0&0&0&3&3&4&5&4&7 \\ 0&0&0&0&0&0&0&1&10&7&5&3 \\ 0&0&0&0&0&0&0&0&5&11&3&7 \\ 0&0&0&0&0&0&0&0&0&2&3&2 \\ 0&0&0&0&0&0&0&0&0&0&12&10 \\ 0&0&0&0&0&0&0&0&0&0&0&4 \end{pmatrix},$$

$$\overline{A}_7 = \begin{pmatrix} 1&8&0&6&3&4&8&4&8&4&11&4 \\ 0&4&7&2&2&8&11&10&9&1&7&11 \\ 0&0&11&4&10&10&10&8&9&9&0&1 \\ 0&0&0&10&0&9&12&3&7&6&11&1 \\ 0&0&0&0&5&3&4&2&9&5&6&12 \\ 0&0&0&0&0&12&12&11&2&12&2&12 \\ 0&0&0&0&0&0&10&2&5&12&10&3 \\ 0&0&0&0&0&0&0&7&7&6&12&12 \\ 0&0&0&0&0&0&0&0&1&0&11&8 \\ 0&0&0&0&0&0&0&0&0&0&12&10 \\ 0&0&0&0&0&0&0&0&0&0&10&0 \\ 0&0&0&0&0&0&0&0&0&0&0&8 \end{pmatrix}, \quad \overline{A}_8 = \begin{pmatrix} 12&3&8&1&2&0&5&5&12&10&9&6 \\ 0&5&8&12&11&5&6&12&1&8&9&10 \\ 0&0&8&2&11&4&7&7&8&12&7&1 \\ 0&0&0&7&10&11&12&11&1&11&1&6 \\ 0&0&0&0&12&4&6&6&10&0&5&7 \\ 0&0&0&0&0&5&0&1&12&11&1&2 \\ 0&0&0&0&0&0&5&9&9&6&0&8 \\ 0&0&0&0&0&0&0&3&10&7&8&12 \\ 0&0&0&0&0&0&0&0&1&12&4&0 \\ 0&0&0&0&0&0&0&0&0&10&5&6 \\ 0&0&0&0&0&0&0&0&0&0&10&11 \\ 0&0&0&0&0&0&0&0&0&0&0&4 \end{pmatrix},$$

$$\overline{A}_9 = \begin{pmatrix} 4&7&4&6&3&4&4&0&4&0&6&12 \\ 0&9&6&7&1&1&6&12&10&3&10&5 \\ 0&0&2&8&0&8&0&2&0&6&0&5 \\ 0&0&0&7&0&12&10&7&0&8&9&3 \\ 0&0&0&0&4&5&10&3&1&7&3&7 \\ 0&0&0&0&0&11&12&5&8&2&3&2 \\ 0&0&0&0&0&0&5&9&10&5&10&7 \\ 0&0&0&0&0&0&0&4&5&12&3&8 \\ 0&0&0&0&0&0&0&0&5&8&6&1 \\ 0&0&0&0&0&0&0&0&0&12&4&8 \\ 0&0&0&0&0&0&0&0&0&0&5&0 \\ 0&0&0&0&0&0&0&0&0&0&0&1 \end{pmatrix}, \quad \overline{A}_{10} = \begin{pmatrix} 1&10&11&0&8&5&8&2&3&9&2&2 \\ 0&3&6&3&10&6&4&2&10&10&8&3 \\ 0&0&8&2&4&7&11&9&9&7&9&3 \\ 0&0&0&3&5&12&5&5&9&9&10&3 \\ 0&0&0&0&12&5&3&0&8&6&12&10 \\ 0&0&0&0&0&10&3&4&9&11&9&12 \\ 0&0&0&0&0&0&1&7&11&8&12&12 \\ 0&0&0&0&0&0&0&2&8&0&3&11 \\ 0&0&0&0&0&0&0&0&3&8&1&3 \\ 0&0&0&0&0&0&0&0&0&0&8&6 \\ 0&0&0&0&0&0&0&0&0&0&1&7 \\ 0&0&0&0&0&0&0&0&0&0&0&7 \end{pmatrix},$$

$$\overline{A}_{11} = \begin{pmatrix} 2&0&7&0&2&7&1&9&9&12&7&0 \\ 0&11&10&4&8&3&11&6&3&10&8&5 \\ 0&0&2&2&3&0&0&6&5&11&10&1 \\ 0&0&0&7&8&6&1&6&2&2&4&10 \\ 0&0&0&0&5&2&5&0&9&0&7&8 \\ 0&0&0&0&0&8&3&7&12&9&7&12 \\ 0&0&0&0&0&0&3&0&1&4&7&9 \\ 0&0&0&0&0&0&0&5&0&9&0&2 \\ 0&0&0&0&0&0&0&0&2&3&5&10 \\ 0&0&0&0&0&0&0&0&0&2&0&3 \\ 0&0&0&0&0&0&0&0&0&0&0&4 \\ 0&0&0&0&0&0&0&0&0&0&0&4 \end{pmatrix}, \quad \overline{A}_{12} = \begin{pmatrix} 1&7&1&9&7&7&8&3&3&9&10&10 \\ 0&1&3&11&5&1&1&3&6&9&7&0 \\ 0&0&5&6&7&11&0&1&4&1&3&7 \\ 0&0&0&9&1&5&0&10&9&10&2&8 \\ 0&0&0&0&12&4&9&10&5&1&10&9 \\ 0&0&0&0&0&10&9&10&1&11&6&0 \\ 0&0&0&0&0&0&8&12&1&0&9&10 \\ 0&0&0&0&0&0&0&12&6&1&10&8 \\ 0&0&0&0&0&0&0&0&7&4&5&12 \\ 0&0&0&0&0&0&0&0&0&11&5&12 \\ 0&0&0&0&0&0&0&0&0&0&12&4 \\ 0&0&0&0&0&0&0&0&0&0&0&8 \end{pmatrix},$$

$\bar{b}_5 = (2, 12, 6, 3, 2, 12, 6, 8, 9, 11, 2, 5)$, $\bar{b}_6 = (2, 6, 4, 1, 2, 10, 8, 0, 1, 8, 0, 12)$,
$\bar{b}_7 = (6, 9, 11, 2, 0, 10, 0, 6, 4, 1, 11, 9)$, $\bar{b}_8 = (5, 10, 2, 4, 6, 8, 10, 2, 4, 12, 2, 4)$,
$\bar{b}_9 = (10, 3, 10, 11, 4, 6, 4, 7, 3, 3, 5, 1)$, $\bar{b}_{10} = (8, 2, 1, 10, 1, 12, 5, 10, 11, 11, 8, 2)$,
$\bar{b}_{11} = (1, 7, 6, 10, 8, 12, 10, 5, 7, 9, 0, 8)$, $\bar{b}_{12} = (3, 3, 5, 1, 1, 8, 11, 0, 3, 10, 12, 6)$,

$$\bar{c}_5 = 11,\ \bar{c}_6 = 1,\ \bar{c}_7 = 1,\ \bar{c}_8 = 4,\ \bar{c}_9 = 1,\ \bar{c}_{10} = 12,\ \bar{c}_{11} = 11,\ \bar{c}_{12} = 4.$$

The above HS scheme corresponds to Rainbow(K; 4, 4, 4) by Proposition 3. The central map $G = (g_5, g_6, g_7, g_8, g_9, g_{10}, g_{11}, g_{12}) : K^{12} \to K^8$ of the corresponding Rainbow is described as

$$g_k(\mathbf{x}) = \mathbf{x}^T A_k \mathbf{x} + b_k \mathbf{x} + c_k.$$

where $A_k \in \mathbb{M}(12, K), b_k \in K^{12}, c_k \in K$ are defined by

$$A_5 = \left(\begin{array}{cccccccc|c} 1&9&0&0&1&0&5&0 & \\ 0&2&0&0&0&2&0&3 & 0 \\ 0&0&1&9&8&0&1&0 & \\ 0&0&0&2&0&10&0&2 & \\ \hline \multicolumn{8}{c|}{0} & 0 \end{array}\right), \quad A_6 = \left(\begin{array}{cccccccc|c} 12&2&0&0&0&1&0&8 & \\ 0&11&0&0&1&0&5&0 & 0 \\ 0&0&1&11&0&8&0&12 & \\ 0&0&0&2&5&0&12&0 & \\ \hline \multicolumn{8}{c|}{0} & 0 \end{array}\right),$$

$$A_7 = \left(\begin{array}{cccccccc|c} 11&0&0&0&8&0&1&0 & \\ 0&4&0&0&0&10&0&2 & 0 \\ 0&0&11&0&12&0&8&0 & \\ 0&0&0&4&0&11&0&10 & \\ \hline \multicolumn{8}{c|}{0} & 0 \end{array}\right), \quad A_8 = \left(\begin{array}{cccccccc|c} 0&0&11&2&0&5&0&1 & \\ 0&0&2&9&8&0&1&0 & 0 \\ 0&0&0&0&0&1&0&8 & \\ 0&0&0&0&1&0&5&0 & \\ \hline \multicolumn{8}{c|}{0} & 0 \end{array}\right),$$

$$A_9 = \left(\begin{array}{cccccccccccc} 5&4&0&5&12&12&5&11&11&2&2&10 \\ 0&10&8&0&12&11&2&3&2&9&3&9 \\ 0&0&5&4&8&2&12&12&11&3&11&2 \\ 0&0&0&10&11&10&12&11&10&4&2&9 \\ 0&0&0&0&12&6&0&5&12&3&4&11 \\ 0&0&0&0&0&11&8&0&3&11&2&5 \\ 0&0&0&0&0&0&12&6&9&2&12&3 \\ 0&0&0&0&0&0&0&11&11&8&3&11 \\ \hline \multicolumn{12}{c}{0} \end{array}\right), \quad A_{10} = \left(\begin{array}{cccccccccccc} 1&10&4&0&6&12&1&8&1&11&8&11 \\ 0&2&0&5&12&12&5&11&11&2&2&10 \\ 0&0&12&3&1&8&7&1&8&11&12&2 \\ 0&0&0&11&5&11&1&1&2&10&2&11 \\ 0&0&0&0&8&11&4&0&8&12&1&9 \\ 0&0&0&0&0&3&0&5&12&3&4&11 \\ 0&0&0&0&0&0&5&2&1&9&5&1 \\ 0&0&0&0&0&0&0&10&4&11&1&10 \\ \hline \multicolumn{12}{c}{0} \end{array}\right),$$

$$A_{11} = \left(\begin{array}{cccccccccccc} 2&0&0&0&2&0&7&2&11&3&11&2 \\ 0&9&0&0&0&9&2&1&10&4&2&9 \\ 0&0&2&0&6&11&2&0&2&11&11&3 \\ 0&0&0&9&11&12&0&9&11&4&10&4 \\ 0&0&0&0&4&0&0&0&9&2&12&3 \\ 0&0&0&0&0&5&0&0&11&8&3&11 \\ 0&0&0&0&0&0&4&0&1&10&9&2 \\ 0&0&0&0&0&0&0&5&10&2&11&8 \\ \hline \multicolumn{12}{c}{0} \end{array}\right), \quad A_{12} = \left(\begin{array}{cccccccccccc} 11&0&2&10&12&5&6&12&5&2&1&11 \\ 0&4&10&4&8&2&12&12&11&3&11&2 \\ 0&0&2&0&6&12&1&8&1&11&8&11 \\ 0&0&0&9&12&12&5&11&11&2&2&10 \\ 0&0&0&0&11&0&3&11&12&4&8&12 \\ 0&0&0&0&0&4&11&6&9&2&12&3 \\ 0&0&0&0&0&0&2&0&8&12&1&9 \\ 0&0&0&0&0&0&0&9&12&3&4&11 \\ \hline \multicolumn{12}{c}{0} \end{array}\right),$$

$b_5 = (1, 6, 4, 3, 5, 5, 12, 7, 0, 0, 0, 0),$ $b_6 = (3, 1, 5, 9, 9, 5, 3, 1, 0, 0, 0, 0),$
$b_7 = (9, 10, 1, 6, 1, 6, 5, 5, 0, 0, 0, 0),$ $b_8 = (8, 4, 3, 1, 10, 12, 9, 5, 0, 0, 0, 0),$
$b_9 = (4, 9, 10, 10, 7, 6, 7, 3, 12, 2, 0, 3),$ $b_{10} = (11, 4, 8, 3, 3, 7, 5, 6, 1, 12, 5, 0),$
$b_{11} = (3, 3, 4, 9, 6, 10, 7, 6, 0, 10, 12, 2),$ $b_{12} = (5, 10, 11, 4, 8, 7, 3, 7, 8, 0, 1, 12),$

$$c_5 = 9,\ c_6 = 1,\ c_7 = 2,\ c_8 = 6,\ c_9 = 8,\ c_{10} = 11,\ c_{11} = 9,\ c_{12} = 0.$$

7 Security Analysis for Attacks against Rainbow

Proposition 6.2 implies that attacks against Rainbow are applicable to HS scheme with $L = K$. In this section, we estimate security conditions against the UOV attack ([13,12]), the MinRank attack ([9,23,3]) and the HighRank attack ([9],[8],[17]), which are well-known attacks against Rainbow without using equation solvers like XL, Gröbner basis algorithm, etc. Our security analysis against these attacks is obtained by combining the results known for these attacks and an analogue (Proposition 3) of Uchiyama-Ogurafs result.

UOV Attack. The UOV attack is effective for the UOV signature scheme [12], and not for Rainbow. In HS scheme, the UOV attack finds the subspace $B^{-1}(\{0\}^{n-r} \times K^r)$ in K^n where B is the matrix expression of the linear part of A_2. The subspace is searched as an invariant subspace of $W_1 W_2^{-1}$ for linear combinations W_1, (invertible) W_2 of the matrices corresponding to the quadratic parts of the components of the public key. From the complexity of the UOV attack [12] and Proposition 3 we have

Proposition 4. *Let $K = GF(2^a)$. The following condition is necessary in order that HS($R; \tilde{n}$) may have a security level of l bits against the UOV attack:*

$$a(n - 2r - 1) + 4 \log_2(r) \geq l, \quad (n = r\tilde{n}).$$

Remark 4. The UOV attack is more efficient in the case of balanced Oil and Vinegar than in the case of general Unbalanced Oil and Vinegar. Therefore, we should not choose $\tilde{n} = 2$ in HS scheme, otherwise, HS scheme corresponds to a balanced Oil and Vinegar scheme.

MinRank Attack. In the MinRank attack, one solves the MinRank problem for rank $2r$. In other words, one finds a $(\lambda_{r+1}, \ldots, \lambda_n) \in K^{n-r}$ such that

$$\operatorname{rank}\left(\sum_{i=r+1}^{n} \lambda_i M_i\right) \leq 2r.$$

where M_i is the symmetric matrix corresponding to the quadratic part of the i-th component of the public key. Once such a matrix is found, one can compute the decomposition

$$K^n = \left(B^{-1}(K^{2r} \times \{0\}^{n-2r})\right) \oplus \left(B^{-1}(\{0\}^{2r} \times K^{n-2r})\right),$$

which helps an adversary to transform the public key into a system of polynomials with a form of a central map of Rainbow. From the complexity of the MinRank attack [23] and Proposition 3 we have

Proposition 5. *Let $K = GF(2^a)$. The following condition is necessary in order that $HS(R;\tilde{n})$ may have a security level of l bits against the MinRank attack:*

$$2ar + \log_2((n-r)(n^2/3 + nr/3 - r^2/6)) \geq l.$$

HighRank Attack. In the HighRank attack, one finds a linear combination, M, of $M_{r+1}, \ldots, M_n$ (as above) such that $\text{rank}(M) = n - r$. The probability that the rank of a random linear combination is equal to $n-r$ is $q^{-(n-r)}$. Once such a value of M is found, one can compute the decomposition

$$K^n = \left(B^{-1}(K^{n-r} \times \{0\}^r)\right) \oplus \left(B^{-1}(\{0\}^{n-r} \times K^r)\right),$$

which helps an adversary to transform the public key into a system of polynomials with a form of central map of Rainbow. Therefore, from the complexity of the HighRank attack [8] and Proposition 3 we have

Proposition 6. *Let $K = GF(2^a)$. The following condition is necessary in order that $HS(R;\tilde{n})$ may have a security level of l bits against the HighRank attack:*

$$ar + \log_2(n^3/6) \geq l.$$

From Proposition 4, 5 and 6, we have

Corollary 1. *In the case of $K = GF(256)$, we must choose a non-commutative ring with dimension more than 10 and $\tilde{n} > 2$ in order that $HS(R;\tilde{n})$ has the 80-bits security level against UOV, MinRank and HighRank attacks.*

8 Efficiency of HS Scheme

Any non-commutative ring R can be embedded in a matrix ring $\mathbb{M}(l, K)$ for some positive integer l. If we can choose a small l, the arithmetic operation of R becomes efficient. The number of field multiplication in solving a linear equations appearing in each layer in the signature generation in our proposed scheme estimated as $O(l^3)$ because the equations is of form of equations with respect to $\mathcal{X} \in \mathbb{M}(l, K)$,

$$\mathcal{A}.\mathcal{X} = \mathcal{B} \quad (\mathcal{A}, \mathcal{B} \in \mathbb{M}(l, K)).$$

On the other hand, that in the corresponding Rainbow estimated as $O(d^3)$ where d is the dimension of R because of Proposition 3. Thus, if $l < d$ is satisfied, the signature generation of our proposed scheme is more efficient than that of the corresponding Rainbow.

8.1 Efficiency in the Case of Group Ring of Dihedral Group

To compare the efficiency of signature generation in HS scheme and the corresponding Rainbow, we prepare dihedral group and its realization. Let n be a positive integer. $M_1 = (a_{ij}), M_2 = (b_{ij}) \in \mathbb{M}(n, K)$ is defined as

$$a_{ij} = \begin{cases} 1 & \text{if } j - i \equiv 1 \pmod{n}, \\ 0 & \text{otherwise}, \end{cases} \qquad b_{ij} = \begin{cases} 1 & \text{if } j + i \equiv 1 \pmod{n}, \\ 0 & \text{otherwise}. \end{cases}$$

We write D_n for the group generated by M_1 and M_2. D_n is isomorphic to the dihedral group with $2n$ elements. $K[D_n]$ denotes the group ring with coefficients in K and associated to D_n, then, it is a non-commutative ring of dimension $2n-1$, realized in $\mathbb{M}(n, K)$. $K[D_n]$ is closed by a transpose operation because D_n is so. Therefore we can use $K[D_n]$ as a base ring in HS scheme. Table 1 compares the efficiency of the signature generation in HS scheme and the corresponding Rainbow. The non-commutative rings used in HS schemes in the table are chosen by $K[D_n]$ where $K = GF(256)$ and $n = 10, 11, 12, 13$. These non-commutative rings satisfy the conditions in Corollary 1. The number of layers in each HS scheme is chosen by 3, and then the corresponding Rainbow of HS$(K[D_n]; 3)$ becomes Rainbow$(K; r, r, r)$ with $r = 2n - 1$ by Proposition 3. We estimate the number of multiplication of $GF(256)$ for efficiency comparison. $\mathrm{M_{sig}}(\mathrm{HS}(R; 3))$ (resp. $\mathrm{M_{sig}}(\mathrm{R}(GF(256); r, r, r))$) stands for the number of multiplications in the signature generation in HS$(R; 3)$ (resp. Rainbow$(GF(256); r, r, r)$). Table 1 shows that the signature generation of HS scheme is about 50% faster than that of the corresponding Rainbow.

Table 1. Efficiency comparison of HS scheme with the corresponding Rainbow (in terms of the number of multiplications in $GF(256)$)

$\mathrm{HS}(R, 3)$	$\mathrm{HS}(K[D_{10}], 3)$	$\mathrm{HS}(K[D_{11}], 3)$	$\mathrm{HS}(K[D_{12}], 3)$	$\mathrm{HS}(K[D_{13}], 3)$
Dimension of R	19	21	23	25
Matrix size	10	11	12	13
$\mathrm{M_{sig}}(\mathrm{HS}(R; 3))$	25353	33233	42581	53521
Corresponding Rainbow $\mathrm{R}(GF(256); r, r, r)$	$\mathrm{R}(19, 19, 19)$	$\mathrm{R}(21, 21, 21)$	$\mathrm{R}(23, 23, 23)$	$\mathrm{R}(25, 25, 25)$
$\mathrm{M_{sig}}(\mathrm{R}(GF(256); r, r, r))$	50198	66766	86618	110050
ratio	50.5%	49.8%	49.2%	48.6%

9 Conclusion

In this paper, we redefine HS scheme as a scheme in the multivariate public key cryptosystem, from a scheme whose security is based on the difficulty of integer factorization. For this proposed scheme, we analysed the security against the attacks of Coppersmith, Stern and Vaudenary for Birational Permutation scheme, the two attacks of Coppersmith for Sato-Araki scheme, and the Min-Rank, HighRank and UOV attacks for Rainbow. The proposed HS scheme can be regarded as a variation of Rainbow using a non-commutative ring. However, if a non-commutative ring used in the proposed scheme is chosen by the group ring associated to dihedral group, the speed of the signature generation can be accelerated by about 50% in comparison with the corresponding Rainbow.

Note that the performance of our designed schemes depend upon the arithmetic operation of adapted non-commutative rings. Therefore, we plan to find a specific non-commutative ring for achieving high processing speeds within our schemes.

Acknowledgements. This work partially supported by JST Strategic Japanese-Indian Cooperative Programme on multidisciplinary Research Field, which combines Information and Communications Technology with Other Fields. The first author is supported by JST A-step feasibility study program, No.AS231Z03613A. The authors would like to thank Jintai Ding and the anonymous reviewers for their helpful comments on the draft manuscript.

References

1. Anshel, I., Anshel, M., Goldfeld, D.: An Algebraic Method for Public-Kkey Cryptography. Math. Res. Lett. 6(3-4), 287–291 (1999)
2. Adleman, L.M., Estes, D.R., McCurley, K.S.: Solving bivariate quadratic congruences in random polynomial time. Math. Comput. 48, 17–28 (1987)
3. Billet, O., Gilbert, H.: Cryptanalysis of Rainbow. In: De Prisco, R., Yung, M. (eds.) SCN 2006. LNCS, vol. 4116, pp. 336–347. Springer, Heidelberg (2006)
4. Coppersmith, D.: Weakness in Quaternion Signatures. In: Wiener, M. (ed.) CRYPTO 1999. LNCS, vol. 1666, pp. 305–314. Springer, Heidelberg (1999)
5. Coppersmith, D., Stern, J., Vaudenay, S.: The security of the birational permutation signature scheme. J. Cryptology 10, 207–221 (1997)
6. Ding, J., Gower, J.E., Schmidt, D.S.: Multivariate Public Key Cryptosystems. In: Advances in Information Security, vol. 25 (2006)
7. Ding, J., Schmidt, D.: Rainbow, a New Multivariable Polynomial Signature Scheme. In: Ioannidis, J., Keromytis, A.D., Yung, M. (eds.) ACNS 2005. LNCS, vol. 3531, pp. 164–175. Springer, Heidelberg (2005)
8. Ding, J., Yang, B.-Y., Chen, C.-H.O., Chen, M.-S., Cheng, C.-M.: New Differential-Algebraic Attacks and Reparametrization of Rainbow. In: Bellovin, S.M., Gennaro, R., Keromytis, A.D., Yung, M. (eds.) ACNS 2008. LNCS, vol. 5037, pp. 242–257. Springer, Heidelberg (2008)
9. Goubin, L., Courtois, N.T.: Cryptanalysis of the TTM Cryptosystem. In: Okamoto, T. (ed.) ASIACRYPT 2000. LNCS, vol. 1976, pp. 44–57. Springer, Heidelberg (2000)
10. Hashimoto, Y., Sakurai, K.: On construction of signature schemes based on birational permutations over noncommutative. Presented at the 1st International Conference on Symbolic Computation and Cryptography (SCC 2008), held in Beijing (April 2008). ePrint `http://eprint.iacr.org/2008/340`
11. Ko, K.H., Lee, S.-J., Cheon, J.H., Han, J.W., Kang, J.-S., Park, C.-S.: New Public-Key Cryptosystem Using Braid Groups. In: Bellare, M. (ed.) CRYPTO 2000. LNCS, vol. 1880, pp. 166–183. Springer, Heidelberg (2000)

12. Kipnis, A., Patarin, J., Goubin, L.: Unbalanced Oil and Vinegar Sschemes. In: Stern, J. (ed.) EUROCRYPT 1999. LNCS, vol. 1592, pp. 206–222. Springer, Heidelberg (1999)
13. Kipnis, A., Shamir, A.: Cryptanalysis of the Oil & Vinegar Signature Scheme. In: Krawczyk, H. (ed.) CRYPTO 1998. LNCS, vol. 1462, pp. 257–266. Springer, Heidelberg (1998)
14. Ong, H., Schnorr, C.P., Shamir, A.: An efficient signature scheme based on quadratic equations. In: Proc. 16th ACM Symp. Theory Comp., pp. 208–216 (1984)
15. Patarin, J.: Cryptanalysis of the Matsumoto and Imai Public Key Scheme of Eurocrypt 1988. In: Coppersmith, D. (ed.) CRYPTO 1995. LNCS, vol. 963, pp. 248–261. Springer, Heidelberg (1995)
16. Petzoldt, A., Bulygin, S., Buchmann, J.: Selecting Parameters for the Rainbow Signature Scheme. In: Sendrier, N. (ed.) PQCrypto 2010. LNCS, vol. 6061, pp. 218–240. Springer, Heidelberg (2010)
17. Petzoldt, A., Bulygin, S., Buchmann, J.: CyclicRainbow – A Multivariate Signature Scheme with a Partially cyclic Public Key. In: Gong, G., Gupta, K.C. (eds.) INDOCRYPT 2010. LNCS, vol. 6498, pp. 33–48. Springer, Heidelberg (2010)
18. Pollard, J.M., Schnorr, C.P.: An efficient solution of the congruence $x^2 + ky^2 \equiv$m (mod n). IEEE Trans. Inf. Theory IT-33, 702–709 (1987)
19. Rai, T.S.: Infinite Gröbner bases and Noncommutative Polly Cracker Cryptosystems. PhD Thesis, Virginia Polytechnique Institute and State Univ. (2004)
20. Satoh, T., Araki, K.: On construction of signature scheme over a certain noncommutative ring. IEICE Trans. Fundamentals E80-A, 702–709 (1997)
21. Shamir, A.: Efficient Signature Schemes Based on Birational Permutations. In: Stinson, D.R. (ed.) CRYPTO 1993. LNCS, vol. 773, pp. 1–12. Springer, Heidelberg (1994)
22. Uchiyama, S., Ogura, N.: Cryptanalysis of the Birational Permutation Signature Scheme over a Non-commutative Ring. JSIAM Lett.ers 2, 85–88 (2010), `http://eprint.iacr.org/2009/245`
23. Yang, B.-Y., Chen, J.-M.: Building Secure Tame-Like Multivariate Public-Key Cryptosystems: The New TTS. In: Boyd, C., González Nieto, J.M. (eds.) ACISP 2005. LNCS, vol. 3574, pp. 518–531. Springer, Heidelberg (2005)

A Extension of HS Scheme

To construct HS scheme, we need the condition of a non-commutative ring R that the realization in matrix algebra of R is closed by a transpose operation. Hashimoto et al. gave an following example of non-commutative rings satisfying the above condition except for quaternion algebras and matrix algebras [10] .

Lemma 1 ([10]). *Let G be a finite group with an embedding $\psi : G \hookrightarrow \mathbb{M}(m, L)$. If $\psi(G)$ is closed by transpose operation, then so is the group algebra $R = L[G](\hookrightarrow \mathbb{M}(m, L))$.*

However, it is difficult to find such G and ψ in general. In this appendix, we consider that the above condition is loosened. First, we introduce a ring with involution.

Definition 1. *We say a map* $* : R \to R$ *for an ring* R *is an involution of* R *if the following conditions are satisfied: For* $a, b \in R$, *(1)* $(a+b)^* = a^* + b^*$, *(2)* $(ab)^* = b^* a^*$, *(3)* $(a^*)^* = a$, *and (4)* $1^* = 1$. *A ring equipped with an involution is called a ring with involution.*

Example 2.

(1) Let R be a subring of a matrix algebra. If R is close by transpose operation, then the transpose is an involution of R.
(2) Let G be a (non-commutative) group. In the group algebra $R = L[G]$, the map defined by the L-linear extension of $G \ni g \mapsto g^{-1} \in G$ becomes an involution of R.

We extend HS scheme to a ring with involution as follows:

Let R be a non-commutative ring over L with an involution $*$. We fix a linear isomorphism

$$\phi : L^r \xrightarrow{\sim} R.$$

Let $\tilde{n}$ be a natural number. The extended HS scheme deploys non-commutative multivariate polynomials as a central map:

$$\tilde{g}_k(x_1, \ldots, x_{\tilde{n}}) = \sum_{i=1}^{k-1} x_i^* \alpha_{ij}^{(k)} x_k + \sum_{1 \le i,j \le k-1} x_i^* \alpha_{ij}^{(k)} x_j + \sum_{1 \le i \le k} \beta_i^{(k)} x_i + \gamma^{(k)} \quad (k = 2, 3, \ldots, \tilde{n}),$$

where $\alpha_{i,j}^{(k)}, \beta_i^{(k)}, \gamma^{(k)} \in R$. The central map of HS scheme is constructed by

$$\tilde{G} = (\tilde{g}_2, \ldots, \tilde{g}_{\tilde{n}}) : R^{\tilde{n}} \to R^{\tilde{n}-1}$$

The key generation, the signature generation and the verification are the same as those of HS scheme described in § 4.2.

Remark 5. 1. This scheme is an extension of HS scheme in § 4.2 because of (1) of Example 2.
2. From (2) of Example 2, we can use any group ring as a base ring of this scheme.
3. The similar statements of security analysis against all attacks dealt with in this paper hold.

Author Index

GPSR Compliance

The European Union's (EU) General Product Safety Regulation (GPSR) is a set of rules that requires consumer products to be safe and our obligations to ensure this.

If you have any concerns about our products, you can contact us on ProductSafety@springernature.com

In case Publisher is established outside the EU, the EU authorized representative is:

Springer Nature Customer Service Center GmbH
Europaplatz 3
69115 Heidelberg, Germany

Batch number: 10859437

Printed by Printforce, the Netherlands